New Tools for Robustness of Linear Systems

New Tools for Robustness of Linear Systems

B. Ross Barmish

**Department of Electrical and
Computer Engineering**
University of Wisconsin

Macmillan Publishing Company
New York
Maxwell Macmillan Canada
Toronto
Maxwell Macmillan International
New York   Oxford   Singapore   Sydney

Macmillan Publishing Company
866 Third Avenue, New York, New York 10022

Macmillan Publishing Company is part
of the Maxwell Communication Group of Companies.

Maxwell Macmillan Canada, Inc.
1200 Eglinton Avenue East
Suite 200
Don Mills, Ontario M3C 3N1

Library of Congress Cataloging in Publication Data

Barmish, B. Ross.
  New Tools for robustness of liner systems/B. Ross Barmish.
    p.   cm.
  Includes bibliographical references and index.
  ISBN 0-02-306055-7
  1. Control theory.   2. Linear systems.   I. Titls.
QA402.3.B346   1994
629.8'312–dc20

                                          93–3909
                                          CIP

Printing: 1 2 3 4 5 6 7 8   Year: 4 5 6 7 8 9 0 1 2 3

Preface

This book is an outgrowth of my longstanding interest in robust control problems involving structured real parametric uncertainty. At the risk of beginning on a controversial note, I believe that it is fair to say that in the robust control field, most research is currently concentrated in five areas. The popular labels for these areas are H_∞, μ, Kharitonov, Lyapunov and QFT. Some colleagues in the field would insist on including L_1 as a sixth area. In terms of the five labels above, the takeoff point for this book is what I believe to be one of the major milestones in the literature relevant to control theory—a 1978 paper in a differential equations journal by the Russian mathematician V. L. Kharitonov; see Kharitonov (1978a).

Kharitonov's paper began to receive attention in the control field in 1983 and provided strong motivation for a decade of furious work by researchers interested in robustness of systems with real parametric uncertainty. I use the word "furious" above because at times, the race for results got rather heated. On numerous occasions, the same result appeared nearly simultaneously in two journals—by different authors, of course. Given this explosive rate of publication, much "smoke" has emerged. The uninitiated reader who wants to become familiar with the new developments faces an enormous pile of papers and may not know which ones to read first. My choice of material for this text implicitly provides my perspective on this matter. One of my main objectives is distillation—taking this large body of new literature, picking out the most important results and simplifying their explanation so as to minimize time investment associated with learning the new techniques. In this regard, many of the proofs are new and given here for the first time.

At the outset, the reader should be aware that the robust control literature does not contain many results "linking" the different areas of research. I am hoping, however, that my exposition will motivate others to undertake efforts aimed at unification of the field; this book is not the "grand unifier." My point of view is as follows: The serious student of robust control might reasonably be expected to take three or four courses in the area. In this sense, my hope is that this book would be a strong competitor for being the text in one of these courses.

After weighing the trade-offs between encyclopedic coverage and pedagogy, I resisted the temptation to let the scope get too broad. I opted to concentrate on trying to write a text which is "technically tight" and yet does not require too high a level of technical sophistication to read. My targeted reader is the beginning graduate student who is familiar with just the basics such as Bode, Nyquist, root locus and elementary state space analysis. For a one-semester course of 13–15 weeks, I would recommend Chapters 1–11 and 14–16. An ambitious instructor might also include selected results from Chapters 12, 13 and 17.

In many places throughout the text, I refer to the value set. Once this rather simple concept is understood, it becomes possible to unify most of the new technical developments emanating from Kharitonov's Theorem. While we may have the illusion that we have been bombarded with dozens of new "lines of proof" over the last decade, the truth of the matter is that most of the seemingly disparate new results can be easily understood with the help of one simple idea—the value set. Granted, I am overstating my case a bit here, but in spirit, I feel that my contention is correct.

At this point, I must note that I have avoided calling the value set concept "new." Value sets arise in many fields, for example, mathematics, economics and optimization. In fact, even in the control literature, value sets appear as early as 1963 in the textbooks of Horowitz and Zadeh and Desoer. What is new in this book is the way the value set is used to unify a large body of literature on robustness of control systems. In fact, one of the greatest challenges in writing this book was taking existing results from the literature and finding new ways to explain them using the value set.

To provide my personal perspective on how this research area came into being, let me begin by noting that Kharitonov's Theorem first came to my attention in 1982 at a workshop in Switzerland organized by Jüergen Ackermann. At that time, I remember sitting next to Manfred Morari and listening to Andrej Olbrot exploit Kharitonov's Theorem to prove a result on delay systems. Given that Kharitonov's Theorem was published in 1978, my immediate reaction to Olbrot's presentation was one of bewilderment. Despite the fact that it was published in a Russian differential equations journal, I could not understand how such an important result had been unheralded in the control community for more than four years. Immediately following the workshop in Switzerland, there was a period of about six months which I spent working with Kris Hollot and Ian Petersen expending considerable effort trying to decide if Kharitonov's cryptic proof was correct. It was.

Apparently, Olbrot was aware of the importance of Kharitonov's Theorem at least one year before the workshop in Switzerland. In a 1981 letter from Olbrot to Ackermann (following a workshop in Bielefeld), the theorem was stated precisely. In his letter, Olbrot also recognized that this result had possible applications to "insensitive stabilization."

Kharitonov's name finally surfaced in the control journals in 1983 in Bialas (1983) and Barmish (1983). While my paper is frequently cited for exposing the power of Kharitonov's Theorem in the "western literature," the paper by Bialas had an equally important role. Although Bialas' attempt to generalize from polynomials to matrices turned out to be incorrect (for example, see Karl, Greschak and Verghese (1984) for a counterexample), his paper served to stimulate researchers to address the following question: To what extent can Kharitonov's strong assumptions on the uncertainty structure be relaxed? An important breakthrough in this direction was the Edge Theorem of Bartlett, Hollot and Huang (1988). This added fuel to the fire just as the flames were beginning to subside.

One final comment to complete this historical perspective: Given that Kharitonov did his seminal work in St. Petersburg, it is amusing that his theorem remained virtually unknown in the former Soviet Union until the late eighties. My understanding is that the result became known to Russian scientists upon visiting the United States.

In the remainder of this preface, I want to gratefully acknowledge the effort and support of a number of individuals. Over a three year period, Zhicheng Shi worked tirelessly on the LaTeX preparation of the text and Hwan-Il Kang served as our "resident expert" on figure preparation. Without their intensive efforts, I would have abandoned this project long ago. I would also like to acknowledge Jerry Hamann's efforts in a few areas. He provided excellent technical commentary on some of the chapters, supported all aspects of the book related to postscript and developed all the style files associated with the page layout. While on the topic of style files, I also owe notes of thanks to Greg Wasilkowski and John Gubner. When I started writing, Greg helped a great deal by allowing us to examine the style files associated with his text on information-based complexity; in the final stages of manuscript preparation, John provided a clever solution of a certain LATEX problem which I thought was insurmountable.

I would also like to thank some of my graduate students who contributed to the overall effort at various stages. I extend my thanks to Saad Saleh for his help during the early part of the writing and to Beng Tak Ting and Medhi Abrishamchian for help with corrections and indexing.

This book could not have been written without the continued support of the National Science Foundation. Through programs administered by Abe Haddad, Michael Polis and Kishan Baheti, it became possible for me to become involved in the area to the point that I felt I had sufficient command of the literature to write a text. Most recent support under NSF Grant ECS-9111570 is gratefully acknowledged.

In many places in the text, there are value set plots which not only help to illuminate the theory but also have a nice aesthetic appeal. These plots were generated by a software program called CONVEX, developed in Italy at Politecnico di Torino and CENS-CNR, by Massimiliano Barberis, Mauro Casales, Diego Cavallera and Roberto Manzin under the direction of Roberto Tempo. This program was then modified at the University of Wisconsin by Carlos Murillo-Sanchez (producing output which is suitable for the production of LATEX picture files).

Over the last few years, one of my primary objectives has been to develop the capability for testing some of the "new tools" for robustness in an industrial environment. In this regard, I am grateful for my collaboration with Centro Ricerche Fiat; most of the material in Chapter 3 is based on this interaction. Far-sighted control engineers like Maurizio Abate are important to the field because they provide the crucial bridge between theory and applications.

About a year ago, I thought the project was nearly complete. However, after soliciting comments on the manuscript from a number of colleagues who volunteered to serve as "readers," I soon realized that there were considerable improvements to be made. In this regard, I extend my thanks to Ted Djaferis, Faryar Jabbari and Lahcen Saydy. Kris Hollot, Ian Petersen, Roberto Tempo, Boris Polyak and Tom Higgins also served as readers, but they need to be mentioned in categories of their own. Kris, Ian and Roberto not only provided me with extensive technical and editorial comments, but also engaged me in lengthy discussions about the "philosophical aspects" of the book. They constantly encouraged me in their capacity as good friends. Over the last few months, Boris Polyak played a very similar role. As far as Tom Higgins is concerned, I am most grateful for our frequent interactions over the last eight years. He is a true scholar who constantly made me aware of results in the literature which were both important and unknown to me.

Throughout the entire course of this project, the support provided by Macmillan Publishing Company was excellent. First and foremost, I wish to express my sincere thanks to John Griffin. In his capacity as Editor, he helped me in a number of ways from start to finish. As the project was drawing to a close, Leo Malek became involved and provided excellent support on many issues associated with final production. Of particular note, Leo connected me with an outstanding copy editor—Lilian Brady.

Finally, my greatest personal note of thanks goes to my immediate family—Marlene, Lara and Sybil. The manuscript could not have been written without their tolerance during my frequent periods of enlistment in the space cadets. This book is dedicated to them.

B. Ross Barmish
Madison, Wisconsin

Contents

New Tools for Robustness of Linear Systems

Part I

Preliminaries

Chapter 1

A Global Overview

Synopsis

The main objective of this chapter is to provide some generalities about the scope of this book. To this end, control problems involving uncertainty are subdivided into different areas via a Problem Tree. This tree is not to be taken too seriously—it only serves as a metaphor for better understanding where the focus of this text lies. A secondary objective of this chapter is to set the stage for the technical exposition to follow. To this end, basic paradigms for robustness are described.

1.1   Introduction

Much of modern control theory addresses problems involving uncertainty. A typical scenario begins with a system to be controlled and a mathematical model which includes uncertain quantities. For example, the mathematical model might involve various physical parameters whose values are specified only within given bounds. In order to provide a global overview of this text, we imprecisely subdivide control problems with uncertainty into three types:

- Adaptive Problems
- Stochastic Problems
- Robustness Problems

This subdivision should not be interpreted too literally. Obviously, many problem formulations with uncertainty do not fit neatly into any of the three categories; e.g., fuzzy control problems and singular perturbation problems. We assume a demarcation between the three problem areas above solely for pedagogical purposes. In addition, the reader familiar with control theory also understands that many control problems with uncertainty may involve ideas from more than one of the three areas above. For example, one can consider a robust adaptive control problem or a stochastic adaptive control problem. Having provided our disclaimers, we stick with our three problem idealization and begin this overview by simply declaring the focal point of this textbook to be a class of robustness problems.

1.2   Robustness Problems

Given that the field of robust control has experienced a large number of breakthroughs over the last two decades, an uninitiated reader might find it quite overwhelming when first attempting to gain some perspective. Since the word "robust" appears in literally dozens of different contexts, it is quite natural to divide the robustness area into a number of different subareas. We indicate such a division via the *Problem Tree* in Figure 1.2.1. As stated in the chapter synopsis, this problem tree is only meant to be interpreted in the sense of a metaphor. The distinction between problem areas can be quite unclear. For example, in Figure 1.2.1, we separate real and complex parametric uncertainty problems knowing full well that in many cases, one encounters a mixture of both types of uncertainty. Similarly, we separate linear and nonlinear problems knowing that we often treat nonlinearities as uncertainties in a linear system. The list of qualifications in Figure 1.2.1 can occupy several pages. Nevertheless, in order to get some perspective on the scope of this book, we ignore these subtleties.

Note that it is possible to more fully develop the branches of the tree. For example, within the class of nonlinear robustness problems, one can further subdivide into many different areas. To create a better focal point, however, we have deliberately pruned various tree branches. When concentrating on the side of the tree dealing with linear time-invariant robustness problems, we are not ruling out solutions which might involve nonlinear control or uncertainties entering nonlinearly into the model. Our point of view is that the mathematical model of the plant includes a "nominal" system which

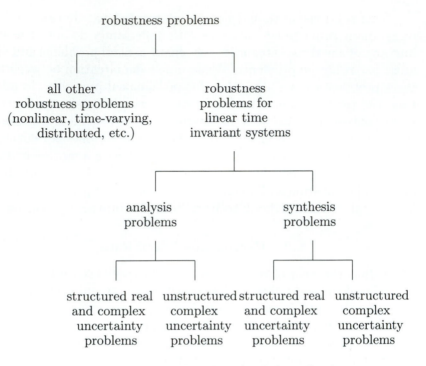

FIGURE 1.2.1 A Problem Tree for Robust Systems

is linear and time-invariant. However, this does not rule out the possibility that uncertainties enter nonlinearly into the plant. For example, consider a transfer function whose description includes uncertain parameters $q_1, q_2, \ldots, q_\ell$. For fixed q_i, the plant is linear and time-invariant but the q_i are permitted to enter transfer function coefficients in a nonlinear manner; in some cases, the nominal plant might be obtained with all q_i set to zero.

At the bottom of the linear time-invariant side of the Problem Tree, we see four types of robustness problems. They involve robustness analysis with structured real and complex uncertainty, robustness analysis with unstructured complex uncertainty, robust synthesis with structured real uncertainty and robust synthesis problems involving unstructured complex uncertainty. In some cases, these uncertainties represent system parameters and in other cases, unmodeled dynamics is the basic concern. We are now prepared to describe the scope of this text. To this end, we first provide some historical context.

1.3   Some Historical Perspective

The primary focal point of this text is robustness analysis problems involving structured real parametric uncertainty. In most cases, the results are easily extended to handle complex uncertainty, and in a few cases, synthesis results are obtained. Suffice it to say, for problems of robust synthesis with real parametric uncertainty, many of the most fundamental research questions are as yet unresolved, particularly at the level of multi-input multi-output (MIMO) systems.

Looking back at the literature dealing with real parametric uncertainty, much impetus for current work is derived from the theory dealing with stability domain characterizations as in Neimark (1949), the contributions to robust synthesis beginning with Horowitz (1963) and the focus on parametric uncertainty and related Popov and Lure-type problems in Siljak (1969). After a hiatus of at least a decade, we see the revitalization of interest in real parametric uncertainty in Ackermann (1980). The approach taken in this text relies heavily on a certain "Zero Exclusion Condition." A version of this condition can be traced back at least as far as Frazer and Duncan (1929). Perhaps it can even be argued that the paper by Frazer and Duncan is one of the pioneering works on robustness of systems.

As far as synthesis with complex uncertainty is concerned, we have recently witnessed some important breakthroughs for classes of synthesis problems with unmodeled dynamics; e.g., one highlight is the H^∞ control problem as formulated by Zames (1981) and elegantly solved via Riccati equations in Doyle, Glover, Khargonekar and Francis (1989). In a sense, these breakthroughs culminate three decades of change which occurred in the field—from preoccupation with optimality in the sixties to emphasis on robustness and MIMO problems in the eighties.

To conclude this section, we mention the large body of literature on robust stabilization in state space pioneered in the seventies by Utkin and Leitmann; e.g., see Utkin (1977) and Leitmann (1979). In retrospect, their robust stabilization theories for systems with time-varying uncertainty which satisfies "matching conditions" was a natural precursor for the more general H^∞ framework of the eighties and nineties.

1.4   Refinement of the Scope

By and large, the results in this book apply most directly to the analysis of the root locations of polynomials whose coefficients depend on uncertain parameters. From a control theoretic point of view, the motivation for studying these new tools is the fact that many robust performance problems for feedback systems can be massaged into equivalent problems involving a polynomial's roots. Using this reduction, the proof of many robustness results given in this book involves exploitation of classical theorems on the geometry of polynomials; e.g., see Marden (1966). In marked contrast to classical literature on polynomials, however, the total emphasis here is on robustness issues.

With the explosive growth of the "new tools literature" following publication of the seminal stability theorem of Kharitonov (1978a), there has been increasing recognition of the fact that many of the new results can actually be applied to a variety of robust performance problems—not just the problem of robust stability. To illustrate this point, imagine that a software package is available which finds the roots of a polynomial. The question is: How might we apply this package to the seemingly unrelated performance problem of checking if a plant has H^∞ norm less than unity? Indeed, we consider a fixed plant expressed as a quotient of polynomials

$$P(s) = \frac{N(s)}{D(s)}$$

with $D(s)$ being stable (all roots in the open left half plane). We assume that this plant is strictly proper and want to determine if the H^∞ norm

$$\|P\|_\infty = \sup_{\omega \geq 0} |P(j\omega)|$$

is less than unity.

To reformulate this problem in a polynomial framework, we make one observation: Since $P(j\omega) \to 0$ as $\omega \to \infty$, continuity of $P(j\omega)$ dictates that the only way that $\|P\|_\infty \geq 1$ can occur is if $|P(j\omega^*)| = 1$ for some frequency $\omega^* \geq 0$. This implies the existence of an angle $0 \leq \theta \leq 2\pi$ such that

$$\frac{N(j\omega^*)}{D(j\omega^*)} = e^{j\theta}.$$

These observations lead us to study the θ-parameterized polynomial

$$p(s, \theta) = N(s) - e^{j\theta}D(s).$$

Now for the punchline: It can be shown that stability of $p(s, \theta)$ for all $\theta \in [0, 2\pi]$ is equivalent to satisfaction of the required inequality

$\|P\|_\infty < 1$; e.g., see Barmish and Khargonekar (1990) for the definition of $p(s,\theta)$ and Chapellat, Dahleh and Bhattacharyya (1990) for the stronger stability result. In summary, the H^∞ problem has been reduced to finding the roots of a family of polynomials parameterized by $\theta \in [0, 2\pi]$.

This example is intended to illustrate one of the possible ways that fundamental polynomial problems arise from control system performance problems. In a robustness context, similar ideas apply with uncertainty entering transfer function coefficients.

1.5   Kharitonov's Theorem: The Spark

Many of the questions addressed in this book are not new—what is new is the machinery introduced and its ability to solve a number of problems which heretofore seemed intractable. The formulation of basic robustness problems and their solution for various special cases goes back a long way. For example, in the early work of Neimark (1949), we see effective techniques for robust stability analysis which work well with a small number of uncertain parameters. In the book by Siljak (1969), special classes of robust stability analysis problems for systems with structured real parametric uncertainty are considered. For problems involving robustness analysis with uncertain parameters entering multilinearly into transfer function coefficients, a powerful tool is the Mapping Theorem given in the book by Zadeh and Desoer (1963). On the synthesis side, we already mentioned the importance of the book by Horowitz (1963) in bringing robustness issues to the fore.

In the mid to late eighties, we see a new "explosion" of research involving structured real parametric uncertainty. In large measure, the reason for this resurgence of interest in the area is the seminal theorem of Kharitonov (1978a); this theorem is the takeoff point for much of the technical exposition in this book. After the control clothing is removed, many of the basic problems which we address can be viewed in the context of an age-old question: How do the roots of a polynomial depend on its coefficients? Although the literature contains a wealth of information on this question (for example, see Gantmacher (1959) and Marden (1966)), the robustness context is missing. In fact, a number of the results in this book can be viewed as robustified versions of classical results.

1.6   The Issue of Uncertainty Structure

After massaging a robustness analysis problem into a polynomial problem with coefficients depending on uncertain parameters, the issue of uncertainty structure arises. In the formalism of the chapters to follow, we deal with a vector of uncertain parameters q and a polynomial which is expressed as

$$p(s, q) = \sum_{i=0}^{n} a_i(q) s^i.$$

The uncertainty structure is manifested via the coefficient functions $a_0(q), \ldots, a_n(q)$. In the idealized framework of Kharitonov (1978a), each component q_i of q enters into only one coefficient. This same "independent" uncertainty structure is also exploited in the important work of Soh, Berger and Dabke (1985), which is described in Chapter 15. In contrast to Kharitonov's framework where a box is used to bound q, Soh, Berger and Dabke use a sphere. The obvious point to note is that an independent uncertainty structure is highly idealized; uncertain parameters of a system generally enter into more than one coefficient of $p(s, q)$, and in many cases, the $a_i(q)$ are often nonlinear functions.

When dealing with uncertainty structure, this text follows a natural progression. First, we explain basic ideas without significant machinery. This is accomplished by studying robust stability for the case of a single uncertain parameter. Subsequently, we deal with independent uncertainty structures as explained above. The next level of complication is the affine linear uncertainty structure. In this case, coefficients depend affine linearly on q and various results on polytopes of polynomials are developed. For affine linear uncertainty structures, the highlight is the Edge Theorem of Bartlett, Hollot and Huang (1988) given in Chapter 9.

To deal with more realistic robust control problems, multilinear and nonlinear uncertainty structures are of paramount importance; e.g., consider $p(s, q)$ above with each coefficient function $a_i(q)$ being multilinear. At this higher level of difficulty, we see a bifurcation in the robustness literature. That is, some authors deal with these more difficult uncertainty structures by restricting their attention to analytically tractable special cases, while other authors resort to mathematical programming. This textbook concentrates on a selected number of analytical results which are available; a highlight is the Mapping Theorem, which is covered in Chapter 14. Although we do not cover the mathematical programming approach in this text, we include the section below as a gateway to this other body of literature.

1.7 The Mathematical Programming Approach

The motivation for a large body of literature is derived from the fact that many robustness problems can be reformulated as mathematical programming problems. Subsequently, one has available a wide variety of software tools to accomplish the required optimization. To illustrate at the simplest of levels, we continue to let q represent an uncertain parameter vector and take Q to be a closed and bounded restraint set for q. We consider a robust stability problem for the polynomial

$$p(s, q) = s^3 + a_2(q)s^2 + a_1(q)s + a_0(q).$$

Namely, determine if all roots of $p(s, q)$ lie in the strict left half plane for all $q \in Q$. The application of the Routh–Hurwitz stability criteria leads to the following conclusion: Robust stability is guaranteed if and only if, for each $q \in Q$, the conditions $a_0(q) > 0$, $a_2(q) > 0$ and $a_1(q)a_2(q) - a_0(q) > 0$ are satisfied. Hence, a solution to the trio of optimization problems

$$\min_{q \in Q} a_0(q); \quad \min_{q \in Q} a_2(q); \quad \min_{q \in Q} [a_1(q)a_2(q) - a_0(q)]$$

leads to a solution of the robust stability problem.

As a second example, we note that mathematical programming problems arise quite naturally in the multivariable stability margin theories of Doyle (1982) and Safonov (1982). In many cases, they take Δ to be a block diagonal uncertainty matrix, and for a fixed matrix M, one seeks a minimum norm Δ such that

$$\det(I + M\Delta) = 0.$$

This is generally called the μ *problem*. Once again, it is straightforward to reformulate this problem as a mathematical program.

The mathematical programming approach to robustness problems has one clear advantage over more analytical approaches. The formulation easily accommodates large classes of robust performance criteria for systems which can have rather complicated uncertainty structures. Many examples illustrating the power of the approach are given in the book by Boyd and Barratt (1990).

On the downside, there are at least two negatives associated with the mathematical programming approach. First, one is faced with the gamut of well-known headaches associated with nonlinear programming—local versus global minima, potentially high computational complexity, etc. To some degree, this issue is being addressed in the literature. For example, in solving the optimization problem associated with μ theory, the underlying control system structure sometimes makes it possible to generate convex bounds which would ordinarily not be available in an arbitrary mathematical program. The systematic exploitation of such bounds leads to "tailored" algorithms. For example, when minimizing some function $f(x)$, the clever exploitation of a convex upper bound $f_{UB}(x)$ might greatly facilitate computation; see Packard (1987) for a good entry point to this literature.

The second negative associated with the mathematical programming approach might be described by the words "loss of insight." When carrying out the required calculations, it is typically unclear how the computed solution depends on the parameters of the controller. One must often resort to repeated trial and error in order to ascertain how changes in various controller parameters affect the performance of the system.

To conclude this section, we note that the problems attacked in this book are all special cases of problems which can be formulated in a mathematical programming context. In contrast to mathematical programming, however, we obtain strong analytical results via the imposition of additional assumptions. Perhaps the prime example illustrating this distinction is Kharitonov's Theorem in Chapter 5; we see that robust stability of interval polynomials can be ascertained by checking only four distinguished extremes.

1.8   The Toolbox Philosophy

The appropriate way to view many of the results in this book is analogous to the way one might view the tools in a toolbox. No single tool is a cure-all for all robustness problems. To provide a crude analogy, when working on a construction project, a wrench is definitely useful when tightening a bolt but is much less useful in driving a nail. In order to make the same point in a control context, consider the classical Nyquist criterion. Although Nyquist analysis has withstood the test of time, no one would argue that the Nyquist criterion by itself solves practical control problems. It is only one of many tools which can be brought to bear. There are many other specific results in modern control theory for which the same argument can be made. For example, the Hurwitz stability criterion and the Small Gain Theorem should rightfully be viewed as two tools which can be brought to bear in appropriate situations.

1.9 The Value Set Concept

To conclude the overview of this book, we briefly mention one piece of technical machinery which we use to unify many results in the literature—the so-called *value set*. An understanding of the value set enables us to appreciate many robustness developments from a single perspective. In many places throughout the book, we state results from the recent literature but provide new proofs—proofs via the value set approach.

We now provide a rough explanation of the value set concept: The main point to note is that we can reformulate many robustness problems in terms of a two-dimensional set which we temporarily call $V(\delta)$; this set lies in the complex plane. Notice that $V(\delta)$ is parameterized via a real scalar δ, which we call a *generalized frequency variable*. Now, as δ increases, the set $V(\delta)$ typically moves around the complex plane. For many problems, it turns out that the *zero exclusion condition*

$$0 \notin V(\delta)$$

for all δ is both necessary and sufficient for satisfaction of the stated robustness specification.

As seen in the sequel, many robustness problems lead to very simple value set geometries which can be exploited to obtain strong analytical results. For example, when dealing with the interval polynomial framework of Kharitonov, the generalized frequency variable δ arises via the substitution $s = j\delta$, and the value set $V(\delta)$ turns out to be a rectangle whose sides are parallel to the real and imaginary axes—a level rectangle.

In large measure, the power of the value set approach is derived from the fact that it is a two-dimensional set, whereas the uncertain parameter set is typically of higher dimension; i.e., although a robustness problem with an ℓ-dimensional uncertain parameter vector q is initially formulated over $\mathbf{R}^\ell$, we need only manipulate the two-dimensional value set $V(\delta)$. Furthermore, since $V(\delta)$ is only two-dimensional, we obtain a second advantage. For cases when $V(\delta)$ is readily constructable, we obtain solutions to robustness problems which lend themselves to implementation in graphics. That is, once we have an analytical description of the value set $V(\delta)$, it is often convenient to simply generate this set on a computer and provide a visual display of its motion with respect to δ.

1.10   Mathematical Model Versus the True System

In the remainder of this chapter, our objective is to create the appropriate "mind set" for the technical exposition to follow. We begin by noting that in classical control theory, one generally begins with a mathematical model of the system and a design is carried out which is aimed at satisfaction of some given set of performance specifications. Examples of such specifications involve prescriptions on damping, overshoot, tracking and frequency response. Once the design is complete, the following question is of critical importance: If we use an inexact mathematical model to derive the controller, will the system perform satisfactorily? In other words, a fundamental concern is that a design based on an inexact mathematical model may result in unacceptable performance when implemented on the true system—the behavior of the true system may be quite different from the behavior predicted by the mathematical model.

To illustrate these ideas, imagine a vehicle dynamics system which is to be designed assuming a certain coefficient of friction $\mu = 0.5$. However, a fundamental concern is that μ may vary depending on whether the road surface is slippery or dry; e.g., consider a rainy day versus a dry day. For example, suppose that 20% variations in the coefficient μ are possible. Hence, the true coefficient of friction lies between 0.4 and 0.6 and the question at hand is whether the system will perform satisfactorily if $\mu = 0.5$ is assumed in the design. This is the issue of robustness with respect to uncertainty in the coefficient μ.

There are two important points to note about the discussion above. First, the uncertainty in μ illustrates only one of the many types of uncertainties one may encounter in a control theoretic context. In addition to structured real parametric uncertainty as illustrated by μ, one may encounter uncertainty due to factors such as unmodeled dynamics, delays and nonlinearities. These factors may also lead to a response of the system which may be radically different from the predictions of the mathematical model. The second point to note is that there is typically more than one uncertain parameter to be considered. For example, for the vehicle system above, one might entertain additional uncertainty in mass loading, various spring constants and damping coefficients.

1.11   Family Paradigm

Throughout this text, we work with bounds on uncertain quantities without assuming a statistical description of any sort. Within this framework, the notion of a *family* $\mathcal{F}$ is fundamental. For example, in the vehicle dynamics problem described above, we adopt the point of view that each admissible value of μ between 0.4 and 0.6 defines a different system. Hence, we have a *family of systems* $\mathcal{F}$ rather than a *fixed* system. In this book, the word "family" is used in a wide variety of contexts; e.g., we refer to a family of polynomials, a family of transfer functions or a family of matrices.

1.12   Robustness Analysis Paradigm

Given a family $\mathcal{F}$ and some property **P**, when we say that $\mathcal{F}$ is *robust*, we mean the following: Every member $f \in \mathcal{F}$ has property **P**. For example, we are often interested in a family of systems $\mathcal{F}$ with **P** being some aspect of performance. Then, when we say that the family $\mathcal{F}$ is *robust*, the understanding is that the performance specification is satisfied for every system in the family; i.e., for all $f \in \mathcal{F}$. There are numerous other possibilities for $\mathcal{F}$ and **P**. For example, $\mathcal{F}$ can be a family of polynomials and **P** might denote stability, or $\mathcal{F}$ might denote a family of transfer functions and **P** can be a specification on the frequency response.

One of the main objectives of this text is to develop machinery which can be used to ascertain whether or not a given family $\mathcal{F}$ is robust. In this regard, families of polynomials and robustness with respect to their root locations occupy much of our attention.

1.12.1   Uncertain System Versus Family of Systems

Throughout the robustness literature, the expression *uncertain system* is often used interchangeably with *family of systems*. For the sake of precision, however, this text makes a distinction between these two expressions. Namely, when we refer to an uncertain system, apriori bounds on uncertain quantities are *not* included. For example, recall the vehicle dynamics system in Section 1.10 with uncertain coefficient of friction μ. When we wish to discuss the model without specifying apriori bounds on μ, we say that we have an uncertain system. That is, we stop short of saying that we have a family of systems because without bounds on the uncertain parameter, our point of view is that the family is incompletely specified. In other words, an uncertain system plus apriori uncertainty bounds defines a family of systems. A similar distinction is made in a number of other situations; e.g., an uncertain polynomial plus uncertainty bounds defines a family of polynomials, or an uncertain transfer function plus uncertainty bounds defines a family of transfer functions.

1.13   Robustness Margin Paradigm

For cases when bounds on the uncertain quantities are not given, we often consider the so-called *Robustness Margin Problem*. The goal is to find the maximal uncertainty bounds under which the performance specification is satisfied. To illustrate, consider the vehicle dynamics problem of Section 1.10 and suppose we know that the coefficient of friction μ is uncertain but we cannot say definitively what variations in μ might be encountered. Now, if the performance specification is satisfied at $\mu = 0.5$, one can consider the following problem: Replace $\mu = 0.5$ by $\mu = 0.5 + \Delta\mu$ and determine how large $\Delta\mu$ can be while preserving satisfaction of the performance specification. This maximal value, call it r_{max}, is called the robustness margin.

EXERCISE 1.13.1 (Multiple Uncertainties):   Consider the vehicle dynamics problem discussed in Section 1.10. Now, however, suppose that there are two uncertain coefficients of friction μ_1 and μ_2. For example, imagine front and rear tires made of different materials. Propose various robustness margin definitions which make sense from an applications point of view.

EXERCISE 1.13.2 (Robustness Analysis Versus Robustness Margin): Provide an interpretation of the following informal statement: The robustness margin paradigm encompasses the robustness analysis paradigm.

EXERCISE 1.13.3 (Recursive Calculation of Robustness Margin): An engineer writes a computer program to solve a robustness analysis problem involving two uncertain friction parameters μ_1 and μ_2 satisfying $\mu_1^2 + \mu_2^2 \leq r$. The user specifies $r \geq 0$, and the program indicates whether or not robustness is guaranteed. The next day, the boss surprises the engineer by indicating that a robustness margin is demanded instead. Using the existing computer code, suggest a recursive procedure for finding the desired robustness margin.

1.14   Robust Synthesis Paradigm

Although this book deals primarily with robustness analysis problems and robustness margin problems, there are a few occasions where the new tools which we describe are readily applicable in a robust synthesis context. A distinguishing feature of the robust synthesis problem is the presence of adjustable design parameters which need to be selected. That is, the family of systems description is expanded to include design parameters which are chosen so as to guarantee that the subsequent robustness analysis succeeds.

From a control theoretic point of view, the adjustable design parameters above are associated with a compensator in a feedback system. In some cases, the number of such parameters is specified. For example, a PI controller $C(s) = K_1 + K_2/s$ is parameterized by the pair of gains (K_1, K_2). The robust design problem is to pick K_1 and K_2 so that performance specifications are met for all admissible values of the uncertain parameters.

More abstractly, suppose that we are given some desired performance specification **P**, a set $\mathcal{C}$ which we call the *class of admissible compensators* and a mapping taking each element $c \in \mathcal{C}$ to a family of systems $\mathcal{F}_c$. Then the *Robust Synthesis Problem* is to pick some $c^* \in \mathcal{C}$ such that every member $f \in \mathcal{F}_{c^*}$ satisfies the given performance specification **P**.

1.15   Testing Sets and Computational Complexity

One of the highlights of this text is the emphasis on *testing sets*. To convey the meaning of a testing set, we provide an example using the notation of Section 1.12. Indeed, let $\mathcal{F}$ be a given family and take **P** to be a desired property representing some robust performance specification. In many instances throughout this text, we identify a finite subset $\mathcal{F}^* = \{f_1^*, f_2^*, \ldots, f_p^*\}$ of $\mathcal{F}$ and prove a result of the following sort: *Property* **P** *holds for all* $f \in \mathcal{F}$ *if and only if Property* **P** *holds for* $f_1^*, f_2^*, \ldots, f_p^*$. When the number of distinguished f_i^* is small, such a result often implies a dramatic reduction in the computational complexity associated with the solution of the robustness problem at hand.

To illustrate the notion of testing sets in a more concrete way, imagine a system with performance specification **P** and two uncertain parameters q_1 and q_2 with bounds $|q_i| \leq 1$ for $i = 1, 2$. Associated with the uncertainty are four extreme combinations $(q_1, q_2) = (-1, -1)$, $(q_1, q_2) = (-1, 1)$, $(q_1, q_2) = (1, -1)$ and $(q_1, q_2) = (1, 1)$ and four associated systems which we designate as f_1^*, f_2^*, f_3^* and f_4^*. Now, from a computational point of view, it is of interest to know whether the set $\mathcal{F}^* = \{f_1^*, f_2^*, f_3^*, f_4^*\}$ serves as a testing set; i.e., testing for satisfaction of specification **P** for the extreme combinations (q_1, q_2) is both necessary and sufficient for robustness.

To conclude, we note that a testing set does not necessarily have to be finite. For example, if the extreme combinations do not suffice, it would be logical to consider the edges of the rectangle associated with the pair (q_1, q_2).

1.16 Conclusion

In summary, one of the primary objectives of this text is to develop robustness criteria which can be executed with "reasonable" computational effort. The first step in this direction is to develop a notational system which facilitates exposition of technical results. This is the focal point of the next chapter.

Notes and Related Literature

NRL 1.1 For parts of the problem tree not covered in this text, good starting references are Rugh (1981), Vidyasagar (1985), Isidori (1985) and Khalil (1992) for nonlinear systems, Astrőm and Wittenmark (1989) for adaptive systems and Kumar and Varaiya (1986) for stochastic systems. A good reference overviewing many aspects of robust control is the text by Weinmann (1991).

NRL 1.2 In Section 1.3, we do not mean to imply that the paper by Doyle, Glover, Khargonekar and Francis (1989) was the first to solve the H^∞ control problem. What distinguishes Doyle, Glover, Khargonekar and Francis (1989) from other work is the elegant mechanization of the solution via Riccati equations. In fact, earlier papers such as Chang and Pearson (1984) and Francis, Helton and Zames (1984) contain solutions which are less accessible to the control community because of their reliance on more abstract interpolation concepts.

NRL 1.3 Ideas which are central to the work of Soh, Berger and Dabke (1985) can be traced back to Fam and Meditch (1978).

NRL 1.4 At the level of multilinear uncertainty structures, impetus for much work along both value set lines and mathematical programming lines was provided by Saeki (1986) in his revival of the Mapping Theorem, which lay dormant since its 1963 publication in the book by Zadeh and Desoer (1963).

NRL 1.5 Formulation and solution of optimization problems obtained from the Hurwitz minors is found in the work of Sideris and Sanchez Pena (1989).

NRL 1.6 From a mathematical point of view, the value set corresponds to the range of a complex-valued function. For example, if $X \subset \mathbf{R}^n$ and $f : X \to \mathbf{C}$, then the range $f(X) = \{f(x) : x \in X\}$ is the set of complex values which can be assumed by f.

NRL 1.7 Value sets associated with rational functions are called *templates* in the original work of Horowitz; e.g., see Horowitz (1963) and Horowitz (1982).

NRL 1.8 The contributions of Dasgupta (1988) and Minnichelli, Anagnost and Desoer (1989) set the stage for a number of ideas in this text. These papers simply explain Kharitonov's Theorem in terms of value set rectangularity and motivate the following question: To what extent is the value set concept useful in the attainment of robustness results for more complicated uncertainty structures beyond those considered by Kharitonov?

NRL 1.9 One of the first demonstrations of the power of Kharitonov's Theorem is given in Barmish (1983); the result of Guiver and Bose (1983) is extended from quartics to polynomials of arbitrary degree.

NRL 1.10 Davison is often credited as being the first to regularly employ the word "robust" in the control literature; e.g., see Davison (1973).

NRL 1.11 In the sixties and seventies, we see a line of research attacking robustness problems via set propagation using the system dynamics. The textbook by Schweppe (1973) consolidates much of this literature. Examples of more work continuing along this line include the papers by Kurzhanskii (1980) and Fogel and Huang (1982).

NRL 1.12 A reconciliation between the mathematical programming and analytical approaches to robustness is rooted in the notion of a *tailored algorithm*. To explain this idea, we consider a robustness problem and evaluate two choices for its solution. *Choice 1*: Reformulate the problem in mathematical programming and "blindly" apply some standard software package hoping that some reasonable engineering solution is obtained. The word "blindly" is used above because apriori results about local versus global extrema are not established and the system structure (underlying to the optimization problem) is not exploited. *Choice 2*: Create a rather specialized mathematical programming code which explicitly takes advantage of system structure or analytical results that may only hold for some idealized version of the problem at hand. The next two notes elaborate on these choices.

NRL 1.13 To provide an example illustrating the distinction between Choices 1 and 2 above, we mention the theory in Chapter 14 dealing with systems having a multilinear uncertainty structure. Making Choice 1 in analyzing such systems amounts to application of a mathematical programming package without specifically exploiting the multilinear structure. On the other hand, if we are aware of the Mapping Theorem, we can exercise Choice 2; i.e., we develop a specialized code which exploits the convex hull approximation which the Mapping Theorem provides. This is what is meant by a *tailored algorithm*.

NRL 1.14 A second example illustrating the notion of a tailored algorithm arises in μ theory. By using convexity properties associated with bounds for the μ problem, one often obtains a better numerical algorithm than would be possible without the exploitation of this information. In summary, it is felt that one fruitful direction of future work involves the integration of the mathematical programming approach and the analytical approach. One incorporates analytical results into the algorithmic steps with the goal of improving the efficiency of computation.

Chapter 2

Notation for Uncertain Systems

Synopsis

This chapter introduces a rather minimal set of notation which is carried throughout the text. Examples are given to illustrate the process of reformulating robustness problems using the given notation.

2.1 Introduction

The primary focal point of this text is robustness problems involving real parametric uncertainty. Our objective in this chapter is to provide a notational system which facilitates mathematical manipulations involving functions of uncertain parameters. In most books on control theory, a transfer function $P(s)$ or a polynomial $p(s)$ has "s" as its only argument. Here, however, we use two argument functions; for example, we write $P(s, q)$ instead of $P(s)$ to emphasize dependence of a transfer function on a vector of uncertain parameters q. This is explained below.

2.2 Notation for Uncertain Parameters

We use the notation q to represent a vector of real *uncertain parameters* with i-th component q_i. We often refer to q simply as the *uncertainty*. If the uncertainty is ℓ-dimensional, it is often convenient to describe q by writing $q = (q_1, q_2, \ldots, q_\ell)$, whereas in other cases, we take q to be a column vector. In either event, we write $q \in \mathbf{R}^\ell$ and it is clear from the context whether an ℓ-tuple or a column vector is intended.

Throughout the text we encounter various *uncertain quantities* which depend on q. To emphasize the dependence on q, we include q as an argument of various functions of interest. For example, as mentioned above, to represent a transfer function with uncertain parameters, we write $P(s, q)$ instead of the usual $P(s)$. If numerator and denominator of this transfer function are of concern, we emphasize the dependence on q by writing

$$P(s, q) = \frac{N(s, q)}{D(s, q)},$$

where $N(s, q)$ and $D(s, q)$ are polynomials in s with coefficients which depend on q. In many cases, we break things down to an even finer level. For example, to denote dependence of the coefficients of $N(s, q)$ and $D(s, q)$ on q, we can write

$$N(s, q) = \sum_{i=0}^{m} a_i(q) s^i$$

and

$$D(s, q) = \sum_{i=0}^{n} b_i(q) s^i.$$

Literally dozens of additional examples illustrating q notation can be drawn from linear systems theory; e.g., if a linear system has a traditional state space representation $\dot{x}(t) = Ax(t)$, we can emphasize the dependence on q by writing

$$\dot{x}(t) = A(q)x(t).$$

Finally, note that we generally append the word "uncertain" to various quantities which depend on q. For example, we refer to an uncertain plant $P(s, q)$, an uncertain polynomial $N(s, q)$ or an uncertain matrix $A(q)$.

On some occasions, it is convenient to introduce a *second vector of uncertain parameters* r which is distinctly different from q. To illustrate, suppose that we wish to differentiate between uncertain parameters which enter the numerator of the plant versus the denominator. In such a situation, we write

$$P(s, q, r) = \frac{N(s, q)}{D(s, r)},$$

where $N(s, q)$ and $D(s, r)$ are uncertain polynomials.

2.3   Subscripts and Superscripts

When working with the uncertain parameter vectors q and r, it is often convenient to include zeroth components q_0 and r_0, respectively. For example, we might focus attention on the uncertain polynomial $p(s, q) = q_0 + q_1 s + q_2 s^2$ with $q \in \mathbf{R}^3$ or the uncertain plant

$$P(s, q, r) = \frac{\displaystyle\sum_{i=0}^{m} q_i s^i}{\displaystyle\sum_{i=0}^{n} r_i s^i}$$

with $q \in \mathbf{R}^{m+1}$ and $r \in \mathbf{R}^{n+1}$.

If we want to designate a vector as being distinguished in some sense, we use superscripts. For example, we might refer to the uncertain parameter vector q^* having i-th component q_i^*.

2.4   Uncertainty Bounding Sets and Norms

For robustness problems, we often assume an apriori bound Q for the vector of uncertain parameters q. We call Q the *uncertainty bounding set*. Motivated by classical engineering considerations, we generally take Q to be a ball in some appropriate norm—usually (but not necessarily) centered at $q = 0$. The two most important norms we consider are ℓ^∞ and ℓ^2. In the ℓ^∞ case, we consider the *max norm*

$$\|q\|_\infty = \max_i |q_i|.$$

We refer to a ball in this norm as a *box*. For example, to describe a box of unit radius with center q^*, we write $\|q - q^*\|_\infty \leq 1$. Often we want to describe such a box via componentwise bounds; e.g., consider

$$Q = \{q \in \mathbf{R}^\ell : q_i^- \leq q_i \leq q_i^+ \text{ for } i = 1, 2, \ldots, \ell\},$$

where q_i^- and q_i^+ are the specified bounds for the i-th component q_i of q.

For the ℓ^2 case, we consider the standard euclidean norm

$$\|q\|_2 = \left(\sum_{i=1}^{\ell} q_i^2 \right)^{\frac{1}{2}}.$$

Hence, a ball of unit radius and center q^* is described by the inequality $\|q - q^*\|_2 \leq 1$ and is referred to as a *sphere*. On a few occasions in this text, we exploit the ℓ^1 norm

$$\|q\|_1 = \sum_{i=1}^{\ell} |q_i|$$

and refer to a ball in this norm as a *diamond*. Analogous to the ℓ^∞ and ℓ^2 cases, the ball of unit radius and center q^* is described by $\|q - q^*\|_1 \leq 1$.

We can also consider weighted versions of any of the norms above. For example, if $w_1, w_2, \ldots, w_\ell$ are positive weights, then we can describe the unit ball in the associated euclidean norm by the inequality

$$\sum_{i=1}^{l} w_i^2 q_i^2 \leq 1$$

and refer to this set as an *ellipsoid*.

2.5   Notation for Families

An uncertain function together with its uncertainty bounding set is called a *family*. Throughout this text, we use a script letter to describe such a family. To illustrate, suppose that we are given an uncertain plant $P(s, q)$ and uncertainty bounding set Q. Then we denote the resulting family of plants by $\mathcal{P} = \{P(\cdot, q) : q \in Q\}$. If $P(s, q) = N(s, q)/D(s, q)$, where $N(s, q)$ and $D(s, q)$ are uncertain polynomials, then we can write $\mathcal{N} = \{N(\cdot, q) : q \in Q\}$ for the family of numerators and use the notation $\mathcal{D} = \{D(\cdot, q) : q \in Q\}$ for the family of denominators. One final example: If $A(q)$ is a matrix having entries which depend on q, we can use the notation $\mathcal{A} = \{A(q) : q \in Q\}$ to describe the resulting family of matrices.

2.6   Uncertain Functions Versus Families

It is important to make a distinction between uncertain functions and families. For example, we differentiate between the uncertain polynomial $p(s, q)$ and a family of polynomials $\mathcal{P} = \{p(\cdot, q) : q \in Q\}$. In other words, the uncertain polynomial $p(s, q)$ in combination with the uncertainty bounding set Q defines the family of polynomials $\mathcal{P}$.

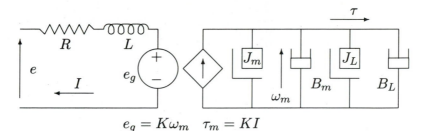

$$e_g = K\omega_m \quad \tau_m = KI$$

FIGURE 2.8.1   The Electromechanical Circuit for Example 2.8.1

2.7   Convention: Real Versus Complex Coefficients

Throughout this book, the standing assumption is that all polynomials have real coefficients. On those occasions when we wish to assume complex coefficients, we state this assumption explicitly. For example, we say, *let $p(s,q)$ be an uncertain polynomial with complex coefficients.* In this regard, our point of view is as follows: In most cases, the results given here are easily modified to obtain complex coefficient analogues. There are, however, some exceptions, and in addition, there are some results which hold for the complex case but not the real case—and vice versa.

2.8   Consolidation of Notation

We now provide examples and exercises illustrating the transcription of robustness problems into the notation of this text.

EXAMPLE 2.8.1 (Torque Control of a DC Motor):   To illustrate the use of q and Q notation, we consider a DC motor driving a viscously damped inertial load as described in Bailey, Panzer and Gu (1988); see Figure 2.8.1. The uncertain parameters in the model come from two sources. Namely, the motor constant K (in volts/rps) and load moment of inertia J_L (in kg-m^3) are imprecisely known; i.e., we take $0.2 \leq K \leq 0.6$ and $10^{-5} \leq J_L \leq 3 \times 10^{-5}$. Associated with the electromechanical circuit in the figure is the transfer function from the armature voltage to the load torque, which is given by

$$P(s) = \frac{K(J_L s + B_L)}{(Ls + R)(J_m s + J_L s + B_m + B_L) + K^2}.$$

Taking uncertain parameters $q_1 = K$ and $q_2 = J_L$ and fixed parameters $J_m = 2 \times 10^{-3}$ kg-m^3 for the motor moment of inertia, $B_m = 2 \times 10^{-5}$ N-m/rps for the motor damping, $L = 10^{-2}$ H for

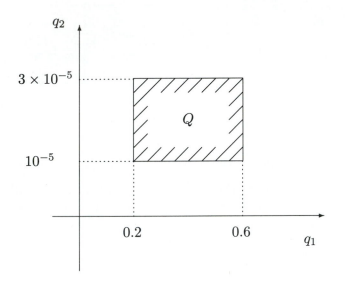

FIGURE 2.8.2   The Bounding Set Q for Example 2.8.1

the armature inductance, $R = 1\Omega$ for the armature resistance and $B_L = 2 \times 10^{-5}$ N-m/rps for the load damping, we obtain a family of transfer functions $\mathcal{P} = \{P(\cdot, q) : q \in Q\}$ described by

$$P(s,q) = \frac{0.5q_1 q_2 s + 10^{-5}q_1}{(10^{-5} + 0.005q_2)s^2 + (0.00102 + 0.5q_2)s + (2 \times 10^{-5} + 0.5q_1^2)}$$

and uncertainty bounds $0.2 \leq q_1 \leq 0.6$ and $10^{-5} \leq q_2 \leq 3 \times 10^{-5}$. In addition, note that the uncertainty bounding set Q is a rectangle in $\mathbf{R}^2$ as shown in Figure 2.8.2.

EXAMPLE 2.8.2 (Performance Specifications):   We consider a family of transfer functions $\mathcal{P} = \{P(\cdot, q) : q \in Q\}$ and turn our attention to the frequency response. For the uncertain transfer function $P(s, q)$, the *uncertain log gain* in decibels is given by

$$LG(\omega, q) = 20 \log |P(j\omega, q)|.$$

Now, a typical performance specification **P** might be as follows: Given a band of frequencies $[\omega_-, \omega_+]$ and two functions of frequency $LG_-(\omega)$ and $LG_+(\omega)$, we seek to guarantee that

$$LG_-(\omega) \leq LG(\omega, q) \leq LG_+(\omega)$$

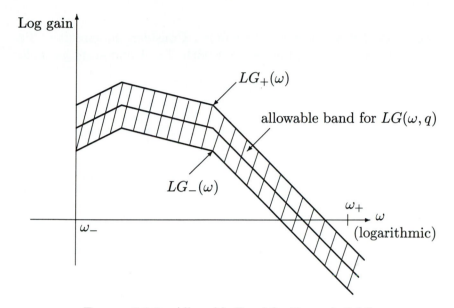

FIGURE 2.8.3   Allowable Band for Example 2.8.2

for all $q \in Q$ and all $\omega \in [\omega_-, \omega_+]$; see Figure 2.8.3. When these performance inequalities are met, we can say that the performance specification **P** *is robustly satisfied.*

EXERCISE 2.8.3 (State Variable Formulation):   Consider the uncertain state variable system described by

$$\dot{x}_1(t) = (1 + q_1)x_1(t) + (q_1 + q_2 + 4)x_2(t);$$
$$\dot{x}_2(t) = (3 + q_2)x_1(t) + (q_1 - q_2 + 3)x_2(t) + (2 + q_1)u(t)$$

with uncertainty bounds $|q_1| \leq 1$ and $|q_2| \leq 1$.
(a) Describe the uncertain matrix $A(q)$ and uncertain vector $b(q)$ associated with the state variable representation

$$\dot{x}(t) = A(q)x(t) + b(q)u(t).$$

(b) Consider the output

$$y(t) = c(q)x(t)$$

for this system. With $c(q) = [0 \quad 1 + q_2]$, find the uncertain transfer function

$$P(s, q) = c(q)[sI - A(q)]^{-1}b(q)$$

for the system.

EXERCISE 2.8.4 (Electrical Circuit):   Consider the circuit of Petersen (1988) shown in Figure 2.8.4 with $\beta = 1$ and state variables

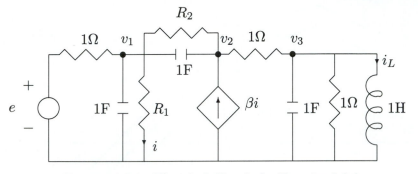

FIGURE 2.8.4   Electrical Circuit for Exercise 2.8.4

$x_1(t) = v_1(t)$, $x_2(t) = v_2(t) - v_1(t)$, $x_3(t) = v_3(t)$ and $x_4(t) = i_L(t)$. Generate state equations which include the two resistances as uncertain parameters. Use $q_1 = 1/R_1$ and $q_2 = 1/R_2$ and show that the uncertain polynomial

$$p(s,q) = s^4 + (3 - 2q_1 + q_2)s^3 + (4 - 7q_1 + 4q_2 - 4q_1q_2)s^2$$
$$+ (5 - 11q_1 - 6q_2 - 8q_1q_2)s + (1 - 3q_1 + 2q_2 - 4q_1q_2)$$

dictates the poles of the system.

EXERCISE 2.8.5 (The Concorde SST):   We consider a model for the Concorde supersonic transport plane given in Dorf (1974) and analyzed in a robustness context in Barmish and Tempo (1991). The plane is flying at MACH 3 with a delta wing configuration and canard surface near the nose. The objective here is to formulate a robustness analysis problem which accounts for uncertainty in the damping and natural frequency of the aircraft transfer function. In the absence of uncertainty, the flight control system is described by the block diagram in Figure 2.8.5. The parameters of the aircraft transfer function vary in accordance with the weight of the plane. We now consider a second order *aircraft transfer function*

$$P(s) = \frac{K_a(10s + 1)}{(s^2/\omega_n^2) + (2\xi s/\omega_n) + 1}$$

subject to the following uncertainty description: K_a varies from 0.02 at medium weight cruise to 0.20 at lightweight cruise with nominal

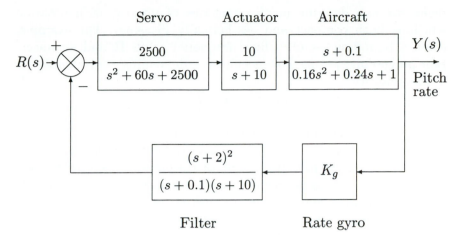

$$\text{FIGURE 2.8.5} \quad \text{Block Diagram for Exercise 2.8.5}$$

value $K_a = 0.1$, the natural frequency ω_n varies by $\pm 10\%$ about the nominal value $\omega_n = 2.5$ and the damping ξ varies by $\pm 10\%$ about the nominal value $\xi = 0.30$. The performance specification $\mathbf{P}$ is to have a closed loop system with a pair of dominant poles at natural frequency close to 2.5 rad/sec and damping close to 0.707.

(a) Find a setting for the rate gyro gain K_g which leads to satisfaction of the performance specification with ω_n, ξ and K_a fixed at their nominal values. Note that this is a simple root locus problem whose solution can be easily obtained using a standard software package.

(b) For the obvious choice $q_1 = K_a$, $q_2 = \xi$ and $q_3 = \omega_n$, compute the aircraft transfer function $P(s,q)$ and describe the uncertainty bounding set Q.

(c) For the less obvious choice $q_1 = K_a - 0.11$, $q_2 = 2\xi\omega_n - 1.5$ and $q_3 = \omega_n^2 - 6.25$, compute the uncertain aircraft transfer function $P(s,q)$ and describe the uncertainty bounding set Q.

(d) Argue that the resulting family of aircraft transfer functions is the same in parts (b) and (c) above.

(e) Is there any reason to prefer the parameterization given in part (c) to the one given in part (b)? Explain.

2.9   Conclusion

The notation in this chapter facilitates presentation of technical results in the sequel. Of particular importance is the fact that our polynomials and rational functions include the extra argument q to emphasize dependence on the uncertainty. Using this extra argument, we can describe many quantities of interest with notation which is both compact and aesthetically pleasing. For example, given a bounding set Q, a fixed frequency $\omega^* \in \mathbf{R}$ and an uncertain polynomial $p(s, q)$ with coefficients depending continuously on q, the quantity

$$\rho_{max} = \max_{q \in Q} |p(j\omega, q)|$$

represents the maximum attainable magnitude.

The next chapter contains no new theoretical concepts. Our primary purpose is to expose some of the issues which arise when setting up robustness problems in a real-world applications context. Hence, the reader who is interested primarily in theoretical fundamentals can proceed directly to Chapter 4 without concern for loss of continuity in the exposition.

Notes and Related Literature

NRL 2.1 The origins of the notational system in this text are rooted in the quadratic stabilizability literature of the seventies. For example, in the work of Leitmann (1979), uncertainty in a state space pair is denoted by writing simply $(\Delta A(r), \Delta B(s))$. To allow for cross coupling between the state and input matrices, the notation $(\Delta A(q), \Delta B(q))$ was introduced; e.g., see Barmish (1985).

NRL 2.2 Since the proof modifications for the complex coefficient case are usually straightforward, this text avoids double citations of the literature; i.e., one citation for the first paper to provide a result and a second citation for a minor extension to the complex case. In fact, a number of the routine extensions to the complex coefficient case are relegated to the exercises.

Chapter 3

Case Study: The Fiat Dedra Engine

Synopsis

To consolidate the concepts and notation of the two preceding chapters, we present a case study involving a model of a Fiat Dedra engine. We also see that a symbolic manipulation program can be quite useful in robustness applications.

3.1 Introduction

In this chapter, we present a case study with two objectives in mind: The first objective is to demonstrate how rather complicated uncertain polynomials and rational functions can arise from systems whose individual parts are quite simple. In contrast to the examples considered in the text thus far, a derivation of the uncertain polynomial governing stability involves so much algebra that only a masochist would attempt it by hand. With the aid of symbolic computation software, however, the long algebraic expressions are easily manipulated. We see that a symbolic manipulation software package such as MACSYMA, Mathematica or SMP can greatly facilitate algebraic manipulations. Our second objective for the case study is to provide an example to which we can refer in later chapters—an example motivated by a real application rather than an example derived from an academic "toy" model.

3.2   Control of the Fiat Dedra Engine

We consider a model for the Fiat Dedra engine given by Abate and di Nunzio (1990) and analyzed in a robustness context by Abate, Barmish, Murillo-Sanchez and Tempo (1992). The focal point is the idle speed control problem which is particularly important for city driving; that is, fuel economy depends strongly on engine performance when idling.

Two common methods to achieve low fuel consumption involve diminishing the idle velocity setpoint and augmenting the air-to-fuel (A:F) ratio. Unfortunately, these methods conflict with other performance criteria. For example, the engine is more likely to stall under torque disturbances when the A:F ratio is high — the so-called lean operating condition. Also, we note that there are internal delays as well as nonlinear effects in the engine which tend to get larger as the idle velocity setpoint diminishes. This hampers global stability of the system. In short, low fuel consumption optimization moves the engine to an operating point which is less stable. This, together with the fact that spark ignition engine models have many uncertain parameters (such as engine temperature, fuel composition, lubricant type and atmospheric pressure), serves as motivation to carry out robustness analysis.

3.3   Discussion of the Engine Model

This section can be skipped by the reader who is not interested in engineering details associated with the derivation of the engine model. In this discussion, we deal with an engine model which has been linearized about the idle operating point. Referring to Figure 3.3.1, the main control inputs are the spark advance $A(s)$ and the throttle valve opening (duty cycle) $D(s)$. The output variables are the manifold pressure $P(s)$ and the engine speed $N(s)$. In addition, the fuel injection command $f(s)$ is viewed as a reference input.

The manifold chamber has a control input $D(s)$ which regulates the incoming air-flow via a bypass valve. Neglecting any small leakage as well as the small fuel flow in the manifold chamber leads to a simple "filling dynamics" equation: The derivative of the manifold pressure $P(t)$ is proportional to a difference of two air mass flows; the incoming air mass flow is denoted by $\dot{M}_a(t)$ and the outgoing flow is denoted by $\dot{M}_b(t)$.

The constant of proportionality K_m depends on factors such as manifold volume, atmospheric pressure, air temperature, gas molec-

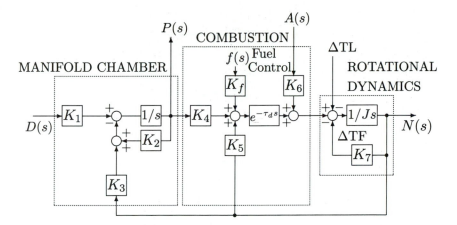

FIGURE 3.3.1   Block Diagram of Fiat Dedra Engine

ular weights and specific heat parameters. We have

$$\dot{P}(t) = K_m[\dot{M}_a(t) - \dot{M}_b(t)].$$

The incoming flow $\dot{M}_a(t)$ is regulated by $D(s)$, whereas the outgoing flow $\dot{M}_b(t)$ is taken to be proportional to the engine speed $N(s)$. This is justified if one thinks of the cylinders as exerting a continuous pumping action, and is actually referred to in mechanical jargon as a *pumping feedback*. Finally, the integration process relating $P(t)$ to $\dot{P}(t)$ is taken to have a time constant $1/K_2$ (dependent on K_m), and the model weights the effects of $D(t)$ and $N(t)$ on $\dot{M}_a(t)$ and $\dot{M}_b(t)$ by K_1 and K_3, respectively. We end up with the expression

$$P(s) = \frac{1}{s + K_2}[K_1 D(s) - K_3 N(s)].$$

As far as the combustion process is concerned, a rather simple torque production model is assumed. This is justified in large measure by the fact that rotational dynamics filter high-frequency effects from the combustion process. Thus, the variables $P(s)$, $N(s)$, $f(s)$ and $A(s)$ are assumed to contribute linearly to the output torque T_e according to the equation

$$T_e(s) = K_6 A(s) + e^{-s\tau_d}[K_4 P(s) + K_5 N(s) + K_f F(s)],$$

where K_4, K_5 and K_f are the constants of proportionality and $e^{-\tau_d s}$ represents the induction-to-power-stroke delay. This delay is associated with the fact that changes in the manifold get reflected in

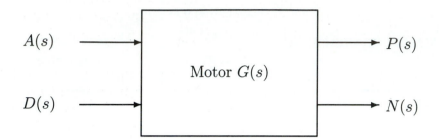

FIGURE 3.3.2   Two-Input/Two-Output Motor Configuration

the combustion chamber only after the cylinder goes from the intake stroke to the power stroke. An approximate expression for τ_d is given in Hazell and Flower (1971): It is

$$\tau_d = \frac{120}{n_c N},$$

where τ_d is the delay in seconds, n_c is the number of independently fired cylinders and N is the idle setpoint speed in rpm. In this case study, we concentrate on the "high idle" setpoint and consider the case when $\tau_d \to 0$. The remaining input in the combustion block is the spark advance; its action is viewed as instantaneous since it acts directly at the cylinders.

The rotational dynamics are described by the basic equation

$$N(s) = \frac{1}{Js + K_7}[T_e(s) - T_L(s)],$$

where J is the moment of inertia, $T_L(s)$ is an external torque load and K_7 is a viscous friction attenuation constant that depends on engine temperature, lubricant type and wear of the engine.

Finally, to complete this discussion, we note that the model assumes a constant air-to-fuel ratio fixed at the stoichiometric value. This leads to setting $K_f = 0$ and a final model with only two inputs $A(s)$ and $D(s)$ and two outputs $P(s)$ and $N(s)$; see Figure 3.3.2. Note that the four transfer functions associated with $G(s)$ are not given in explicit form. We now concentrate on this point.

3.4   Transfer Function Matrix for the Engine

The objective in this section is to derive a transfer function matrix representing the engine model. To this end, we write

$$\begin{bmatrix} P(s) \\ N(s) \end{bmatrix} = \begin{bmatrix} g_{11}(s) & g_{12}(s) \\ g_{21}(s) & g_{22}(s) \end{bmatrix} \begin{bmatrix} A(s) \\ D(s) \end{bmatrix}$$

and exploit Figure 3.3.1 to obtain the $g_{ij}(s)$ above. Indeed, four lengthy but straightforward applications of Mason's Rule leads to

$$g_{11}(s) = \frac{P(s)}{A(s)} \Big|_{D(s)=0}$$

$$= \frac{-K_3 K_6}{Js^2 + (JK_2 + K_7 - K_5)s + (K_3 K_4 + K_2 K_7 - K_2 K_5)};$$

$$g_{12}(s) = \frac{P(s)}{D(s)} \Big|_{A(s)=0}$$

$$= \frac{K_1[Js + (K_7 - K_5)]}{Js^2 + (JK_2 + K_7 - K_5)s + (K_3 K_4 + K_2 K_7 - K_2 K_5)};$$

$$g_{21}(s) = \frac{N(s)}{A(s)} \Big|_{D(s)=0}$$

$$= \frac{K_6(s + K_2)}{Js^2 + (JK_2 + K_7 - K_5)s + (K_3 K_4 + K_2 K_7 - K_2 K_5)};$$

$$g_{22}(s) = \frac{N(s)}{D(s)} \Big|_{A(s)=0}$$

$$= \frac{K_1 K_4}{Js^2 + (JK_2 + K_7 - K_5)s + (K_3 K_4 + K_2 K_7 - K_2 K_5)}.$$

For this engine model, the gains $K_1, K_2, \ldots, K_7$ and the inertia J are viewed as the uncertain parameters which affect the robustness of the system.

3.5   Uncertain Parameters in the Engine Model

To conform with our standard q notation for uncertain parameters, we take $q_1 = K_1$, $q_2 = K_2$, $q_3 = K_3$, $q_4 = K_4$, $q_5 = K_7 - K_5$, $q_6 = K_6$ and $q_7 = J$. We take advantage of the fact that K_5 and K_7 always enter $G(s)$ as a difference; this reduces the number of uncertain parameters. To summarize, the transfer function matrix of the engine is a 2×2 matrix $G(s, q)$ with entries

$$g_{11}(s, q) = \frac{-q_3 q_6}{q_7 s^2 + (q_2 q_7 + q_5)s + (q_3 q_4 + q_2 q_5)};$$

$$g_{12}(s, q) = \frac{q_1 (q_7 s + q_5)}{q_7 s^2 + (q_2 q_7 + q_5)s + (q_3 q_4 + q_2 q_5)};$$

$$g_{21}(s, q) = \frac{q_6 (s + q_2)}{q_7 s^2 + (q_2 q_7 + q_5)s + (q_3 q_4 + q_2 q_5)};$$

$$g_{22}(s, q) = \frac{q_1 q_4}{q_7 s^2 + (q_2 q_7 + q_5)s + (q_3 q_4 + q_2 q_5)}.$$

In accordance with Fiat specifications, there are various operating conditions of interest which can be expressed in terms of the uncertain parameters q_i. The description of these operating points is given in Chapter 10 when we return to this example to demonstrate robustness analysis techniques.

3.6   Discussion of the Controller Model

The controller used here was originally designed by linear quadratic methods; see Abate and Di Nunzio (1990). As seen in Figure 3.6.1, the controller has inputs $P_0(s) - P(s)$ and $N_0(s) - N(s)$, where $P(s)$ is the motor pressure, $P_0(s)$ is the reference for motor pressure, $N(s)$ is the motor speed and $N_0(s)$ is the reference for motor speed. The two outputs of the controller are taken to be the analog signals representing the spark advance a and the duty cycle d. Two salient features of the controller are as follows: The integral action on the speed error is used for steady-state purposes and the "derivator" in the spark advance path guarantees no steady-state offset. The elimination of this offset is a design requirement.

We describe the controller by a 2×2 transfer function model; i.e., we write

$$\begin{bmatrix} A(s) \\ D(s) \end{bmatrix} = H(s) \begin{bmatrix} p(s) \\ n(s) \end{bmatrix}.$$

To find the entries $h_{ij}(s)$ of $H(s)$, let k_{ij} denote the (i, j)-th entry of the 2×4 gain matrix K_C in Figure 3.6.1. From the block diagram,

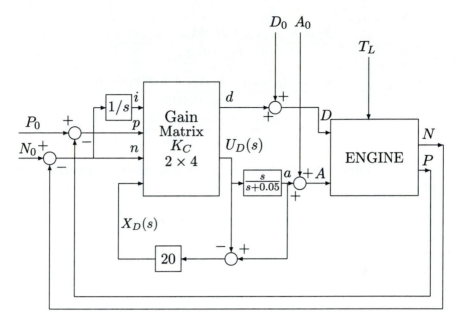

FIGURE 3.6.1   Controller for the Fiat Dedra Engine

we obtain the relationships

$$D(s) = \frac{k_{11}}{s}n(s) + k_{12}p(s) + k_{13}n(s) + k_{14}X_D(s);$$

$$U_D(s) = \frac{k_{21}}{s}n(s) + k_{22}p(s) + k_{23}n(s) + k_{24}X_D(s)$$

$$= \frac{s+0.05}{s}A(s);$$

$$X_D(s) = -\frac{1}{s}A(s).$$

We now solve for $A(s)$ and $D(s)$ in terms of $p(s)$ and $n(s)$. A lengthy but straightforward computation leads to transfer function matrix entries

$$h_{11}(s) = \frac{k_{22}s}{s+k_{24}+0.05};$$

$$h_{12}(s) = \frac{k_{23}s + k_{21}}{s+k_{24}+0.05};$$

$$h_{21}(s) = \frac{k_{12}s + (k_{12}k_{24} + 0.05k_{12} - k_{14}k_{22})}{s + k_{24} + 0.05};$$

$$h_{22}(s) = \frac{k_{13}s + (k_{13}k_{24} - k_{14}k_{23} + 0.05k_{13} + k_{11})}{s + k_{24} + 0.05}$$

$$+ \frac{(k_{11}k_{24} - k_{14}k_{21} + 0.05k_{11})}{s(s + k_{24} + 0.05)}.$$

Finally, to complete the description of the controller, we provide the gain matrix

$$K_C = \begin{bmatrix} 0.0081 & 0.1586 & 0.0872 & -0.1202 \\ 0.0187 & 0.0848 & 0.1826 & -0.0224 \end{bmatrix}.$$

Now, by substituting for the k_{ij} above, we obtain the entries $h_{ij}(s)$ in the controller transfer function matrix $H(s)$. We obtain

$$h_{11}(s) = \frac{0.0848s}{s + 0.0276};$$

$$h_{12}(s) = \frac{0.1826s + 0.0187}{s + 0.0276};$$

$$h_{21}(s) = \frac{0.1586s + 0.0145703}{s + 0.0276};$$

$$h_{22}(s) = \frac{0.0872s^2 + 0.0324552s + 0.0024713}{s^2 + 0.0276s}.$$

3.7   The Closed Loop Polynomial

From linear systems theory (for example, see Chen (1984)), closed loop stability considerations lead us to study the numerator of the uncertain rational function

$$R(s, q) = \det[I + G(s, q)H(s)].$$

Now, using the expressions found for $g_{ij}(s, q)$ and $h_{ij}(s)$ above, a lengthy computation is required to obtain the desired closed loop polynomial. This is readily accomplished via symbolic computation. The end result is the desired numerator polynomial written in the form $p(s,q) = \sum_{i=0}^{7} a_i(q)s^i$. For the reader interested in verification, the extremely lengthy formulas for the $a_i(q)$ are given in Appendix A. The conclusion to be drawn from the appendix is that without a symbolic computation package, the task of generating a closed form for $p(s,q)$ would be monumental.

3.8   Is Symbolic Computation Really Needed?

As demonstrated above and in examples and exercises in Chapter 2, the uncertain functions which we wish to study are often derived from more basic uncertainty descriptions. For example, when a state variable system is described via an uncertain matrix $A(q)$, then calculation of an explicit expression for the characteristic polynomial $p(s,q) = \det(sI - A(q))$ involves considerable algebra. In fact, for practical applications of interest, the derivation of explicit expressions for the required uncertain polynomials can involve literally hundreds of terms—often thousands. A dramatic illustration of this point is provided by our case study involving a Fiat Dedra engine. As seen in Appendix A, the uncertain polynomial of interest is quite complicated.

This leads us to consider the following question: In order to carry out a robustness analysis, do we really need to obtain explicit expressions for the uncertain functions of interest? For example, in analyzing the stability of a family of matrices $\{A(q) : q \in Q\}$ via its uncertain characteristic polynomial $p(s,q)$, do we really need to carry out the determinant calculations symbolically in (s,q)? The reader who is unconvinced about the importance of this question is urged to consider an uncertain 4×4 matrix $A(q)$ with a simple uncertainty structure—say each entry of $A(q)$ depends linearly on three uncertain parameters q_1, q_2 and q_3 and the objective is to derive a closed form for the characteristic polynomial.

In principle, the application of the robustness tools in the chapters to follow do not require explicit expressions for the uncertain functions of interest. What matters is the ability to carry out repeated evaluations. To illustrate, suppose that we are interested in performing a robustness analysis involving the $n \times n$ uncertain matrix $A(q)$ and the tool being applied requires evaluation of the characteristic polynomial $p(s,q) = \det(sI - A(q))$ for

$$(s,q) = (s_1, q^1), (s_2, q^2), \ldots, (s_N, q^N).$$

Notice that evaluation of $p(s,q)$ for $(s,q) = (s_i, q^i)$ does not necessarily imply that an explicit expression for $p(s,q)$ is needed. For

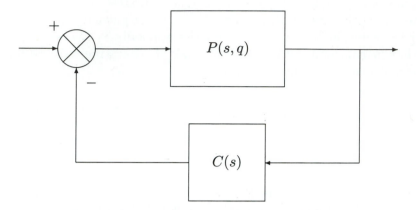

FIGURE 3.9.1   Standard Feedback Configuration

each (s_i, q^i) combination, one can compute det $(s_i I - A(q^i))$ without going through the intermediate step of symbolically computing the coefficients of $p(s, q)$. This argument, however, should not be erroneously construed to mean that a symbolic manipulation package is not helpful. It may well be the case that the required determinant calculations above are "expensive" for certain combinations of n and N. For such cases, an overall computational saving might result by working with a symbolic representation for $p(s, q)$. In other words, even if the initial computational cost of symbolic evaluation is high, the savings accrued when evaluating $p(s_i, q^i)$ may render the overall effort worthwhile.

3.9   Symbolic Computation with Transfer Functions

Recall that the starting point in many robustness problems is an uncertain plant described by $P(s, q) = N(s, q)/D(s, q)$, where $N(s, q)$ and $D(s, q)$ are uncertain polynomials. Suppose that this plant is connected in a feedback configuration with a compensator denoted by $C(s) = N_C(s)/D_C(s)$, where $N_C(s)$ and $D_C(s)$ are fixed polynomials; see Figure 3.9.1. Now, in order to provide a closed form for the closed loop transfer function

$$P_{CL}(s,q) = \frac{P(s,q)}{1 + P(s,q)C(s)} = \frac{N(s,q)D_C(s)}{N(s,q)N_C(s) + D(s,q)D_C(s)},$$

a symbolic computation in (s, q) is required. In some cases, only part of $P_{CL}(s, q)$ might need to be computed. For example, when studying stability of the feedback system above, the uncertain polynomial

$$p(s, q) = N(s, q)N_C(s) + D(s, q)D_C(s)$$

is the symbolically computed quantity of interest.

EXERCISE 3.9.1 (From Control System to Polynomial):   Recall the torque control problem for the DC motor in Example 2.8.1 and consider a controller given by $C(s) = K_1 + K_2/s$ and numerical values given by $J_m = 2 \times 10^{-3}$, $B_m = 2 \times 10^{-5}$, $L = 10^{-2}$, $R = 1$ and $B_L = 2 \times 10^{-5}$. Obtain explicit formulas for the coefficients of the uncertain polynomial $p(s, q)$ which determines the closed loop poles. Express this uncertain polynomial in the form

$$p(s, q) = a_2(q)s^2 + a_1(q)s + a_0(q)$$

and note that the formulas for the $a_i(q)$ are parameterized in the two gains K_1 and K_2.

3.10   Conclusion

This chapter completes Part I of this text. We are now well prepared to proceed toward the attainment of technical results. After introducing the notion of robust stability in Chapter 4, we reach the first mountaintop in Chapter 5—Kharitonov's seminal theorem for robust stability of interval polynomials. Through understanding of the key ideas associated with Kharitonov's simple but elegant interval polynomial framework, it becomes possible to introduce a number of important technical ideas which are essential in the later chapters.

Notes and Related Literature

NRL 3.1 A more detailed discussion on the derivation of the engine model is given by Powell (1979) and Dobner (1980) where much emphasis is placed on obtaining a complete nonlinear formulation. Work aimed at finding a more simple linear model can be found in the papers by Dobner and Fruechte (1983) and Powell, Cook and Grizzle (1987).

NRL 3.2 The problems encountered in idle speed engine control are explained more extensively by Washino, Nishiyama and Ohkubo (1986), Nishimura and Ishii (1986), Yamagushi, Takizawa, Sanbuichi and Ikeura (1986), and Ando and Motomochi (1987).

NRL 3.3 The paper by Olbrot and Powell (1989) provides a nice analysis of the stability problem associated with idle speed control.

Part II

From Robust Stability to the Value Set

Chapter 4

Robust Stability with a Single Parameter

Synopsis

This chapter introduces the robust stability problem and provides some basic results for a family of polynomials or matrices whose description involves only one uncertain parameter. The highlight of the chapter is the eigenvalue criterion of Bialas.

4.1 Stability and Robust Stability

In this chapter, we begin exposition of technical results by concentrating on the robust stability problem. We consider the special case obtained with only one uncertain parameter. There are a number of reasons for restricting attention to this specialized one-parameter setting before proceeding to the general case. First, from a pedagogical point of view, it is felt that dealing with only one parameter facilitates understanding of the more general analysis to follow—common sense takes us a long way when only a single-parameter is involved. Second, for the case of a single parameter, we often obtain stronger results than in a more general setting. Finally, perhaps the most important reason is the fact that there is an important class of robust stability problems with many uncertain parameters which can be reduced to the single parameter case. For example, this class arises when the uncertainty bounding set Q is a box and the coefficients of an uncertain polynomial $p(s, q)$ depend affine linearly on q.

Under these conditions, the Edge Theorem (see Chapter 9) enables us to reduce the multiple-parameter problem to a finite number of single-parameter problems.

4.2   Basic Definitions and Examples

For the sake of completeness, we now provide two basic definitions.

DEFINITION 4.2.1 (Stability):   A fixed polynomial $p(s)$ is said to be *stable* if all its roots lie in the strict left half plane.

DEFINITION 4.2.2 (Robust Stability):   A given family of polynomials $\mathcal{P} = \{p(\cdot, q) : q \in Q\}$ is said to be *robustly stable* if, for all $q \in Q$, $p(s, q)$ is stable; that is, for all $q \in Q$, all roots of $p(s, q)$ lie in the strict left half plane.

EXAMPLE 4.2.3 (Commonsense Analysis):   A family of first order plants, characterized by uncertainty in the location of a simple pole, is described by $P(s, q) = 1/(s - q)$ and uncertainty bounding $|q| \leq 2$. When the system is compensated with a unity feedback $C(s) = 1$, we obtain the closed loop polynomial $p(s, q) = s + 1 - q$. For this system, $p(s, q)$ has a single root $s_1(q) = -1 + q$. Clearly, the resulting family of polynomials $\mathcal{P}$ is *not* robustly stable because for $q \geq 1$, $s_1(q)$ lies in the right half plane. More generally, with uncertainty bound $|q| \leq r$, it is easy to see that $\mathcal{P}$ is robustly stable if and only if $r < 1$.

EXAMPLE 4.2.4 (Slightly More Complicated):   For the second order uncertain polynomial $p(s, q) = s^2 + (2 - q)s + (3 - q)$, coefficient positivity considerations lead us to conclude that robust stability is guaranteed if and only if the uncertainty bounding set Q is contained in $(-\infty, 2)$. For illustrative purposes, we also study robust stability using the quadratic formula. Indeed, we obtain two roots,

$$s_{1,2}(q) = \begin{cases} (-1 + q/2) \pm j(\sqrt{8 - q^2}/2) & \text{if } 0 \leq q \leq 2\sqrt{2}; \\ (-1 + q/2) \pm (\sqrt{q^2 - 8}/2) & \text{if } 2\sqrt{2} \leq q \leq 4, \end{cases}$$

which are plotted for $Q = (0, 4]$ in Figure 4.2.1. From the figure, it is obvious that the resulting family of polynomials $\mathcal{P} = \{p(\cdot, q) : q \in Q\}$ is not robustly stable.

EXERCISE 4.2.5 (Robustness Margin):   Consider the uncertain

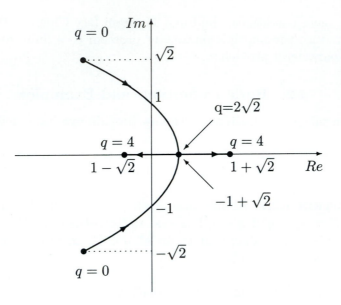

FIGURE 4.2.1   Root Locations for Example 4.2.4

polynomial

$$p(s, q) = s^3 + (2 - q)s^2 + (3 - q)s + 4$$

without any specified uncertainty bound for q. Letting $Q_r = [-r, r]$, show that the robustness margin

$$r_{max} = \sup\{r : p(s, q) \text{ is stable for all } q \in Q_r\}$$

is given by $r_{max} \approx 0.43$.

4.3   Root Locus Analysis

If $p(s, q)$ is an uncertain polynomial with a single uncertain parameter entering affine linearly into the coefficients, we see below that the robust stability problem can be analyzed by classical root locus and Nyquist methods. The key idea, introduced in many undergraduate texts, involves creation of a *fictitious plant*.

EXAMPLE 4.3.1 (Root Locus):   We consider the uncertain polynomial $p(s, q) = s^2 + (2 - q)s + (3 - q)$ in Example 4.2.4 and notice that $p(s, q)$ can be written as $p(s, q) = (s^2 + 2s + 3) - q(s + 1)$. Hence,

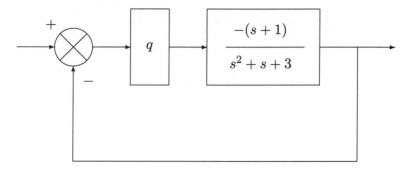

FIGURE 4.3.1   Unity Feedback System for Example 4.3.1

if we consider a *fictitious plant*

$$P(s) = -\frac{s+1}{s^2 + 2s + 3}$$

with gain q and unity feedback as shown in Figure 4.3.1, robust stability analysis is accomplished by generation of a classical root locus. If the uncertainty bounding set Q is an interval $[q^-, q^+]$, then we restrict out attention to the portion of the root locus corresponding to gain q between q^- and q^+. Note that this unity feedback system is fictitious in the following sense: The underlying system which gives rise to $p(s, q)$ may be quite different from the one depicted in Figure 4.3.1. For example, the polynomial $p(s, q)$ above arises when analyzing stability of the unity feedback system with uncertain plant

$$P(s, q) = \frac{(2-q)s}{s^2 + (3-q)}.$$

Notice that the fictitious plant $P(s)$ does not correspond to an evaluation $P(s, q)$ for some specific $q \in Q$.

4.4   Generalization of Root Locus

We now generalize the root locus ideas introduced via an example in the preceding section. Namely, suppose that $p(s, q)$ has degree n for all $q \in Q$ and uncertain coefficients which are affine linear functions of a single uncertain parameter q; i.e., the coefficient of s^i in $p(s, q)$ has the form

$$a_i(q) = \alpha_i q + \beta_i,$$

where α_i and β_i are real. In this section and the next, we view the degree requirement on $p(s, q)$ as a mathematical condition; further

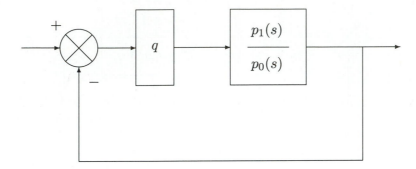

FIGURE 4.4.1   A Unity Feedback System for the Fictitious Plant

interpretation of this "invariant degree" condition is relegated to Section 4.6.

In view of the assumed form for $a_i(q)$, the uncertain polynomial $p(s, q)$ admits a decomposition of the form

$$p(s, q) = p_0(s) + q p_1(s),$$

where $p_0(s)$ and $p_1(s)$ are fixed polynomials. We can now study robust stability using a root locus plot for the fictitious plant

$$P(s) = \frac{p_1(s)}{p_0(s)}$$

with unity feedback and gain q; see Figure 4.4.1. Clearly, a necessary and sufficient condition for robust stability is that the distinguished portion of the root locus corresponding to $q \in Q$ remains within the strict left half plane.

For the more general case involving more complicated coefficient dependence on q, we can still view the robust stability problem in a root locus context; i.e., while sweeping q between q^- and q^+, one computes roots $s_1(q), s_2(q), \ldots, s_n(q)$ of $p(s, q)$. Again, a necessary and sufficient condition for robust stability is that each of these root branches remains within the strict left half plane. Notice, however, that this interpretation is not as nice as the one obtained with affine linear coefficient dependence on q; i.e., if the dependence on q is nonlinear, there is no feedback system associated with the problem and the classical rules of thumb for root locus generation do not apply.

4.5   Nyquist Analysis

We now describe an approach to robust stability analysis based on Nyquist plots rather than root locus plots. As in the preceding section, we consider the family of polynomials $\mathcal{P}$ which is described by $p(s,q) = p_0(s) + qp_1(s)$ and $Q = [q^-, q^+]$. For simplicity, we take $q^- = 1$ and $q^+ = 1$ and make the apriori assumptions that $p_0(s)$ is stable and $p(s,q)$ has degree n for all $q \in Q$. Since stability of $p_0(s)$ is necessary for robust stability of $\mathcal{P}$, the stability condition we imposed is nonrestrictive; recall that the discussion of the fixed degree requirement on $p(s,q)$ is relegated to the next section.

EXERCISE 4.5.1 (Relationship with Nyquist Plot): With the setup and notation above, consider the fictitious plant $P(s) = p_1(s)/p_0(s)$ and argue that the family of polynomials $\mathcal{P}$ is robustly stable if and only if

$$P(j\omega) \notin (-\infty, -1] \cup [1, \infty)$$

for all frequencies $\omega \in \mathbf{R}$. In other words, robust stability of $\mathcal{P}$ is equivalent to an appropriately constructed Nyquist plot not intersecting a forbidden portion of the real axis. Now, derive a similar result for the general uncertainty bounding set $Q = [q^-, q^+]$.

EXERCISE 4.5.2 (Root Locus and Nyquist): For the uncertain plant

$$P(s,q) = \frac{s^2 + (4+q)s + (3+q)}{s^4 + (3+q)s^3 + (5+q)s^2 + (2+q)s + 4}$$

with unity feedback compensator and uncertainty bound $|q| \leq 1$, analyze robust stability by generating appropriate root locus and Nyquist plots. Verify that both methods lead to the same conclusion. Subsequently, consider a variable uncertainty bound $Q_r = [-r, r]$ and use your graphical output to compute a robustness margin

$$r_{max} = \sup\{r : p(s,q) \text{ is stable for all } q \in Q_r\}$$

for the closed loop system.

4.6   The Invariant Degree Concept

In the preceding two sections, we assumed that $p(s,q)$ has degree n for all $q \in Q$. Similarly, in almost all chapters to follow, we impose an "invariant degree" assumption whenever convenient. The objective of this section is to demonstrate that such an assumption is rather benign. In a feedback control context, we see that an invariant degree assumption amounts to appropriate properness of the loop function $P(s,q)C(s)$. We also explain the technical reason for imposition of invariant degree conditions.

DEFINITION 4.6.1 (Invariant Degree):   A family of polynomials given by $\mathcal{P} = \{p(\cdot, q) : q \in Q\}$ is said to have *invariant degree* if the following condition holds: Given any $q^1, q^2 \in Q$, it follows that

$$\deg p(s, q^1) = \deg p(s, q^2).$$

If, for all $q \in Q$, $\deg p(s,q) = n$, then we call $\mathcal{P}$ a *family of n-th order polynomials*. Finally, if $\mathcal{P}$ does not have invariant degree, we say that *degree dropping* occurs.

REMARKS 4.6.2 (Highest Order Coefficient):   If we begin with the uncertain polynomial $p(s,q) = \sum_{i=0}^{n} a_i(q)s^i$, notice that $\mathcal{P}$ has invariant degree if and only if $a_n(q) \neq 0$ for all $q \in Q$. We see below that the invariant degree assumption is intimately related to properness of feedback control problems.

EXERCISE 4.6.3 (Invariant Degree in Feedback Systems):   Consider a family of plants $\mathcal{P}$ described by

$$P(s,q) = \frac{N(s,q)}{D(s,q)}$$

with uncertainy bound $q \in Q$ and $N(s,q)$ and $D(s,q)$ being uncertain polynomials with $D(s,q)$ monic. Now, given a compensator

$$C(s) = \frac{N_C(s)}{D_C(s)}$$

which is connected in a feedback configuration, prove that if the uncertain loop function $P(s,q)C(s)$ is strictly proper, then the resulting family of closed loop polynomials described by

$$p(s,q) = N(s,q)N_C(s) + D(s,q)D_C(s)$$

and $q \in Q$ has invariant degree. For the case when $P(s,q)$ and $C(s)$ are both proper but not strictly proper, and $D(s,q)$ is not necessarily monic, let $a_n(q)$, $b_n(q)$, c_m and d_m denote the highest order coefficients of $N(s,q)$, $D(s,q)$, $N_C(s)$ and $D_C(s)$, respectively.

Now, assuming $a_n(q)$ and $b_n(q)$ depend continuously on q and Q is closed and bounded, argue that the resulting family of closed loop polynomials has invariant degree if and only if

$$\min_{q \in Q} |a_n(q)c_m + b_n(q)d_m| > 0.$$

REMARKS 4.6.4 (Technical Role of Invariant Degree):   We consider an uncertain polynomial $p(s, q)$, and, for illustrative purposes, suppose that $Q = [0, 1]$ is the uncertainty bounding set. Assuming $p(s, 0)$ is stable, we imagine q increasing from 0 to 1 and obtain a root locus which begins in the strict left half plane at the roots of $p(s, 0)$. If no degree dropping occurs, the number of branches of this root locus remains constant. Hence, the only way that we can experience a transition from stability to instability is by having one or more branches cross the imaginary axis. On the other hand, if $p(s, q)$ experiences degree dropping, then as q varies from 0 to 1, the branches of the root locus can "leapfrog" from the strict left half plane into the strict right half plane without crossing the imaginary axis. This phenomenon is illustrated below.

EXAMPLE 4.6.5 (Leapfrogging):   Consider the uncertain polynomial $p(s, q) = qs^2 - s - 1$ with uncertainty bounding set $Q = [0, 1]$. Observe that $p(s, 0) = -s - 1$ has a single left half plane root $s = -1$ and $p(s, 1) = s^2 - s - 1$ has a strict right half plane root at $s = (1 + \sqrt{5})/2$. However, we claim that this family of polynomials has no imaginary axis roots. This is apparent by observing that with $s = j\omega$, we have $p(j\omega, q) = -1 - q\omega^2 - j\omega \neq 0$ for all $q \in Q$ and all $\omega \in \mathbf{R}$; i.e., $p(j\omega, q)$ is nonvanishing on the imaginary axis.

EXERCISE 4.6.6 (Robustness Margin with Degree Dropping):   To further demonstrate the subtleties involved in robust stability analysis with degree dropping, consider the uncertain polynomial given by $p(s, q) = (2 + q_1)s^2 + (5 + q_1 q_2)s + (3 + q_1 + q_2)$ with uncertainty bounding set Q_r described by $|q_i| \leq r$ for $i = 1, 2$. Letting the *robustness margin for degree k* be given by

$$r_{max,k} = \sup\{r : \deg p(s, q) = k \text{ for all } q \in Q_r\},$$

find $r_{max,0}$, $r_{max,1}$ and $r_{max,2}$.

4.7   Eigenvalue Criteria for Robustness

In this section and the next four, we concentrate on technical ideas due to Bialas (1985) with refinements due to Fu and Barmish (1988). Indeed, we consider the uncertain polynomial $p(s, q) = p_0(s) + qp_1(s)$ with an uncertainty bounding set $Q = [q^-, q^+]$. Now, however, instead of using root locus or Nyquist analysis as in Sections 4.3 and 4.5, we provide a solution to the robust stability and robustness margin problems which is more in the spirit of a closed form. To this end, we use the coefficients of $p_0(s)$ and $p_1(s)$ to construct a special matrix whose eigenvalues tell us what we want to know about robust stability. First, however, we require some definitions.

DEFINITION 4.7.1 (Subfamilies):   Consider the uncertain polynomial $p(s, q) = p_0(s) + qp_1(s)$ with $p_0(s)$ assumed stable and the uncertainty bounding set $Q = [q^-, q^+]$ with $q^- \leq 0$ and $q^+ \geq 0$. We define the *subfamilies*

$$\mathcal{P}(q^+) = \{p(\cdot, q) : 0 \leq q \leq q^+\}$$

and

$$\mathcal{P}(q^-) = \{p(\cdot, q) : q^- \leq q \leq 0\}$$

of the original polynomial family $\mathcal{P} = \{p(\cdot, q) : q \in Q\}$.

DEFINITION 4.7.2 (Maximal Stability Interval):   Associated with the subfamily $\mathcal{P}(q^+)$ is the *right-sided robustness margin*

$$q^+_{max} = \sup\{q^+ : \mathcal{P}(q^+) \text{ is robustly stable}\},$$

and associated with the subfamily $\mathcal{P}(q^-)$ is the *left-sided robustness margin*

$$q^-_{min} = \inf\{q^- : \mathcal{P}(q^-) \text{ is robustly stable}\}.$$

Subsequently, we call

$$Q_{max} = (q^-_{min}, q^+_{max})$$

the *maximal interval* for robust stability.

DEFINITION 4.7.3 (The Hurwitz Matrix):   For a fixed polynomial

$$p(s) = a_n s^n + a_{n-1} s^{n-1} + \cdots + a_1 s + a_0$$

with $a_n > 0$, the $n \times n$ array

$$H(p) = \begin{bmatrix} a_{n-1} & a_{n-3} & a_{n-5} & \cdots & & \cdots \\ a_n & a_{n-2} & a_{n-4} & \cdots & & \cdots \\ 0 & a_{n-1} & a_{n-3} & a_{n-5} & \cdots & \\ 0 & a_n & a_{n-2} & a_{n-4} & \cdots & \\ 0 & 0 & 0 & a_n & & \ddots \\ 0 & 0 & 0 & 0 & \cdots & a_0 \end{bmatrix}$$

is called the *Hurwitz Matrix* associated with $p(s)$.

REMARKS 4.7.4 (Hurwitz Stability Criterion):   We recall the classical Hurwitz stability criterion: A polynomial $p(s)$ above is stable if and only if all principal minors of $H(p)$ are positive. For example, the first principal minor is $\Delta_1 = a_{n-1}$, the second principal minor is

$$\Delta_2 = \det \begin{bmatrix} a_{n-1} & a_{n-3} \\ a_n & a_{n-2} \end{bmatrix}$$

and the last principal minor is $\Delta_n = \det H(p)$.

DEFINITION 4.7.5 ($\lambda_{max}^+(M)$ and $\lambda_{min}^-(M)$):   Given an $n \times n$ matrix M, we define $\lambda_{max}^+(M)$ to be the maximum positive real eigenvalue of M. When M does not have any positive real eigenvalues, we take $\lambda_{max}^+(M) = 0^+$. Similarly, we define $\lambda_{min}^-(M)$ to be the minimum negative real eigenvalue of M. When M does not have any negative real eigenvalues, we take $\lambda_{min}^-(M) = 0^-$.

THEOREM 4.7.6 (Eigenvalue Criterion):   *Consider the uncertain polynomial $p(s,q) = p_0(s) + qp_1(s)$ with $p(s,0) = p_0(s)$ stable and having positive coefficients and* $\deg p_0(s) > \deg p_1(s)$. *Then the maximal interval for robust stability is described by*

$$q_{max}^+ = \frac{1}{\lambda_{max}^+(-H^{-1}(p_0)H(p_1))}$$

and

$$q_{min}^- = \frac{1}{\lambda_{min}^-(-H^{-1}(p_0)H(p_1))},$$

where, for the purpose of conformability of matrix multiplication, $H(p_1)$ is an $n \times n$ matrix obtained by treating $p_1(s)$ as an n-th order polynomial.

4.8   Machinery for the Proof

The proof of Theorem 4.7.6 is established with the aid of three technical lemmas. No proof is given for the first two of these lemmas because they are standard results which can be found in books such as Gantmacher (1959) and Marden (1966). The first lemma provides Orlando's formula and the second lemma is the classical result on continuous dependence of the roots of a polynomial on parameters. This continuity result is used many times in later chapters—both explicitly and implicitly.

LEMMA 4.8.1 (Orlando's Formula): *Consider a fixed polynomial*

$$p(s) = a_n s^n + a_{n-1} s^{n-1} + \cdots + a_1 s + a_0$$

with $a_n > 0$, roots $s_1, s_2, \ldots, s_n$ and Hurwitz matrix $H(p)$. Then

$$det\ H(p) = (-1)^{n(n-1)/2} a_n^{n-1} a_0 \prod_{1 \leq i < k \leq n} (s_i + s_k).$$

LEMMA 4.8.2 (Continuous Root Dependence): *Consider the family of polynomials $\mathcal{P}$ described by $p(s,q) = \sum_{i=0}^{n} a_i(q) s^i$ and $q \in Q$. Assume that $\mathcal{P}$ has invariant degree and coefficient functions $a_0(q), a_1(q), a_2(q), \ldots, a_n(q)$ which depend continuously on q. Then the roots of $p(s,q)$ vary continuously with respect to $q \in Q$. That is, there exist continuous mappings $s_i : Q \to \mathbf{C}$ for $i = 1, 2, \ldots, n$ such that $s_1(q), s_2(q), \ldots, s_n(q)$ are the roots of $p(s,q)$.*

LEMMA 4.8.3 (Nonsingularity Invariance): *Consider the family of polynomials $\mathcal{P}$ described by $p(s,q) = p_0(s) + q p_1(s)$ and uncertainty bounding set $Q = [q^-, q^+]$. Assume that $p_0(s)$ has positive coefficients, is stable and $\deg p_0(s) > \deg p_1(s)$. Then the subfamily $\mathcal{P}(q^+)$ is robustly stable if and only if $H(p(\cdot, q))$ is nonsingular for all $q \in [0, q^+]$. Similarly, the subfamily $\mathcal{P}(q^-)$ is robustly stable if and only if $H(p(\cdot, q))$ is nonsingular for all $q \in [q^-, 0]$.*

PROOF: We establish the result for $\mathcal{P}(q^+)$ and note that the proof for $\mathcal{P}(q^-)$ is identical. Proceeding with necessity, we assume that $\mathcal{P}(q^+)$ is robustly stable and simply observe that nonsingularity of

$H(p(\cdot, q))$ for all $q \in [0, q^+]$ follows from the classical Hurwitz Stability Criterion; see Remarks 4.7.4. That is, the last principal minor is the determinant of $H(p(\cdot, q))$, which must be nonvanishing.

To establish sufficiency, we assume that $H(p(\cdot, q))$ is nonsingular for all $q \in [0, q^+]$ and must prove that $\mathcal{P}(q^+)$ is robustly stable. Proceeding by contradiction, suppose $p(s, q^*)$ is unstable for some $q^* \in [0, q^+]$. In accordance with Lemma 4.8.2, there is a continuously varying root $s_{i*}(q)$ which is in the open left half plane when $q = 0$ (recall $p_0(s)$ is stable) and in the closed right half plane when $q = q^*$. By continuity of $s_{i*}(q)$, there exists some $\hat{q} \in (0, q^*]$ such that $s_{i*}(\hat{q})$ lies on the imaginary axis. This situation is illustrated pictorially in Figure 4.8.1; the crossing of the axis is apparent.

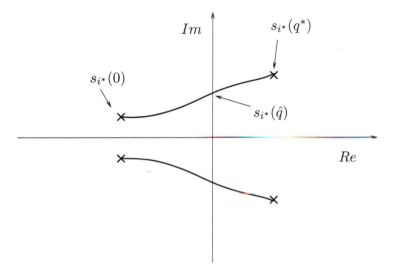

FIGURE 4.8.1   Continuous Root Path Crossing Imaginary Axis

To complete the proof, we claim that $H(p(\cdot, \hat{q}))$ is singular; this is the contradiction we seek. This claim is easily established using Orlando's formula in Lemma 4.8.1. Indeed, there are two possibilities. The first possibility is that $s_i(\hat{q}) = 0$. In this case, $a_0(\hat{q}) = 0$, which forces $\det H(p(\cdot, \hat{q})) = 0$ via Orlando's formula. The second possibility is that $s_{i*}(\hat{q}) \neq 0$. Since the roots of $p(s, \hat{q})$ appear in conjugate pairs and $s_{i*}(\hat{q})$ is purely imaginary, there must exist a different root $s_{k*}(\hat{q})$ such that $s_{k*}(\hat{q}) = -s_{i*}(\hat{q})$. This implies that the term $s_{i*}(\hat{q}) + s_{k*}(\hat{q})$ vanishes in Orlando's formula, which again forces $\det H(p(\cdot, \hat{q})) = 0$. ∎

4.9   Proof of the Theorem

To prove Theorem 4.7.6, we derive the formula for q_{max}^+ and simply note that the derivation of q_{min}^- runs along identical lines. Indeed, for fixed $q^+ \geq 0$, it follows from Lemma 4.8.3 that $\mathcal{P}(q^+)$ is robustly stable if and only if $H(p(\cdot, q))$ is nonsingular for all $q \in [0, q^+]$. Noting that

$$H(p(\cdot, q)) = H(p_0 + qp_1) = H(p_0) + qH(p_1),$$

it follows that q_{max}^+ is the largest value of q^+ such that

$$\det[H(p_0) + qH(p_1)] \neq 0$$

for all $q \in (0, q^+)$. Now, since $p_0(s)$ is stable, $H(p_0)$ is invertible and we can multiply by $H^{-1}(p_0)/q$ above and characterize q_{max}^+ by the condition

$$\det \left[\frac{1}{q}I + H^{-1}(p_0)H(p_1) \right] \neq 0$$

for all $q \in (0, q_{max}^+)$. There are now two cases to consider.

Case 1: The matrix $-H^{-1}(p_0)H(p_1)$ has no positive real eigenvalues. In this case, there is no $q > 0$ leading to a zero determinant above. Hence, we obtain $q_{max}^+ = +\infty$.

Case 2: The matrix $-H^{-1}(p_0)H(p_1)$ has positive real eigenvalues $0 < \lambda_1^+ \leq \lambda_2^+ \leq \cdots \leq \lambda_k^+$. Hence, the largest value of $q > 0$ leading to a nonvanishing determinant above is $q_{max}^+ = 1/\lambda_k^+$.

Combining Cases 1 and 2 now yields the formula

$$q_{max}^+ = \frac{1}{\lambda_{max}^+(-H^{-1}(p_0)H(p_1))}. \quad ∎$$

EXERCISE 4.9.1(Left-Sided and Right-Sided Robustness Margins): Find the left-sided and right-sided robustness margins for the uncertain polynomial $p(s, q) = s^4 + (6 + q)s^3 + 4s^2 + (10 + q)s + 8$.

4.10   Convex Combinations and Directions

In the derivation of analytical results, it is often more convenient to describe $p(s, q) = p_0(s) + qp_1(s)$, the one-parameter uncertain polynomial, using the notion of *convex combinations*. With $Q = [q^-, q^+]$, the associated family of polynomials has *extreme points* $p(s, q^-)$ and

$p(s, q^+)$. Furthermore, given any $q \in Q$, we can view $p(s, q)$ as a point on a line segment joining $p(s, q^-)$ and $p(s, q^+)$ in the space of polynomials; we express $p(s, q)$ as a *convex combination* of $p(s, q^-)$ and $p(s, q^+)$ by taking

$$\lambda = \frac{q^+ - q}{q^+ - q^-}$$

and writing

$$\tilde{p}(s, q) = \lambda p(s, q^-) + (1 - \lambda) p(s, q^+).$$

Conversely, for every $\lambda \in [0, 1]$, there corresponds some $q \in [q^-, q^+]$ such that $\tilde{p}(s, \lambda) = p(s, q)$. Given this isomorphism between $q \in Q$ and $\lambda \in [0, 1]$, it is purely a matter of convenience whether we work with the original family of polynomials or we work with an equivalent family $\tilde{\mathcal{P}} = \{\tilde{p}(\cdot, \lambda) : \lambda \in [0, 1]\}$ defined by

$$\tilde{p}(s, \lambda) = \lambda \tilde{p}_0(s) + (1 - \lambda) \tilde{p}_1(s),$$

where $\tilde{p}_0(s)$ and $\tilde{p}_1(s)$ are fixed polynomials. That is, $\mathcal{P} = \tilde{\mathcal{P}}$. Of course, the fixed polynomials associated with $\tilde{\mathcal{P}}$ are not the same ones associated with $\mathcal{P}$.

EXERCISE 4.10.1 (Representation Using a Direction):   For the family of polynomials described by $p(s, q) = p_0(s) + q p_1(s)$ and $Q = [q^-, q^+]$, let $f(s) = p_0(s) + q^- p_1(s)$ and $g(s) = (q^+ - q^-) p_1(s)$ and define

$$p(s, \lambda) = f(s) + \lambda g(s).$$

Prove that the family of polynomials $\mathcal{P} = \{p(\cdot, \lambda) : \lambda \in [0, 1]\}$ is the same as the original family. *Hint*: If $q \in Q$, let

$$\lambda = \frac{q - q^-}{q^+ - q^-}$$

and consider $\lambda \in [0, 1]$.

4.11   The Theorem of Bialas

To consolidate the technical ideas associated with Theorem 4.7.6 and the convex combination representation above, we relegate the theorem of Bialas (1985) to an exercise. Note that the proof for the exercise below is established via a straightforward mimic of the proof of Theorem 4.7.6. However, care must be taken in making the distinction between the robustness margin problem and the robust stability problem.

EXERCISE 4.11.1 (Bialas (1985)):   Consider the family of polynomials $\mathcal{P}$ described by $p(s, \lambda) = \lambda p_0(s) + (1 - \lambda)p_1(s)$ and $\lambda \in [0, 1]$, where $p_0(s)$ and $p_1(s)$ are fixed polynomials with $p_0(s)$ stable with positive coefficients and $n = \deg p_0(s) > \deg p_1(s)$. Prove that $\mathcal{P}$ is robustly stable if and only if the matrix $H^{-1}(p_0)H(p_1)$ has no purely real nonpositive eigenvalues.

4.12   The Matrix Case

In this section, we outline the steps required for generalization of Theorem 4.7.6 to the matrix case. To this end, we consider an uncertain matrix of the form

$$A(q) = A_0 + qA_1,$$

where A_0 and A_1 are fixed $n \times n$ matrices. Assuming A_0 is stable, we define left-sided and right-sided robustness margins as in the polynomial case; i.e., for the subfamilies of matrices given by

$$\mathcal{A}(q^+) = \{A(q) : 0 \le q \le q^+\}$$

and

$$\mathcal{A}(q^-) = \{A(q) : q^- \le q \le q^+\},$$

we define

$$q_{max}^+ = \sup\{q^+ : \mathcal{A}(q^+) \text{ is robustly stable}\}$$

and

$$q_{max}^- = \inf\{q^+ : \mathcal{A}(q^-) \text{ is robustly stable}\}$$

with the understanding that a family of matrices $\mathcal{A}$ is deemed *stable* if every $A \in \mathcal{A}$ has all its roots in the strict left half plane. Equivalently, the polynomial

$$p_A(s) = \det(sI - A)$$

is stable for all $A \in \mathcal{A}$.

We are now prepared to expose the "secret" associated with the matrix case: We convert the robust stability problem for the $n \times n$ matrix family $\mathcal{A}$ into a robust nonsingularity problem for an $n^2 \times n^2$

matrix family $\mathcal{A}^+$. Furthermore, a typical uncertain matrix $A^+(q)$ in $\mathcal{A}^+$ has the form

$$A^+(q) = A_0^+ + qA_1^+,$$

where A_0^+ and A_1^+ are fixed $n^2 \times n^2$ matrices. Once this step is carried out, the line of proof used in Theorem 4.7.6 immediately applies. That is, the nonsingularity arguments given above the matrix $H(p_0) + qH(q_1)$ are now replaced by identical arguments involving $A_0^+ + qA_1^+$. We now describe the mechanics associated with the transformation of a stability problem into a nonsingularity problem.

DEFINITION 4.12.1 (Kronecker Operations): Suppose that A and B are square matrices of dimensions n_1 and n_2, respectively, and let a_{ij} denote the (i,j)-th entry of A. Then, the *Kronecker product* $A \otimes B$ is the square matrix of dimension $n_1 n_2$ with (i,j)-th block $a_{ij}B$. The *Kronecker sum* $A \oplus B$ also has dimension $n_1 n_2$ and is given by the formula

$$A \oplus B = A \otimes I_{n_2} + I_{n_1} \otimes B,$$

where I_k denotes the $k \times k$ identity matrix. Finally, the *Kronecker difference* $A \ominus B$ is given by the formula

$$A \ominus B = A \oplus (-B).$$

REMARKS 4.12.2 (Special Case): In the results to follow, A is an $n \times n$ matrix and we often encounter $A \oplus A$. Using the formula

$$A \oplus A = A \otimes I_n + I_n \otimes A,$$

it is straightforward to see that $A \oplus A$ is the $n^2 \times n^2$ matrix having (i,j)-th block $A + a_{ij}I_n$ for $i = j$ and $a_{ij}I_n$ for $i \neq j$.

EXAMPLE 4.12.3 (Kronecker Operations): Suppose,

$$A = \begin{bmatrix} 1 & 2 \\ 3 & 4 \end{bmatrix}; \quad B = \begin{bmatrix} 5 & 6 \\ 7 & 8 \end{bmatrix}.$$

Then $A \otimes B$ and $A \oplus B$ are 4×4 matrices. Moreover,

$$A \otimes B = \begin{bmatrix} 5 & 6 & 10 & 12 \\ 7 & 8 & 14 & 16 \\ 15 & 18 & 20 & 24 \\ 21 & 24 & 28 & 38 \end{bmatrix} ;$$

$$A \oplus B = \begin{bmatrix} 1 & 0 & 2 & 0 \\ 0 & 1 & 0 & 2 \\ 3 & 0 & 4 & 0 \\ 0 & 3 & 0 & 4 \end{bmatrix} + \begin{bmatrix} 5 & 6 & 0 & 0 \\ 7 & 8 & 0 & 0 \\ 0 & 0 & 5 & 6 \\ 0 & 0 & 7 & 8 \end{bmatrix} = \begin{bmatrix} 6 & 6 & 2 & 0 \\ 7 & 9 & 0 & 2 \\ 3 & 0 & 9 & 6 \\ 0 & 3 & 7 & 12 \end{bmatrix}$$

are obtained via a straightforward computation.

REMARKS 4.12.4 (Eigenvalues Associated with Kronecker):   In the sequel, it is useful to exploit some well-known facts about the eigenvalues of Kronecker products and sums. Indeed, let $A \in \mathbf{R}^{n_1 \times n_1}$ and $B \in \mathbf{R}^{n_2 \times n_2}$ and take $\lambda_i(A)$ and $\lambda_i(B)$ to be the i-th eigenvalue of A and B, respectively. Then the eigenvalues of $A \otimes B$ consist of the $n_1 n_2$ products of the form $\lambda_{i_1}(A)\lambda_{i_1}(B)$ and the eigenvalues of $A \oplus B$ consist of the $n_1 n_2$ sums of the form $\lambda_{i_1}(A) + \lambda_{i_1}(B)$.

EXERCISE 4.12.5 (Eigenvalue Computation):   With A and B as in Example 4.12.3, use the remark above to find the eigenvalues of $A \oplus B$ and $A \otimes B$. Observe that the solution of the two required eigenvalue problems are easily carried out by hand because A and B are 2×2 matrices. Without exploitation of the remark, however, two much more difficult 4×4 eigenvalue problems need to be solved.

EXERCISE 4.12.6 (Matrix Problem Is More General):   Argue that the matrix problem of finding q_{min}^- and q_{max}^+ is a generalization of the polynomial problem. *Hint*: Argue that the embedding of polynomial $p_0(s) + q p_1(s)$ into a companion canonical form leads to the desired matrix uncertainty structure $A(q) = A_0 + q A_1$. Furthermore, show that this matrix uncertainty structure has a characteristic polynomial of the form

$$p(s, q) = p_0(s) + q p_1(s) + q^2 p_2(s) + \cdots + q^n p_n(s),$$

where $p_0(s)$, $p_1(s)$, ..., $p_n(s)$ are fixed.

EXERCISE 4.12.7 (From Stability to Nonsingularity): Consider the family of $n \times n$ matrices $\mathcal{A} = \{A(q) : q \in Q\}$ with $Q = [q^-, q^+]$ being a prescribed interval and A_0 assumed stable. Letting

$$\mathcal{A}^+ = \{A(q) \oplus A(q) : q \in Q\},$$

prove that $\mathcal{A}$ is robustly stable if and only if $\mathcal{A}^+$ is robustly nonsingular; i.e., every $A^+ \in \mathcal{A}^+$ is nonsingular. *Hint*: Specialize the eigenvalue characterization in Remarks 4.12.4 to $A(q) \oplus A(q)$.

REMARKS 4.12.8 (Linkage and Formulae): We are now prepared to derive the desired formulas for the left-sided and right-sided sided robustness margins q_{max}^+ and q_{min}^- for the family of matrices $\mathcal{A}$. Indeed, in view of the reduction of the stability problem to a nonsingularity problem described above, we simply mimic the arguments used to prove Theorem 4.7.6; see Section 4.9. With $(A_0 \oplus A_0) + q(A_1 \oplus A_1)$ replacing $H(p_0) + qH(p_1)$, we arrive at the formulas

$$q_{max}^+ = \frac{1}{\lambda_{max}^+(-(A_0 \oplus A_0)^{-1}(A_1 \oplus A_1))};$$

$$q_{min}^- = \frac{1}{\lambda_{min}^-(-(A_0 \oplus A_0)^{-1}(A_1 \oplus A_1))}.$$

EXERCISE 4.12.9 (Consolidation): A state variable model for an A4D jet fighter is given by

$$\dot{x}(t) = \begin{bmatrix} -0.0605 & -32.37 & 0.0 & 32.2 \\ -0.00014 & -1.475 & 1.0 & 0.0 \\ -0.0111 & -34.72 & -2.793 & 0.0 \\ 0.0 & 0.0 & 1.0 & 0.0 \end{bmatrix} x(t) + \begin{bmatrix} 0.0 \\ -0.1064 \\ -33.8 \\ 0.0 \end{bmatrix} u(t)$$

$$= Ax(t) + Bu(t),$$

where the state $x(t)$ has components $x_1(t)$, which represents the forward velocity, $x_2(t)$, which represents the angle of attack, $x_3(t)$, which represents the pitching velocity and $x_4(t)$, which represents the pitch angle. The output of the system $y(t)$ is the forward velocity $x_1(t)$ and the input $u(t)$ is the elevator angle. For this system, a Kalman filter design is used for stabilization. The controller is given by

$$u(t) = [\, 0.9630 \quad -14.0959 \quad 1.1138 \quad 11.7908\,]\hat{x}(t),$$

where $\hat{x}(t)$ is the state estimate obtained from the equation

$$\dot{\hat{x}}(t) = (A + BK - LC)\hat{x}(t) + Ly(t)$$

with output matrix $C = [1\ 0\ 0\ 0]$, state feedback vector

$$K = [\, 0.9639 \quad -14.0959 \quad 1.1138 \quad 19.79\,]$$

and transpose of the observer gain vector

$$L^T = [9.1417 \quad -0.1723 \quad 1.6171 \quad 1.1261\,]^T.$$

To study robust stability with respect to changes in the longitudinal static stability derivative, we replace the $(3,2)$ entry of A by $a_{32}(q) = -34.72 + q$; hence we write $A(q)$ instead of A. For the controller observer system with closed loop matrix

$$A_{CL} = \begin{bmatrix} A(q) & BK \\ LC & A(q) + BK - LC \end{bmatrix},$$

find the maximal interval of q for robust stability using the formulas for q_{min}^- and q_{max}^+ above.

4.13   Introduction to Robust $\mathcal{D}$-Stability

In this section, we extend robust stability concepts to allow for a more general root location region. For example, if $\mathcal{P}$ is a family of second order plants, a typical robustness specification might be as follows: Each plant in $\mathcal{P}$ should have a damping ratio no larger than some prescribed ξ and a degree of stability of at least σ. Letting $D(s,q)$ denote the uncertain denominator polynomial, this specification is tantamount to a constraint that for each $q \in Q$, the roots of $D(s,q)$ lie in a region $\mathcal{D}$ of the sort given in Figure 4.13.1. In each of these two examples, we insist that the system has a certain robust $\mathcal{D}$-stability property; see definition to follow. By dealing with general $\mathcal{D}$ regions in later chapters, we obtain a unified theory; e.g., the

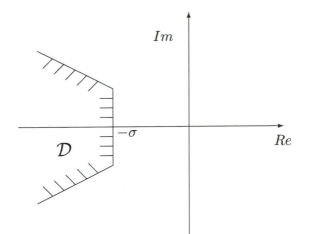

FIGURE 4.13.1 $\mathcal{D}$ Region for Damping and Degree of Stability Constraints

same theory applies to both continuous and discrete-time systems. In the continuous case, it is convenient to take $\mathcal{D}$ to be a subset of the left half plane, whereas in discrete-time $\mathcal{D}$ is taken to be a subset of the unit disc. We now provide some basic definitions.

DEFINITION 4.13.1 ($\mathcal{D}$-Stability):   Let $\mathcal{D} \subseteq \mathbf{C}$ and take $p(s)$ to be a fixed polynomial. Then $p(s)$ is said to be $\mathcal{D}$-stable if all its roots lie in the region $\mathcal{D}$.

DEFINITION 4.13.2 (Robust $\mathcal{D}$-Stability):   A family of polynomials $\mathcal{P} = \{p(\cdot, q) : q \in Q\}$ is said to be *robustly $\mathcal{D}$-stable* if, for all $q \in Q$, $p(s, q)$ is $\mathcal{D}$-stable; i.e., all roots of $p(s, q)$ lie in $\mathcal{D}$. For the special case when $\mathcal{D}$ is the open left half plane, $\mathcal{P}$ is simply said to be *robustly stable* and for the special case when $\mathcal{D}$ is the open unit disc, $\mathcal{P}$ is said to be *robustly Schur stable*.

EXERCISE 4.13.3 (A Subtlety):   For high order control systems, a typical specification might be as follows: The closed loop polynomials should have a pair of "dominant roots" in circles of given radius $\epsilon > 0$ centered at $-\alpha \pm j\beta$, and all remaining roots having real part less than $-\sigma$, where $\sigma > 0$. This leads us to consider a $\mathcal{D}$ region as in Figure 4.13.2. In a robustness context, if deg $p(s, q) = n$ for each $q \in Q$, we require 2 roots in $\mathcal{D}_1 \bigcup \mathcal{D}_2$ and $n - 2$ roots in $\mathcal{D}_3$. Is this

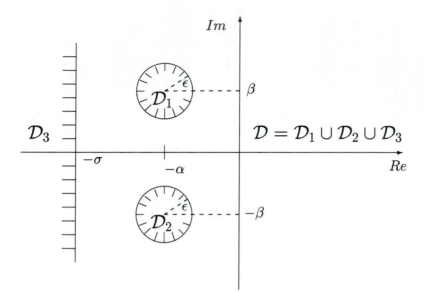

FIGURE 4.13.2 $\mathcal{D}$ Region for Dominant Roots and Degree of Stability

a robust $\mathcal{D}$-stability specification? Explain.

4.14   Robust $\mathcal{D}$-Stability Generalizations

The main objective of this section is to point out that for polynomials with one uncertain parameter, the root locus and Nyquist analyses of Sections 4.3 and 4.5 admit robust $\mathcal{D}$-stability generalizations in a rather obvious way. Indeed, let $\mathcal{D}$ be a desired root location region and take $\mathcal{P} = \{p(\cdot, q) : q \in Q\}$ to be a family of polynomials described by

$$p(s, q) = p_0(s) + q p_1(s)$$

with $p_0(s)$ and $p_1(s)$ being fixed polynomials and uncertainty bounding set $Q = [q^-, q^+]$. We assume that $\mathcal{P}$ has invariant degree. Now, arguing as in Section 4.3, the root locus plot for

$$P(s) = \frac{p_1(s)}{p_0(s)}$$

tells us everything we wish to know about robust $\mathcal{D}$-stability of $\mathcal{P}$; i.e., $\mathcal{P}$ is robustly stable if and only if the distinguished portion of the root locus plot for $p(s)$, corresponding to plant gain $q \in [q^-, q^+]$, remains in $\mathcal{D}$. As far as a Nyquist-like criterion is concerned, slightly more thought is required; see the exercise below.

EXERCISE 4.14.1 (Nyquist-Like Result for Robust $\mathcal{D}$-Stability):
Consider the family of polynomials $\mathcal{P}$ above with associated fictitious
plant $P(s)$. In addition, assume $p_0(s)$ is $\mathcal{D}$-stable and let $\mathcal{D} \subset \mathbf{C}$ be
open with boundary $\partial \mathcal{D}$. Prove that $\mathcal{P}$ is robustly $\mathcal{D}$-stable if and
only if for all $d \in \partial \mathcal{D}$, $p(d)$ does not meet a subset of the real axis
determined by q^- and q^+. Characterize this "forbidden" subset.

4.15   Extreme Point Results

Thus far, we have demonstrated that we can readily obtain solutions
to robust stability problems with a single uncertain parameter entering affine linearly into a polynomial's coefficients. Given that our
criteria involve root locus plots, Nyquist plots and eigenvalue computation, the following question is natural to ask: Can we guarantee
robust stability more simply by checking stability of the extreme
points? The exercise below answers this question in the negative for
the general case.

EXERCISE 4.15.1 (Stability of Extreme Points Do Not Suffice):
With $f(s) = 10s^3 + s^2 + 6s + 0.57$ and $g(s) = s^2 + 2s + 1$ as specified
in Bialas and Garloff (1985), consider the family of polynomials $\mathcal{P}$
which is described by $p(s, \lambda) = f(s) + \lambda g(s)$ and $\lambda \in [0, 1]$. Verify
that the two extremes $p(s, 0) = f(s)$ and $p(s, 1) = g(s)$ are stable
but $p(s, \lambda)$ is unstable for a range of $\lambda \in [0, 1]$; e.g., the intermediate
polynomial $p(s, 0.5)$ is unstable.

REMARKS 4.15.2 (Conditions for Extremality):   In view of the
exercise above, we conclude that additional conditions must be imposed on the pair $(f(s), g(s))$ in order to infer robust stability of
the entire family from stability of the extremes. In fact, the issue of
extreme point results is a main focal point in many of the chapters
to follow. As we develop new machinery, increasingly sophisticated
extreme point results are obtained. To appreciate the need for additional machinery, the ambitious reader might attempt the exercise
below from first principles.

EXERCISE 4.15.3 (For the Undergraduate):   Let $f(s)$ be a stable
polynomial of degree two or more and assume that $f(s) + (s + 1)$ is
stable. Now prove that

$$p(s, \lambda) = f(s) + \lambda(s + 1)$$

is stable for all $\lambda \in [0, 1]$.

4.16   Conclusion

In this chapter, we concentrated on one uncertain parameter and obtained three different solutions to the robust stability problem—a solution involving root locus plots, a solution involving Nyquist plots and a solution involving eigenvalues. For the more general robust $\mathcal{D}$-stability problem, the extension of the root locus and Nyquist concepts was immediate. However, no extension of the eigenvalue criterion of Theorem 4.7.6 was given. Although such an extension exists, we do not provide it until much later in the text because it requires a much more abstract level of presentation. Although not recommended, it is possible to proceed directly from this point in the text to Chapter 17 and read the $\mathcal{D}$-stability generalization; that is, the tools in Chapters 6–16 are not instrumental to the proofs given in Chapter 17.

Notes and Related Literature

NRL 4.1 If we consider complex rather than real uncertainty, there is a simple connection between the ideas in this chapter and the classical Small Gain Theorem. Indeed, suppose that q is a complex uncertain parameter with uncertainty bound $|q| \leq 1$. Take $p_0(s)$ to be a stable complex coefficient polynomial and $p_1(s)$ is another complex coefficient polynomial with $\deg p_1(s) < \deg p_0(s)$. Then, with $p(s, q) = p_0(s) + qp_1(s)$, it is easy to show that the resulting family of polynomials $\mathcal{P} = \{p(\cdot, q) : q \in Q\}$ is robustly stable if and only if the fictitious plant $P(s) = p_1(s)/p_0(s)$ has H^∞ norm less than unity. This is the same condition we obtain using the classical feedback interconnection associated with the Small Gain Theorem. That is, given the plant $P(s)$ with complex uncertainty q in the feedback path, the quantity $1/||P||_\infty$ indicates how large $|q|$ can be before instability is encountered.

NRL 4.2 The survey paper by Brewer (1978) provides a detailed review of Kronecker operations.

NRL 4.3 For the matrix case, there are many other transformations (besides the Kronecker sum) taking the robust stability problem into a robust nonsingularity problem. For example, with $m = n(n+1)/2$, there exist many possible linear mappings $T : \mathbf{R}^{n \times n} \to \mathbf{R}^{m \times m}$ which serve the same function as the Kronecker sum. For such cases, the eigenvalue problem associated with computation of q_{min}^- and q_{max}^+ can be significantly smaller; see Fu and Barmish (1988) for details.

Chapter 5

The Spark: Kharitonov's Theorem

Synopsis

This chapter is devoted to the seminal theorem of Kharitonov. The technical ideas underlying the proof serve as a pedagogical stepping stone for development of the more general value set concept which unifies many results in later chapters. In fact, the Kharitonov rectangle which we introduce is actually a value set corresponding to a rather specialized uncertainty structure.

5.1   Introduction

The main result in this chapter, Kharitonov's Theorem, addresses a rather specialized problem—robust stability of an interval polynomial family. The elegance of the solution immediately sets one's thought processes in motion; i.e., seeing such a dramatic breakthrough for the robust stability problem for interval polynomials, one cannot help but wonder what powerful results are possible for more general robustness problems. In a sense, most of the chapters to follow are testimonials to the new way of thinking which comes from the proof of Kharitonov's Theorem.

5.2   Independent Uncertainty Structures

In this section, we introduce the independent uncertainty structure. Results for this highly specialized structure should not be viewed as an end in itself. With this simpler theory under our belts, however, we are prepared to deal with more general polytopic and multilinear uncertainty structures in the chapters to follow.

Perhaps the most compelling motivation for the study of independent uncertainty structures is derived from the following scenario: An engineer generates a fixed model for a control system and obtains the associated characteristic polynomial $p(s)$. Although the presence of parametric uncertainty is acknowledged, the dependence on q is complicated and highly nonlinear. Despite the fact that the uncertainty structure is too complicated to analyze mathematically, it is still important to know something about the degree of robustness. In such cases, a sound argument can be made for imposition of an independent uncertainty structure. For example, using an independent uncertainty structure, we can use the theory in this chapter to determine what percentage variations in the coefficients of polynomial $p(s)$ can be tolerated.

It is also worth noting that in many cases, a more complicated uncertainty structure admits a certain type of overbounding by an independent uncertainty structure. Hence, once we have results for the independent case, we often obtain sufficient conditions for the more complicated case at hand. To illustrate, after overbounding a complicated uncertainty structure by an independent uncertainty structure, one might compute a robustness margin of 13% when the true robustness margin is 16%. It can be argued that the conservatism resulting from overbounding is not critical when the performance specification is still met.

DEFINITION 5.2.1 (Independent Uncertainty Structure): An uncertain polynomial

$$p(s, q) = \sum_{i=0}^{n} a_i(q)s^i$$

is said to have an *independent uncertainty structure* if each component q_i of q enters into only one coefficient.

EXERCISE 5.2.2 (Independent Uncertainty Structure): Does the uncertain polynomial

$$p(s, q) = s^3 + (q_1 + 4q_2 + 6)s^2 + (q_1 - 3q_4)s + (q_0 + 5)$$

have an independent uncertainty structure? Explain.

5.3 Interval Polynomial Family

In this section, we define interval polynomial families and the concept of lumping. By lumping, we mean combining uncertainties so as to obtain a description of the same family of polynomials involving a smaller number of uncertain parameters.

DEFINITION 5.3.1 (Interval Polynomial Family): A family of polynomials $\mathcal{P} = \{p(\cdot, q) : q \in Q\}$ is said to be an *interval polynomial family* if $p(s, q)$ has an independent uncertainty structure, each coefficient depends continuously on q and Q is a box. For brevity, we often drop the word "family" and simply refer to $\mathcal{P}$ as an *interval polynomial*.

EXAMPLE 5.3.2 (Simple Interval Polynomial): An interval polynomial family $\mathcal{P}$ arises from the uncertain polynomial described by
$$p(s, q) = (5 + q_4)s^4 + (3 + q_3)s^3 + (2 + q_2)s^2 + (4 + q_1)s + (6 + q_0)$$
with uncertainty bounds $|q_i| \leq 1$ for $i = 0, 1, 2, 3, 4$.

EXAMPLE 5.3.3 (Some Coefficients Fixed): Notice that the definition of interval polynomial does not rule out the possibility that some coefficients of $p(s, q)$ are fixed rather than uncertain; e.g., consider $p(s, q) = (5 + q_4)s^4 + 3s^3 + (2 + q_2)s^2 + (4 + q_1)s + 6$ with a given box Q for the uncertainty bounding set.

EXAMPLE 5.3.4 (Lumping Interval Polynomials): The uncertainty representation often involves a certain type of redundancy. For example, if $p(s, q) = s^3 + (5 + q_2 + 2q_3)s^2 + (6 + 2q_1 + 5q_4)s + (3 + q_0)$ and bounds $|q_i| \leq 0.5$ for $i = 0, 1, 2, 3, 4$, one can "lump" the uncertainty as follows: Define new uncertain parameters $\tilde{q}_2 = 5 + q_2 + 2q_3$, $\tilde{q}_1 = 6 + 2q_1 + 5q_4$ and $\tilde{q}_0 = 3 + q_0$, a new uncertainty bounding set $\tilde{Q}$ by $2.5 \leq \tilde{q}_0 \leq 3.5$, $2.5 \leq \tilde{q}_1 \leq 9.5$ and $3.5 \leq \tilde{q}_2 \leq 6.5$ and a new uncertain polynomial $\tilde{p}(s, \tilde{q}) = s^3 + \tilde{q}_2 s^2 + \tilde{q}_1 s + \tilde{q}_0$. We call $\tilde{\mathcal{P}} = \{\tilde{p}(\cdot, \tilde{q}) : \tilde{q} \in \tilde{Q}\}$ a *lumped version* of the original family $\mathcal{P}$ and leave it to the reader to verify that $\tilde{\mathcal{P}} = \mathcal{P}$.

EXERCISE 5.3.5 (Lumping with More Complicated Dependence): The objective of this exercise is to demonstrate that lumping is possible with more complicated dependence on q. To this end, consider an interval polynomial family $\mathcal{P}$ described by

$$p(s, q) = (5 + e^{q_1} \cos q_2)s^2 + (\sin(q_3 + q_4) + 4)s + (q_5 q_6^2 + e^{q_7})$$

and $|q_i| \leq 1$ for $i = 1, 2, \ldots, 7$. Provide a characterization of a lumped version $\tilde{\mathcal{P}}$ of $\mathcal{P}$.

EXERCISE 5.3.6 (A Lumping Theorem): This exercise generalizes on the one above. Indeed, consider an interval polynomial family $\mathcal{P} = \{p(\cdot, q) : q \in Q\}$ with $p(s, q)$ having coefficients depending continuously on q. Prove that there exists a second interval polynomial family $\tilde{\mathcal{P}} = \{\tilde{p}(\cdot, \tilde{q}) : \tilde{q} \in \tilde{Q}\}$ with $\tilde{p}(s, \tilde{q})$ of the form $\tilde{p}(s, \tilde{q}) = \sum_{i=0}^{n} \tilde{q}_i s^i$ and, moreover, $\tilde{\mathcal{P}} = \mathcal{P}$.

5.4 Shorthand Notation

In view of the discussion of lumping above, we henceforth work with an uncertain polynomial of the form

$$p(s, q) = \sum_{i=0}^{n} q_i s^i$$

when dealing with an interval family. Such a family is completely described by the shorthand notation

$$p(s, q) = \sum_{i=0}^{n} [q_i^-, q_i^+] s^i$$

with $[q_i^-, q_i^+]$ denoting the bounding interval for the i-th component of uncertainty q_i. In the context of this convenient abuse of notation, we can refer to $p(s, q)$ as an *interval polynomial*.

5.5 The Kharitonov Polynomials

In order to describe Kharitonov's Theorem for robust stability, we first define four fixed polynomials associated with an interval polynomial family $\mathcal{P}$. In the definition below, note that the polynomials are fixed in the sense that only the bounds q_i^- and q_i^+ enter into the description but not the q_i themselves. We also emphasize that the number of polynomials is four—independent of the degree of $p(s, q)$. That is, four is a magic number.

DEFINITION 5.5.1 (The Kharitonov Polynomials): Associated with the interval polynomial $p(s, q) = \sum_{i=0}^{n} [q_i^-, q_i^+] s^i$ are the four fixed *Kharitonov polynomials*

$$K_1(s) = q_0^- + q_1^- s + q_2^+ s^2 + q_3^+ s^3 + q_4^- s^4 + q_5^- s^5 + q_6^+ s^6 + \cdots;$$

$$K_2(s) = q_0^+ + q_1^+ s + q_2^- s^2 + q_3^- s^3 + q_4^+ s^4 + q_5^+ s^5 + q_6^- s^6 + \cdots;$$
$$K_3(s) = q_0^+ + q_1^- s + q_2^- s^2 + q_3^+ s^3 + q_4^+ s^4 + q_5^- s^5 + q_6^- s^6 + \cdots;$$
$$K_4(s) = q_0^- + q_1^+ s + q_2^+ s^2 + q_3^- s^3 + q_4^- s^4 + q_5^+ s^5 + q_6^+ s^6 + \cdots.$$

EXAMPLE 5.5.2 (Construction of Kharitonov Polynomials): The Kharitonov polynomials are easily constructed by inspection. To illustrate, the four Kharitonov polynomials corresponding to the interval polynomial

$$p(s, q) = [1, 2]s^5 + [3, 4]s^4 + [5, 6]s^3 + [7, 8]s^2 + [9, 10]s + [11, 12]$$

are

$$K_1(s) = 11 + 9s + 8s^2 + 6s^3 + 3s^4 + s^5;$$
$$K_2(s) = 12 + 10s + 7s^2 + 5s^3 + 4s^4 + 2s^5;$$
$$K_3(s) = 12 + 9s + 7s^2 + 6s^3 + 4s^4 + s^5;$$
$$K_4(s) = 11 + 10s + 8s^2 + 5s^3 + 3s^4 + 2s^5.$$

5.6   Kharitonov's Theorem

We now present the celebrated theorem of Kharitonov (1978a) and also illustrate its application. The proof of the theorem is relegated to the next two sections.

THEOREM 5.6.1 (Kharitonov (1978a)):   *An interval polynomial family $\mathcal{P}$ with invariant degree is robustly stable if and only if its four Kharitonov polynomials are stable.*

EXAMPLE 5.6.2 (Application of Kharitonov's Theorem):   For the interval polynomial

$$p(s, q) = [0.25, 1.25]s^3 + [2.75, 3.25]s^2 + [0.75, 1.25]s + [0.25, 1.25],$$

the four Kharitonov polynomials are

$$K_1(s) = 0.25 + 0.75s + 3.25s^2 + 1.25s^3;$$
$$K_2(s) = 1.25 + 1.25s + 2.75s^2 + 0.25s^3;$$
$$K_3(s) = 1.25 + 0.75s + 2.75s^2 + 1.25s^3;$$
$$K_4(s) = 0.25 + 1.25s + 3.25s^2 + 0.25s^3.$$

Using the classical Hurwitz criterion, it is easy to verify that all four Kharitonov polynomials above are stable. Hence, we conclude that the interval polynomial family is robustly stable.

EXERCISE 5.6.3 (Application of Kharitonov's Theorem):   Consider the interval polynomial family which is given in Example 5.5.2. Is it robustly stable?

5.7   Machinery for the Proof

For some readers, there is a temptation to skip sections containing technical proofs. For the case of Kharitonov's Theorem, however, the author's advice is to continue reading. The ideas introduced in this section and the next are at the heart of many generalizations presented in later chapters. In addition, the geometrical ideas in the proof suggest ideas for computer-aided analysis. Most notably, the proof makes use of the so-called Kharitonov rectangle. This rectangle is in fact a special type of "value set" which plays a major role in later chapters.

5.7.1   The Kharitonov Rectangle

In this subsection, we consider an elementary geometry problem: Given an interval polynomial $p(s,q) = \sum_{i=0}^{n}[q_i^-, q_i^+]s^i$ and a fixed frequency $\omega = \omega_0$, describe the set of possible values that $p(j\omega_0, q)$ can assume as q ranges over the box Q. More formally, we want to describe the subset of the complex plane given by

$$p(j\omega_0, Q) = \{p(j\omega_0, q) : q \in Q\}.$$

We call $p(j\omega_0, Q)$ the *Kharitonov rectangle* at frequency $\omega = \omega_0$. To justify this name, we now prove that $p(j\omega_0, Q)$ is a rectangle with vertices which are obtained by evaluating the four *fixed* Kharitonov polynomials $K_1(s)$, $K_2(s)$, $K_3(s)$ and $K_4(s)$ at $s = j\omega_0$; i.e., the vertices of $p(j\omega_0, Q)$ are precisely the $K_i(j\omega_0)$.

To establish rectangularity, we examine the real and imaginary parts of $p(j\omega_0, q)$. Indeed, we first observe that

$$Re\, p(j\omega_0, q) = \sum_{i\ even} q_i(j\omega_0)^i = q_0 - q_2\omega_0^2 + q_4\omega_0^4 - q_6\omega_0^6 + q_8\omega_0^8 - \cdots$$

and

$$Im\, p(j\omega_0, q) = \frac{1}{j}\sum_{i\ odd} q_i(j\omega_0)^i = q_1\omega_0 - q_3\omega_0^3 + q_5\omega_0^5 - q_7\omega_0^7 + q_9\omega_0^9 - \cdots.$$

Notice that no q_i which enters $Re\ p(j\omega_0, q)$ enters $Im\ p(j\omega_0, q)$ and vice versa. In view of this decoupling between real and imaginary parts, the set $p(j\omega_0, Q)$ consists of all complex numbers z such that

$$Re\ z = q_0 - q_2\omega_0^2 + q_4\omega_0^4 - q_6\omega_0^6 + q_8\omega_0^8 - \cdots$$

for some admissible $q \in Q$ and

$$Im\ z = q_1\omega_0 - q_3\omega_0^3 + q_5\omega_0^5 - q_7\omega_0^7 + q_9\omega_0^9 - \cdots$$

for some admissible $q \in Q$.

We now argue that the set of all generatable pairs $(Re\ z, Im\ z)$ above is a rectangle which is obtained by finding the minimum and maximum values of $Re\ p(j\omega_0, q)$ and $Im\ p(j\omega_0, q)$ with respect to $q \in Q$. Indeed, since each q_i enters only one coefficient of $p(s, q)$, for $Re\ p(j\omega_0, q)$, we can minimize or maximize each term individually to obtain

$$\min_{q \in Q} Re\ p(j\omega_0, q) = q_0^- - q_2^+\omega_0^2 + q_4^-\omega_0^4 - q_6^+\omega_0^6 + q_8^-\omega_0^8 - \cdots$$
$$= Re\ K_1(j\omega_0)$$

and

$$\max_{q \in Q} Re\ p(j\omega_0, q) = q_0^+ - q_2^-\omega_0^2 + q_4^+\omega_0^4 - q_6^-\omega_0^6 + q_8^+\omega_0^8 - \cdots$$
$$= Re\ K_2(j\omega_0).$$

As far as $Im\ p(j\omega_0, q)$ is concerned, one must pay attention to the sign of ω_0 in deciding whether to use q_i^- or q_i^+ when minimizing or maximizing. Keeping this issue in mind, for $\omega_0 \geq 0$, we obtain

$$\min_{q \in Q} Im\ p(j\omega_0, q) = q_1^-\omega_0 - q_3^+\omega_0^3 + q_5^-\omega_0^5 - q_7^+\omega_0^7 + \cdots$$

and similarly, for $\omega_0 < 0$,

$$\min_{q \in Q} Im\ p(j\omega_0, q) = q_1^+\omega_0 - q_3^-\omega_0^3 + q_5^+\omega_0^5 - q_7^-\omega_0^7 + \cdots.$$

Combining these two cases, we arrive at

$$\min_{q \in Q} Im\ p(j\omega_0, q) = \begin{cases} Im\ K_3(j\omega_0) & \text{if } \omega_0 \geq 0; \\ Im\ K_4(j\omega_0) & \text{if } \omega_0 < 0. \end{cases}$$

For the maximization problem, the same type of reasoning leads to

$$\max_{q \in Q} Im\ p(j\omega_0, q) = \begin{cases} Im\ K_4(j\omega_0) & \text{if } \omega_0 \geq 0; \\ Im\ K_3(j\omega_0) & \text{if } \omega_0 < 0. \end{cases}$$

Thus far, our arguments indicate that $p(j\omega_0, Q)$ is bounded by the rectangle given in Figure 5.7.1; i.e., if $z \in p(j\omega_0, Q)$ and $\omega_0 \geq 0$,

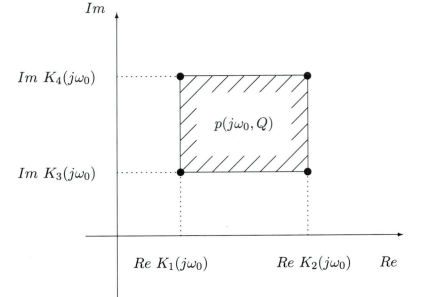

FIGURE 5.7.1   The Kharitonov Rectangle for $\omega_0 \geq 0$

then

$$Re\ K_1(j\omega_0) \leq Re\ z \leq Re\ K_2(j\omega_0);$$
$$Im\ K_3(j\omega_0) \leq Im\ z \leq Im\ K_4(j\omega_0).$$

To complete the argument, we now claim that this bounding rectangle is precisely equal to $p(j\omega_0, Q)$. That is, every value in this rectangle is realizable by some $q \in Q$. Indeed, by viewing $Re\ p(j\omega_0, q)$ as a mapping of $(q_0, q_2, q_4, \ldots)$ to $\mathbf{R}$ and $Im\ p(j\omega_0, q)$ as a mapping from $(q_1, q_3, q_5, \ldots)$ to $\mathbf{R}$, a simple intermediate value argument guarantees that for each z satisfying the two inequalities above, there exists some uncertainty $q_z \in Q$ such that $p(j\omega_0, q_z) = z$. In summary, the set $p(j\omega_0, Q)$ is precisely the rectangle depicted in Figure 5.7.1.

We now relate the vertices of the rectangle $p(j\omega_0, Q)$ to the Kharitonov polynomials:

$$\text{Southwest Vertex} = Re\ K_1(j\omega_0) + jIm\ K_3(j\omega_0)$$

$$= Re \ K_1(j\omega_0) + j Im \ K_1(j\omega_0)$$
$$= K_1(j\omega_0);$$

$$\text{Northeast Vertex} = Re \ K_2(j\omega_0) + j Im \ K_4(j\omega_0)$$
$$= Re \ K_2(j\omega_0) + j Im \ K_2(j\omega_0)$$
$$= K_2(j\omega_0);$$

$$\text{Southeast Vertex} = Re \ K_2(j\omega_0) + j Im \ K_3(j\omega_0)$$
$$= Re \ K_3(j\omega_0) + j Im \ K_3(j\omega_0)$$
$$= K_3(j\omega_0);$$

$$\text{Northwest Vertex} = Re \ K_1(j\omega_0) + j Im \ K_4(j\omega_0)$$
$$= Re \ K_4(j\omega_0) + j Im \ K_4(j\omega_0)$$
$$= K_4(j\omega_0).$$

This leads to our final depiction of the *Kharitonov rectangle* given in Figure 5.7.2. The key point to note is that each vertex is associated

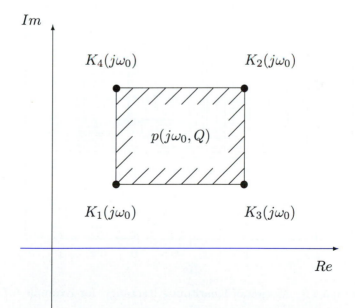

FIGURE 5.7.2   Simplified Kharitonov Rectangle for $\omega_0 \geq 0$

with a unique Kharitonov polynomial.

EXERCISE 5.7.2 (Kharitonov Rectangle for $\omega_0 < 0$ and $\omega_0 = 0$): Sketch the Kharitonov rectangle $p(j\omega_0, Q)$ for $\omega_0 < 0$ with vertices carefully labeled. For $\omega_0 = 0$, notice that $p(j\omega_0, Q) = [q_0^-, q_0^+]$.

REMARKS 5.7.3 (Motion of Kharitonov Rectangle): Thus far, the discussion of the Kharitonov rectangle has been in the context of a frozen frequency $\omega = \omega_0$. We now entertain the notion of sweeping the frequency. Indeed, we begin at $\omega = 0$ and imagine ω increasing. This results in motion of the Kharitonov rectangle. That is, we have a rectangle moving around the complex plane with vertices $K_i(j\omega)$ obtained by evaluation of the Kharitonov polynomials. Generally, the dimensions of this rectangle vary with the frequency ω.

EXAMPLE 5.7.4 (Illustration of Motion):   For the interval polynomial

$$p(s, q) = [0.25, 1.25]s^3 + [2.75, 3.25]s^2 + [0.75, 1.25]s + [0.25, 1.25]$$

which we analyzed in Example 5.6.2, we illustrate the motion of the Kharitonov rectangle $p(j\omega, Q)$ in Figure 5.7.3 for twenty frequencies

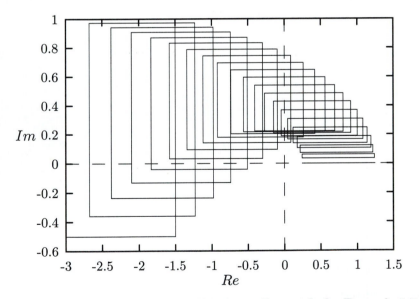

FIGURE 5.7.3   Motion of Kharitonov Rectangle for Example 5.7.4

evenly spaced between $\omega = 0$ and $\omega = 1$. Notice that this rectangle begins at $\omega = 0$ as an interval on the positive real axis and then moves from the first to the second quadrant as ω is increased.

5.7.5   Angle Considerations

In this subsection, we review some basic facts about the angle of a polynomial as a function of frequency. We include the proof of the well-known lemma below because the underlying ideas are useful in later chapters. In a control setting, the lemma below is often credited to Mikhailov (1938).

LEMMA 5.7.6 (Monotonic Angle Property):   *Suppose that $p(s)$ is a stable polynomial. Then the angle of $p(j\omega)$ is a strictly increasing function of $\omega \in \mathbf{R}$. Furthermore, as ω varies from 0 to $+\infty$, $\angle p(j\omega)$ experiences an increment of $n\pi/2$.*

PROOF: First, we write $p(s) = K \prod_{i=1}^{n}(s - z_i)$, where $K \in \mathbf{R}$ and $Re\ z_i < 0$ for $i = 1, 2, \ldots, n$. The angle of $p(j\omega)$ is given by

$$\angle p(j\omega) = \angle K + \sum_{i=1}^{n} \angle (j\omega - z_i).$$

With $\theta_i(\omega) = \angle (j\omega - z_i)$ and the aid of Figure 5.7.4, we make the

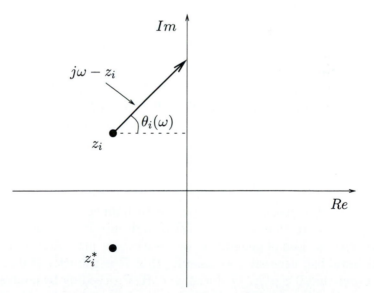

FIGURE 5.7.4 $\theta_i(\omega)$ is a Strictly Increasing Function of ω

following observations, noting that z_i lies in the strict left half plane:

If z_i is purely real, then as ω varies from 0 to $+\infty$, $\theta_i(\omega)$ is strictly increasing and experiences a net increment of $\pi/2$. If z_i is complex, we work with z_i in combination with its conjugate z_i^*. Now, as ω increases from 0 to $+\infty$, the corresponding angles $\theta_i(\omega)$ are strictly increasing and contribute a net increment total of π. The proof of the lemma is completed by summing over the $\theta_i(\omega)$. ∎

EXERCISE 5.7.7 (More General Angle Considerations):   Suppose $p(s)$ is an n-th order polynomial with n_1 roots in the strict left half plane and n_2 roots in the strict right half plane. Assume $n_1 + n_2 = n$ and show that as ω varies from 0 to $+\infty$, $\measuredangle p(j\omega)$ experiences a total change in angle of $(n_1 - n_2)\pi/2$. Also modify the result to allow for the case when $p(s)$ has some roots on the imaginary axis.

5.7.8   The Zero Exclusion Condition

In this subsection, we introduce the Zero Exclusion Condition. The technical ideas associated with this condition arise time and time again throughout the remainder of this text. Since we are currently working within the framework of interval polynomials, the lemma below is not stated in full generality; the most general version which we provide is given in Theorem 7.4.2. In addition to facilitating the proof of Kharitonov's Theorem, the lemma below is also of practical use because it suggests a simple test for robust stability which is easy to implement in graphics.

LEMMA 5.7.9 (Zero Exclusion Condition):   *Suppose that an interval polynomial family $\mathcal{P} = \{p(\cdot, q) : q \in Q\}$ has invariant degree and at least one stable member $p(s, q^0)$. Then $\mathcal{P}$ is robustly stable if and only if $z = 0$ is excluded from the Kharitonov rectangle at all nonnegative frequencies; i.e.,*

$$0 \notin p(j\omega, Q)$$

for all frequencies $\omega \geq 0$.

PROOF: We first justify the restriction to nonnegative frequencies. To this end, note that $z \in p(j\omega, Q)$ if and only if $z^* \in p(-j\omega, Q)$. Hence, without loss of generality, we restrict our attention to $\omega \geq 0$.

To establish necessity, we assume that $\mathcal{P}$ is robustly stable and must prove that $0 \notin p(j\omega, Q)$ for all $\omega \in \mathbf{R}$. Proceeding by contradiction, suppose that $0 \in p(j\omega^*, Q)$ for some frequency $\omega^* \in \mathbf{R}$. Then $p(j\omega^*, q^*) = 0$ for some $q^* \in Q$; i.e., the polynomial $p(s, q^*)$ has a root at $s = j\omega^*$ which contradicts robust stability of $\mathcal{P}$.

To establish sufficiency, we assume that $0 \notin p(j\omega, Q)$ for all $\omega \in \mathbf{R}$ and must show that $\mathcal{P}$ is robustly stable. Proceeding by contradiction, if $\mathcal{P}$ is not robustly stable, then $p(s, q^1)$ is unstable for some $q^1 \in Q$. Now, for $\lambda \in [0, 1]$, let

$$\tilde{p}(s, \lambda) = p(s, \lambda q^1 + (1 - \lambda)q^0)$$

and notice that $\tilde{p}(s, \lambda) \in \mathcal{P}$ because $\lambda q^1 + (1 - \lambda)q^0 \in Q$. Moreover, for $\lambda = 0$, $\tilde{p}(s, 0) = p(s, q^0)$ has all roots in the strict left half plane and for $\lambda = 1$, $\tilde{p}(s, 1) = p(s, q^1)$ has at least one root in the closed right half plane. Since the roots of $\tilde{p}(s, \lambda)$ depend continuously on λ (Lemma 4.8.2), there exists a $\lambda^* \in [0, 1]$ such that $\tilde{p}(s, \lambda^*)$ has a root on the imaginary axis. Equivalently, $p(j\omega^*, \lambda^* q^1 + (1 - \lambda^*)q^0) = 0$ for some $\omega^* \in \mathbf{R}$. This implies that $0 \in p(j\omega^*, Q)$, which is the contradiction we seek. ∎

REMARKS 5.7.10 (Real Versus Complex Coefficients):   When working with the Zero Exclusion Condition for the complex coefficient case, we can no longer restrict attention to $\omega \geq 0$; i.e., we cannot exploit the fact that $z \in p(j\omega, Q)$ if and only if $z^* \in p(-j\omega, Q)$. In this case, the lemma above requires a minor modification: Under the standing hypotheses, $\mathcal{P}$ is robustly stable if and only if $0 \notin p(j\omega, Q)$ for all $\omega \in \mathbf{R}$. This arises in Chapter 6 when we consider the complex coefficient version of Kharitonov's Theorem.

5.8   Proof of Kharitonov's Theorem

The proof of necessity is trivial; i.e., if $\mathcal{P}$ is robustly stable, it follows that the four Kharitonov polynomials are stable because $K_i(s) \in \mathcal{P}$ for $i = 1, 2, 3, 4$. To establish sufficiency, we assume that the four Kharitonov polynomials are stable and must prove that $\mathcal{P}$ is robustly stable. Proceeding by contradiction, suppose that $\mathcal{P}$ is not robustly stable. Using the standard notation $p(s, q) = \sum_{i=0}^{n}[q_i^-, q_i^+]s^i$, we consider two cases.

Case 1: $0 \in [q_0^-, q_0^+]$. Recalling the invariant degree assumption, it must be true that q_n^- and q_n^+ have the same sign. Without loss of generality, say that the signs of q_n^- and q_n^+ are positive. Then it follows that at least one of the four Kharitonov polynomials, call it $K_{i*}(s)$, has coefficient of s^n, which is positive, and coefficient of s^0, which is nonpositive. This contradicts the assumed stability of $K_{i*}(s)$ because a stable polynomial must have nonzero coefficients which all have the same sign.

Case 2: $0 \notin [q_0^-, q_0^+]$. Since $p(j0, Q) = [q_0^-, q_0^+]$ is the Kharitonov rectangle at $\omega = 0$, we have $0 \notin p(j0, Q)$. On the other hand, since $\mathcal{P}$ is not robustly stable, we know by the Zero Exclusion Condition (Lemma 5.7.9) that $0 \in p(j\omega^*, Q)$ for some $\omega^* \in \mathbf{R}$. Now, using the fact that $0 \notin p(j0, Q)$, the continuous motion of the vertices $K_i(j\omega)$ of $p(j\omega, Q)$ guarantees that there must be some frequency $\hat{\omega} > 0$ for which $z = 0$ pierces the boundary of the rectangle $p(j\hat{\omega}, Q)$. Without loss of generality, assume that this piercing occurs on the southern boundary of $p(j\hat{\omega}, Q)$ as shown in Figure 5.8.1. Also, note

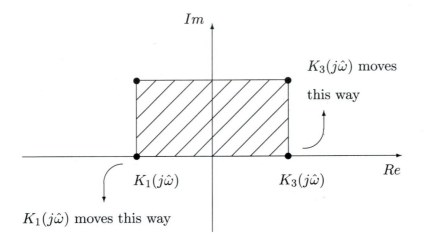

FIGURE 5.8.1   Piercing the Boundary of the Kharitonov Rectangle

that $z = 0$ cannot be coincident with $K_1(j\hat{\omega})$ or $K_3(j\hat{\omega})$ because $K_1(s)$ and $K_3(s)$ are assumed stable. To complete the proof, we exploit continuity of the $K_i(j\omega)$ and the Monotonic Angle Property (Lemma 5.7.6). Namely, for $\delta\hat{\omega} > 0$ suitably small, it follows that

$$0^\circ < \measuredangle K_3(j(\hat{\omega} + \delta\hat{\omega})) < 90^\circ$$

and

$$180^\circ < \measuredangle K_1(j(\hat{\omega} + \delta\hat{\omega})) < 270^\circ.$$

We now have the contradiction which we seek because simultaneous satisfaction of the two angle inequalities above makes it impossible for the southern boundary of the rectangle $p(j(\hat{\omega} + \delta\hat{\omega}), Q)$ to remain parallel to the real axis. ∎

5.9   Formula for the Robustness Margin

For an interval polynomial family, by combining the results of this chapter with those of Chapter 4, we obtain the robustness margin formulas of Fu and Barmish (1988). To this end, we describe an n-th order interval polynomial family with stable nominal $p_0(s)$ and variable uncertainty bound $r \geq 0$ by writing

$$p_r(s, q) = p_0(s) + r \sum_{i=0}^{n-1} [-\epsilon_i, \epsilon_i] s^i.$$

We view the $\epsilon_i \geq 0$ above as scale factors which determine the aspect ratios of the uncertainty bounding set Q_r. Letting $\mathcal{P}_r$ denote the resulting family of polynomials, our objective is to provide a formula for the robustness margin

$$r_{max} = \sup\{r : \mathcal{P}_r \text{ is robustly stable}\}.$$

To obtain the desired formula, we first argue that Kharitonov's Theorem enables us to reduce the robustness margin problem to four separate problems for the uncertain polynomials $\{p_0(s) + qp_{1,i}(s)\}_{i=1}^{4}$, where

$$p_{1,1}(s) = -\epsilon_0 - \epsilon_1 s + \epsilon_2 s^2 + \epsilon_3 s^3 - \epsilon_4 s^4 - \epsilon_5 s^5 + \epsilon_6 s^6 + \cdots;$$

$$p_{1,2}(s) = \epsilon_0 + \epsilon_1 s - \epsilon_2 s^2 - \epsilon_3 s^3 + \epsilon_4 s^4 + \epsilon_5 s^5 - \epsilon_6 s^6 - \cdots;$$

$$p_{1,3}(s) = \epsilon_0 - \epsilon_1 s - \epsilon_2 s^2 + \epsilon_3 s^3 + \epsilon_4 s^4 - \epsilon_5 s^5 - \epsilon_6 s^6 + \cdots;$$

$$p_{1,4}(s) = -\epsilon_0 + \epsilon_1 s + \epsilon_2 s^2 - \epsilon_3 s^3 - \epsilon_4 s^4 + \epsilon_5 s^5 + \epsilon_6 s^6 - \cdots.$$

Now, applying Theorem 4.7.6 and taking the worst case with respect to $i = 1, 2, 3, 4$, we arrive at the formula

$$r_{max} = \min_{i \leq 4} \frac{1}{\lambda_{max}^{+}(-H^{-1}(p_0)H(p_{1,i}))}.$$

5.10   Robust Stability Testing via Graphics

The Zero Exclusion Condition (see Lemma 5.7.9) suggests a simple graphical procedure for checking robust stability—watch the motion of the Kharitonov rectangle $p(j\omega, Q)$ as ω varies from 0 to $+\infty$ and determine by inspection if the condition $0 \notin p(j\omega, Q)$ is satisfied. This raises the following question: Can we find some finite precomputable *cutoff frequency* $\omega_c > 0$ such that $0 \notin p(j\omega, Q)$ for all $\omega \geq \omega_c$? That is, can we terminate the frequency sweep at the frequency $\omega = \omega_c$?

The existence of ω_c is easily established using the invariant degree condition. Indeed, suppose that $p(s, q) = \sum_{i=0}^n [q_i^-, q_i^+] s^i$ and, without loss of generality, assume that $q_i^- > 0$ for $i = 0, 1, \ldots, n$. Then given any $q \in Q$, it is easy to see that for $\omega \geq 0$,

$$|p(j\omega, q)| \geq q_n^- \omega^n - \sum_{i=0}^{n-1} q_i^+ \omega^i.$$

Since the right-hand side tends to $+\infty$ as $\omega \to +\infty$, it follows that for any prescribed $\beta > 0$ there exists an $\omega_c > 0$ such that $|p(j\omega, q)| \geq \beta$ for all $\omega > \omega_c$. Hence, $0 \notin p(j\omega, Q)$ for all $\omega > \omega_c$.

In fact, we can easily compute an appropriate ω_c. For example, one can take ω_c to be the largest real root of the polynomial

$$f(\omega) = q_n^- \omega^n - \sum_{i=1}^{n-1} q_i^+ \omega^i.$$

Other possibilities for estimating ω_c (often less conservatively) are suggested from classical bounds on the roots of a polynomial. For example, in Marden (1966), it is seen that the roots of a fixed positive coefficient polynomial $p(s) = \sum_{i=0}^n a_i s^i$ lie in a disc of radius

$$R = 1 + \frac{\max\{a_0, a_1, \ldots, a_{n-1}\}}{a_n}.$$

Hence, for the interval polynomial $p(s, q)$, with $q_n^- > 0$, it follows that an appropriate cutoff frequency is given by

$$\omega_c = 1 + \frac{\max\{q_0^+, q_1^+, \ldots, q_{n-1}^+\}}{q_n^-}.$$

EXAMPLE 5.10.1 (Illustration of Graphics Method):   We consider the interval polynomial family $\mathcal{P} = \{p(\cdot, q) : q \in Q\}$ described by

$$p(s, q) = s^6 + [3.95, 4.05]s^5 + [3.95, 4.05]s^4 + [5.95, 6.05]s^3$$
$$+ [2.95, 3.05]s^2 + [1.95, 2.05]s + [0.45, 0.55].$$

In accordance with Lemma 5.7.9, the first step in the graphical test for robust stability requires that we guarantee that at least one polynomial in $\mathcal{P}$ is stable. Using the midpoint of each interval above, we

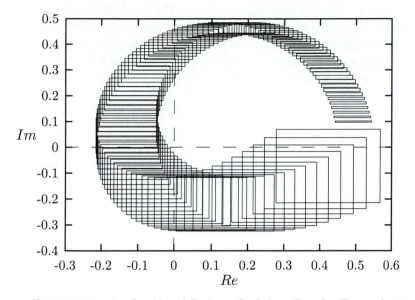

FIGURE 5.10.1   Graphical Robust Stability Test for Example 5.10.1

obtain $p(s, q^0) = s^6 + 4s^5 + 4s^4 + 6s^3 + 3s^2 + 2s + 0.5$, whose roots are $s_1 \approx -3.2681$, $s_{2,3} \approx -0.1328 \pm 0.9473j$, $s_{4,5} \approx -0.0731 \pm 0.7190j$ and $s_6 \approx -0.3201$.

Next, in accordance with the discussion of cutoff frequencies above, we compute the largest real root of the test polynomial $f(\omega)$; that is, with

$$f(\omega) = \omega^6 - 4.05\omega^5 - 4.05\omega^4 - 6.05\omega^3 - 3.05\omega^2 - 2.05\omega - 0.55,$$

we obtain $\omega_c \approx 5.1023$ as an acceptable cutoff frequency for the required Kharitonov rectangle plot. In Figure 5.10.1, we provide a "zoom" of the required plot using 100 evenly spaced frequencies in the critical range $0 \leq \omega \leq 1$. For ω in this range, the Kharitonov rectangle makes its closest approach to $z = 0$. Since $0 \notin p(j\omega, Q)$, we conclude that the family of polynomials $\mathcal{P}$ is robustly stable.

5.11   Overbounding via Interval Polynomials

As mentioned in the introduction to this chapter, the independent uncertainty structure is restrictive because uncertain parameters typically enter into more than one coefficient. For such "dependent" uncertainty structures, we consider two alternatives: The first alternative is to develop more general results; this is the topic of later chapters. The second alternative is the so-called *overbounding method*, which is described below. One warning, however, is in order: Although the overbounding method is easy to use, it may lead to unduly conservative results; i.e., we only obtain sufficient conditions for robustness. In short, associated with overbounding is a trade-off between ease of use and degree of conservatism.

In the remainder of this section, we no longer require the polynomials $p(s, q)$ to have an independent uncertainty structure, and, in addition, Q is not necessarily a box. We begin with the uncertain polynomial $p(s, q) = \sum_{i=0}^{n} a_i(q)s^i$ and an uncertainty bounding set Q which is closed and bounded. Assuming the coefficient functions $a_i(q)$ depend continuously on q, we define the bounds

$$\overline{q}_i^- = \min_{q \in Q} a_i(q)$$

and

$$\overline{q}_i^+ = \max_{q \in Q} a_i(q)$$

and simply observe that the family of polynomials $\overline{\mathcal{P}}$ described by

$$\overline{p}(s, \overline{q}) = \sum_{i=0}^{n} [\overline{q}_i^-, \overline{q}_i^+] s^i$$

is a superset of $\mathcal{P}$. Therefore, any robustness property which holds for the interval polynomial family $\overline{\mathcal{P}}$ must hold for $\mathcal{P}$. In particular, robust stability of $\overline{\mathcal{P}}$ implies robust stability of $\mathcal{P}$. Note, however, that the converse is not true. These points are illustrated via the examples below.

EXAMPLE 5.11.1 (Success of Overbounding):   Consider the family of polynomials $\mathcal{P}$ described by

$$p(s, q) = s^4 + (5 + 0.2q_1q_2 + 0.1q_1 - 0.1q_2)s^3 + (6 + 3q_1q_2 - 4q_2)s^2 + (6 + 6q_1 - 8q_2)s + (0.5 - 3q_1q_2)$$

and uncertainty bound $|q_i| \leq 0.25$ for $i = 1, 2$. The objective is to determine whether $\mathcal{P}$ is robustly stable. To this end, we compute bounds

$$\overline{q}_0^- = \min_{q \in Q} a_0(q) = \min_{-0.25 \leq q_i \leq 0.25} (0.5 - 3q_1q_2) = 0.3125;$$

$$\overline{q}_0^+ = \max_{q \in Q} a_0(q) = \max_{-0.25 \leq q_i \leq 0.25} (0.5 - 3q_1q_2) = 0.6875;$$

$$\bar{q}_1^- = \min_{q \in Q} a_1(q) = \min_{-0.25 \le q_i \le 0.25} (6 + 6q_1 - 8q_2) = 2.5;$$

$$\bar{q}_1^+ = \max_{q \in Q} a_1(q) = \max_{-0.25 \le q_i \le 0.25} (6 + 6q_1 - 8q_2) = 9.5.$$

Similar computations yield $\bar{q}_2^- = 4.8125$, $\bar{q}_2^+ = 7.1875$, $\bar{q}_3^- = 4.9475$ and $\bar{q}_3^+ = 5.0375$. Hence, an interval polynomial family $\overline{\mathcal{P}}$ used for overbounding is described by

$$\bar{p}(s, \bar{q}) = s^4 + [4.9475, 5.0375]s^3 + [4.8125, 7.1875]s^2$$
$$+ [2.5, 9.5]s + [0.3125, 0.6875].$$

By applying Kharitonov's Theorem to the overbounding family $\overline{\mathcal{P}}$ above, it is straightforward to verify that the four Kharitonov polynomials are stable. Hence, from the robust stability of $\overline{\mathcal{P}}$, we conclude that the original family $\mathcal{P}$ must also be robustly stable.

EXERCISE 5.11.2 (Failure of Overbounding):   In this exercise, the objective is to illustrate how overbounding can fail. To this end, consider the family of polynomials $\mathcal{P}$ given in Wei and Yedavalli (1989); i.e., the family $\mathcal{P}$ is described by

$$p(s, q) = s^4 + s^3 + 2qs^2 + s + q$$

with uncertainty bounding set $Q = [1.5, 4]$. Argue that $\mathcal{P}$ is robustly stable but the overbounding family

$$\bar{p}(s, \bar{q}) = s^4 + s^3 + [3, 8]s^2 + s + [1.5, 4]$$

has an unstable Kharitonov polynomial.

5.12   Conclusion

In a sense, Kharitonov's Theorem raises more questions than it answers. To illustrate the type of questions suggested by Kharitonov's Theorem, we consider the robust Schur stability problem for an interval polynomial family $\mathcal{P}$: Indeed, if the associated four Kharitonov polynomials have all their roots in the interior of the unit disc, does it follow that $\mathcal{P}$ is robustly Schur stable? If not, does it suffice to test polynomials associated with all the vertices of Q? More generally, for what type of root location regions does a Kharitonov-like extreme point result hold? The list of possible questions seems endless. In Chapter 13, we characterize classes of $\mathcal{D}$ regions for which $\mathcal{D}$-stability of the polynomials associated with the extreme points of the Q box implies robust $\mathcal{D}$-stability of the associated interval polynomial family.

Notes and Related Literature

NRL 5.1 The paper by Faedo (1953) appears to have provided important motivation for Kharitonov's work.

NRL 5.2 Kharitonov's original proof is based on the Hermite–Biehler Theorem; e.g., see Gantmacher (1959). Indeed, consider a polynomial $p(s)$ decomposed into even and odd parts $p(s) = p_{even}(s^2) + s p_{odd}(s^2)$. Then, according to the Hermite–Biehler Theorem, $p(s)$ is stable if and only if $p_{even}(x)$ and $p_{odd}(x)$ have highest order coefficients of the same sign and negative real distinct interlacing roots; e.g., if polynomial $p(s)$ has odd degree and $x_{e,1} < x_{e,2} < \cdots < x_{e,m}$ and $x_{o,1} < x_{o,2} < \cdots < x_{o,m}$ are the roots of $p_{even}(x)$ and $p_{odd}(x)$, respectively, then $x_{o,1} < x_{e,1} < x_{o,2} < x_{e,2} < \cdots < x_{o,m} < x_{e,m}$. The key idea behind the original proof of Kharitonov's Theorem is as follows: Given an interval polynomial family $\mathcal{P}$, one creates "root intervals" for the even and odd parts. The endpoints of these intervals are associated with the Kharitonov polynomials. Subsequently, it is argued that satisfaction of the root interlacing condition for each Kharitonov polynomial implies satisfaction of the root interlacing condition for the entire family. The Hermite–Biehler line of attack is not pursued in this text because we want to explain as many results as possible within the unifying framework of value sets. The Kharitonov rectangle is in fact an example illustrating the more general value set concept of Chapter 7.

NRL 5.3 The key ideas underlying our proof of Kharitonov's Theorem come from Dasgupta (1988) and Minnichelli, Anagnost and Desoer (1989). More specifically, we note that Dasgupta (1988) exposes the rectangular geometry of $p(j\omega, Q)$ and Minnichelli, Anagnost and Desoer (1989) exploits rectangularity and the Zero Exclusion Condition to obtain a simple proof of the theorem.

NRL 5.4 The paper by Frazer and Duncan (1929) appears to be the first to use the Zero Exclusion Condition in a robust stability context.

NRL 5.5 For more complicated uncertainty structures, Wei and Yedavalli (1989) propose a transformation technique in lieu of overbounding. Their approach involves applying a q-dependent linear transformation to the even or odd parts of $p(s, q)$. As a simple illustration, take $p(s) = p_{even}(s^2) + s p_{odd}(s^2)$ and suppose, $R(q)$ and $I(q)$ are positive functions of q. Defining the transformed polynomial $\tilde{p}(s, q) = R(q) p_{even}(s^2) + s I(q) p_{odd}(s^2)$, it is easy to show that robust stability remains invariant and in some cases, a reduction of conservatism may result. The potential for further research involving such methods is illustrated by the family $\mathcal{P}$ in Exercise 5.11.2. A robust stability test based on overbounding by an interval polynomial is inconclusive but multiplication of the even part by $1/q$ and the odd part by unity leads to an interval polynomial whose robust stability is easily verified by Kharitonov's Theorem.

NRL 5.6 There are a number of papers in the literature involving transformations aimed at facilitating robust stability analysis. For example, using the shifted circles in Petersen (1989), one can deal with the so-called Delta transform for a discrete-time system; for similar extreme point results involving Delta transformation, see also Soh (1991). The paper by Vaidyanathan (1990) provides another example of a transformation used for discrete-time problems.

NRL 5.7 Some alternatives to the technique described in Section 5.11 are given in papers by Djaferis (1991) and Pujara (1990). These papers describe different overbounding families which are sometimes useful.

NRL 5.8 Rather than working with the original coefficients, one can consider a bounding box B in the space of Markov parameters. By breaking an n-th order $p(s)$ into its even and odd parts as $p(s) = p_{even}(s^2) + s p_{odd}(s^2)$, a continued fraction expansion for $p_{odd}(x)/p_{even}(x)$ leads to the set of Markov parameters; see Gantmacher (1959). If $n = 2m$, we obtain parameters $(b_0, b_1, \ldots, b_{2m-1})$, and if $n = 2m - 1$, we obtain $(b_{-1}, b_0, \ldots, b_{2m-1})$. With this representation, robust stability is guaranteed if and only if two distinguished polynomials are stable. For example, if $n = 2m$ and the box B is described by $b_i^- \leq b_i \leq b_i^+$ for $i = 0, 1, 2, \ldots, 2m - 1$, the first distinguished polynomial has Markov parameters $(b_0^-, b_1^+, b_2^-, \ldots, b_{2m-1}^+)$ and the second distinguished polynomial has Markov parameters $(b_0^+, b_1^-, b_2^+, \ldots, b_{2m-1}^-)$; see Hollot (1989) for further elaboration. Of course, a fundamental limitation of these results is that the relationship between the Markov parameters and the original parameters is generally quite complicated. This complication motivates interesting research problems involving system identification for robust control.

NRL 5.9 We mention a body of work aimed at generalization of Kharitonov's Theorem to *scattering Hurwitz polynomials*. For example, in papers by Bose (1988), Kim and Bose (1988) and Basu (1989), the uncertain polynomial $p(s, q)$ is replaced by a multivariate uncertain polynomial $p(s_1, s_2, \ldots, s_n, q)$ and interval bounds on the coefficients are imposed.

Chapter 6

Embellishments of Kharitonov's Theorem

Synopsis

For interval polynomial families, this chapter provides a number of extensions, refinements and alternatives to Kharitonov's Theorem. Of particular note is the Tsypkin–Polyak plot for easy visualization of robustness margins.

6.1 Introduction

Once one is familiar with the technical ideas associated with the Kharitonov rectangle, it becomes possible to develop many extensions and refinements of the results in Chapter 5. This chapter concentrates on an important subset of these extensions and refinements. In particular, when considering the robust stability problem for low order interval polynomials. We see that fewer than four Kharitonov polynomials need only be tested. The chapter also includes extensions of Kharitonov's Theorem for problems involving degree dropping and problems involving complex coefficients. Finally, two "alternative" robust stability tests are described. The first involves plotting a scalar function of frequency and the second involves a Nyquist-like plot in the complex plane.

6.2   Low Order Interval Polynomials

The objective in this section is to establish the fact that less than four Kharitonov polynomials are needed for robust stability testing when an interval polynomial has degree five or less; this is the result of Anderson, Jury and Mansour (1987).

THEOREM 6.2.1 (Simplified Kharitonov Theorem for Degree $n = 3$): *Consider an interval polynomial family $\mathcal{P}$ with invariant degree $n = 3$ and lowest order coefficient bound $q_0^- > 0$. Then $\mathcal{P}$ is robustly stable if and only if the single Kharitonov polynomial $K_3(s)$ is stable.*

PROOF: As in the proof of Kharitonov's Theorem, necessity follows immediately from the fact that $K_3(s) \in \mathcal{P}$. To establish sufficiency, we now assume that $K_3(s)$ is stable and must prove that $\mathcal{P}$ is robustly stable. In view of the Zero Exclusion Condition (Lemma 5.7.9), we must show that $0 \notin p(j\omega, Q)$ for all $\omega \geq 0$. Indeed, as ω ranges from 0 to $+\infty$, the Monotonic Angle Property (Lemma 5.7.6) indicates that $\measuredangle K_3(j\omega)$ increases monotonically from 0 to $3\pi/2$. A typical trajectory of $K_3(j\omega)$, obtained by varying ω from 0 to $+\infty$, is shown in Figure 6.2.1.

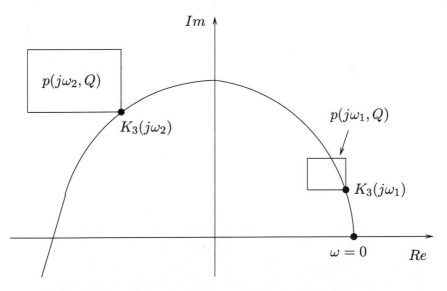

FIGURE 6.2.1   Typical Trajectory of $K_3(j\omega)$ for $\omega \geq 0$

The proof is now completed by noting that all points in the Kharitonov rectangle $p(j\omega, Q)$ lie northwest of $K_3(j\omega)$ for $\omega \geq 0$. Therefore, if $\measuredangle K_3(j\omega)$ increases monotonically from 0 to $3\pi/2$, avoiding $z = 0$, it is impossible for zero to enter $p(j\omega, Q)$ at any frequency $\omega \geq 0$. Hence, $\mathcal{P}$ is robustly stable. ∎

EXERCISE 6.2.2 (The Role of $q_0^- > 0$):   Does Theorem 6.2.1 hold without the assumption $q_0^- > 0$? Explain.

EXERCISE 6.2.3 (Result for $n = 4$ and $n = 5$):   Consider an interval polynomial family $\mathcal{P}$ with invariant degree $n = 4$ and lowest order coefficient bound $q_0^- > 0$.
(a) Prove that $\mathcal{P}$ is robustly stable if and only if the two Kharitonov polynomials $K_2(s)$ and $K_3(s)$ are stable.
(b) Now, establish a three-polynomial result for an interval polynomial family with invariant degree $n = 5$. Note that the statement of the result no longer requires the assumption $q_0^- > 0$.

6.3   Extensions with Degree Dropping

The analysis of singular control problems provides motivation for asking if Kharitonov's Theorem can be extended to accommodate degree dropping. To illustrate how degree dropping arises, suppose that one begins with the standard singular system description

$$E(\epsilon)\dot{x}(t) = Ax(t)$$

with $x(t) \in \mathbf{R}^n$ and matrix $E(\epsilon)$ being singular for some values of the real parameter ϵ. For example, it is often the case that $E(\epsilon)$ is singular for $\epsilon = 0$. We observe that the characteristic polynomial

$$p(s, \epsilon) = \det(sE(\epsilon) - A)$$

can exhibit degree dropping for values of ϵ rendering $E(\epsilon)$ singular. As a simple illustration, if $E(\epsilon) = \epsilon$ and $A = -1$, the polynomial $p(s, \epsilon) = \epsilon s + 1$ drops in degree from one to zero when $\epsilon = 0$.

We consider $n \geq 2$ and first dispose with trivial cases for which $p(s, q) = \sum_{i=0}^{n} [q_i^-, q_i^+] s^i$ can drop in degree with coefficients of opposite sign for some member of the family. For example, if $q_n^+ > 0$ and $q_{n-1}^- < 0$, then the polynomial

$$p^*(s) = q_n^+ s^n + q_{n-1}^- s^{n-1} + \sum_{i=0}^{n-2} q_i^+ s^i$$

is a member of the family but cannot be stable because its coefficients do not all have the same sign.

In view of the argument above, henceforth we concentrate on the case when $q_i^- \geq 0$ for $i = 0, 1, \ldots, n$. First, we consider the case when the degree can drop by three or more.

EXERCISE 6.3.1 (Degree Drop of Three or More):   Consider the interval polynomial $p(s, q) = \sum_{i=0}^{n} [q_i^-, q_i^+] s^i$ with degree $n \geq 4$, $q_n^- = q_{n-1}^- = q_{n-2}^- = 0$ and $q_i^+ > 0$ for $i = 0, 1, 2, 3, 4$. Prove that this family of polynomials is not robustly stable; this is the result of Meerov (1947). *Hint*: With $f(s) = \sum_{i=1}^{n} q_i^+ s^i$ and $g(s) = \sum_{i=0}^{n-3} q_i^+ s^i$, relate the robust stability problem to a root locus problem for the fictitious plant $P(s) = g(s)/f(s)$. Now study the asymptotes associated with this root locus problem.

EXERCISE 6.3.2 (Degree Drop of One):   In this exercise, we examine some of the issues addressed in Mori and Kokame (1992). The starting point is an interval polynomial family $\mathcal{P}$ described by $p(s, q) = \sum_{i=0}^{n} [q_i^-, q_i^+] s^i$.
(a) Suppose that $n \leq 4$ and assume a degree drop of at most one; say $q_n^- = 0$, $q_n^+ > 0$ and $q_{n-1}^- > 0$. Show that robust stability of $\mathcal{P}$ is equivalent to stability of the four Kharitonov polynomials.
(b) For degree drops of one at most and $n \geq 5$, show that robust stability $\mathcal{P}$ is equivalent to stability of six polynomials—the four full order Kharitonov polynomials plus the two auxiliary polynomials

$$K_5(s) = q_{n-1}^- s^{n-1} + q_{n-2}^- s^{n-2} + q_{n-3}^+ s^{n-3} + q_{n-4}^+ s^{n-4} + \cdots$$

and

$$K_6(s) = q_{n-1}^+ s^{n-1} + q_{n-2}^- s^{n-2} + q_{n-3}^- s^{n-3} + q_{n-4}^+ s^{n-4} + \cdots$$

associated with degree dropping.

6.4   Interval Plants with Unity Feedback

We now develop an extension of Kharitonov's Theorem to a class of unity feedback control systems. To this end, a definition is required.

DEFINITION 6.4.1 (Interval Plants):   An *interval plant family* $\mathcal{P}$ is described by

$$P(s, q, r) = \frac{N(s, q)}{D(s, r)}$$

with an uncertain numerator polynomial $N(s, q) = \sum_{i=0}^{m} q_i s^i$, uncertain denominator polynomial $D(s, r) = \sum_{i=0}^{n} r_i s^i$ and boxes Q and

R as uncertainty bounding sets for q and r, respectively; i.e., $\mathcal{P}$ is a quotient of interval polynomial families. For notational simplicity, we can write

$$P(s,q,r) = \frac{\displaystyle\sum_{i=0}^{m}[q_i^-,q_i^+]s^i}{\displaystyle\sum_{i=0}^{n}[r_i^-,r_i^+]s^i}.$$

We use the notation $\mathcal{P} = \{P(\cdot,q,r) : q \in Q; r \in R\}$ and often refer to $\mathcal{P}$ simply as an *interval plant*.

REMARKS 6.4.2 (Lumping for Interval Plants): For a proper interval plant defined by the condition $n \geq m$, observe that if unity feedback is used, we can lump uncertainties in the closed loop polynomial; i.e.,

$$\begin{aligned}
p(s,q,r) &= \sum_{i=0}^{m}[q_i^-,q_i^+]s^i + \sum_{i=0}^{n}[r_i^-,r_i^+]s^i \\
&= \sum_{i=0}^{m}[q_i^- + r_i^-, q_i^+ + r_i^+]s^i + \sum_{i=m+1}^{n}[r_i^-,r_i^+]s^i.
\end{aligned}$$

In other words, the imposition of unity feedback preserves the interval polynomial structure. This fact is slightly generalized in the exercise below.

EXERCISE 6.4.3 (Pure Gain Compensator): In this exercise, the result of Ghosh (1985) is established. Indeed, consider an interval plant with pure gain compensator $C(s) = K$ and give conditions on the uncertainty bounds under which the closed loop polynomial has invariant degree. Under such conditions, consider $K \geq 0$ and $K < 0$ as separate cases and argue that the family of closed loop polynomials is robustly stable if and only if four Kharitonov polynomials associated with this family are stable. Describe the four *Kharitonov plants* associated with the four Kharitonov polynomials. Notice that the sign of K is instrumental to the selection of the four relevant plants.

EXERCISE 6.4.4 (Interval Plant Calculation): Consider a unity feedback control system with interval plant

$$P(s,q) = \frac{[0.75, 1.25]s + [0.75, 1.25]}{s^3 + [2.75, 3.25]s^2 + [8.75, 9.25]s + [0.75, 9.25]}.$$

(a) Using Theorem 6.2.1, determine if the resulting family of closed loop polynomials is robustly stable.

(b) Suppose that instead of using a unity feedback, a compensator $C(s) = 1/s$ is used. Use the result in Exercise 6.2.3 to determine if the resulting family of closed loop polynomials is robustly stable.

EXERCISE 6.4.5 (Nonunity Feedback): Consider the interval plant family of Hollot and Yang (1990) described by

$$P(s,q) = \frac{q}{(s+0.1)(s+0.2)}$$

and $Q = [1, 5000]$ with compensator

$$C(s) = \frac{(s+3)(s+4)}{s(s+25)(s+75)}$$

connected in a feedback configuration. Using the uncertain closed loop polynomial

$$p(s,q) = N(s,q)N_C(s) + D(s,q)D_C(s),$$

verify that the family $\mathcal{P} = \{p(\cdot, q) : q \in Q\}$ has invariant degree, stable extremes $p(s, 1)$ and $p(s, 5000)$ but $p(s, 2)$ is unstable.

6.5   Frequency Sweeping Function $H(\omega)$

This embellishment of Kharitonov's Theorem can be contrasted to the graphical test for robust stability given in Section 5.10. Recall that the graphical test for robust stability involves checking zero exclusion from the Kharitonov rectangle as ω is varied from 0 to $+\infty$. In this section, we see that we can study robust stability by checking for positivity of a specially constructed scalar function of frequency $H(\omega)$. Hence, instead of generating two-dimensional Kharitonov rectangles, we can examine the plot of the scalar function $H(\omega)$ to determine if the family of polynomials $\mathcal{P}$ is robustly stable. The theorem below can be viewed as a frequency domain alternative to Kharitonov's Theorem.

THEOREM 6.5.1 (Barmish (1989)): *Let $\mathcal{P}$ be an interval polynomial family with invariant degree, at least one stable member and associated Kharitonov polynomials $K_1(s)$, $K_2(s)$, $K_3(s)$ and $K_4(s)$. Then, with*

$$H(\omega) = \max\{Re\ K_1(j\omega), -Re\ K_2(j\omega), Im\ K_3(j\omega), -Im\ K_4(j\omega)\},$$

it follows that $\mathcal{P}$ *is robustly stable if and only if*

$$H(\omega) > 0$$

for all frequencies $\omega \geq 0$.

PROOF: Letting $p(j\omega, Q)$ denote the Kharitonov rectangle as defined in Section 5.7.1, recall the Zero Exclusion Condition (Lemma 5.7.9) indicates that $\mathcal{P}$ is robustly stable if and only if $0 \notin p(j\omega, Q)$ for all frequencies $\omega \geq 0$. Now, we refer to the Kharitonov rectangle in Figure 5.7.2 and note that the argument to follow does not require $p(j\omega, Q)$ to be in the first quadrant. Indeed, at frequency $\omega \geq 0$, zero is excluded from the Kharitonov rectangle $p(j\omega, Q)$ if and only if one or more of the following conditions holds: The point $z = 0$ lies to the left of the western boundary of $p(j\omega, Q)$ in the sense that $Re\ K_1(j\omega) > 0$; the point $z = 0$ lies to the right of the eastern boundary of $p(j\omega, Q)$ in the sense that $Re\ K_2(j\omega) < 0$; the point $z = 0$ lies below the southern boundary of $p(j\omega, Q)$ in the sense that $Im\ K_3(j\omega) > 0$; the point $z = 0$ lies above the northern boundary of $p(j\omega, Q)$ in the sense that $Im\ K_4(j\omega) < 0$. Equivalently, $0 \notin p(j\omega, Q)$ if and only if $Re\ K_1(j\omega) > 0$ or $-Re\ K_2(j\omega) > 0$ or $Im\ K_3(j\omega) > 0$ or $-Im\ K_4(j\omega) > 0$. The "or" of these four conditions is equivalent to the requirement that

$$\max\{Re\ K_1(j\omega), -Re\ K_2(j\omega), Im\ K_3(j\omega), -Im\ K_4(j\omega)\} > 0.$$

That is, $0 \notin p(j\omega, Q)$ for all $\omega \geq 0$ if and only if $H(\omega) > 0$ for all frequencies $\omega \geq 0$. ∎

EXAMPLE 6.5.2 (Plotting $H(\omega)$):   To illustrate the use of $H(\omega)$, we consider the interval polynomial

$$p(s, q) = [0.75, 1.25]s^3 + [2.75, 3.25]s^2 + [0.75, 1.25]s + [0.75, 1.25]$$

which was already analyzed using Kharitonov's Theorem in Example 5.6.2. To apply Theorem 6.5.1, we first verify that the family of polynomials has at least one stable member. Indeed, with $q_0 = q_1 = q_3 = 1.25$ and $q_2 = 3.25$, stability is trivially verified. Next, by a straightforward substitution into the $H(\omega)$ formula above, we obtain

$$H(\omega) = \max\{0.75 - 3.25\omega^2, -1.25 + 2.75\omega^2, 0.75\omega + 1.25\omega^3, -1.25\omega - 0.75\omega^3\}.$$

A plot of $H(\omega)$ versus ω is indicated in Figure 6.5.1. Since this function remains positive for all $\omega \geq 0$, Theorem 6.5.1 guarantees

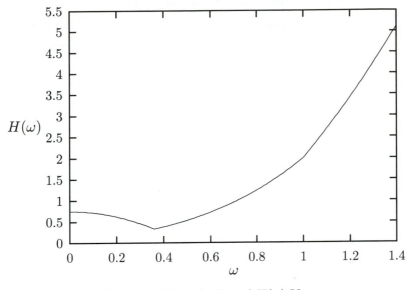

FIGURE 6.5.1   A Plot of $H(\omega)$ Versus ω

robust stability. Note that we do not carry the frequency sweep all the way out to $\omega = +\infty$ because we can guarantee apriori that $H(\omega) > 0$ beyond some cutoff frequency $\omega = \omega_c$; see the discussion in Section 5.10 and the exercise below.

EXERCISE 6.5.3 (High-Frequency Behavior):   Given an interval polynomial family with invariant degree, prove that the condition

$$\lim_{\omega \to \infty} H(\omega) = +\infty$$

is satisfied.

EXERCISE 6.5.4 (Relationship with Kharitonov Rectangle):   For complex numbers $z \in \mathbf{C}$, consider the classical *max norm* defined by

$$\|z\|_\infty = \max\{|Re\ z|, |Im\ z|\}.$$

Under the hypotheses of Theorem 6.5.1, prove that if $0 \notin p(j\omega, Q)$ at some frequency $\omega \geq 0$, then

$$H(\omega) = \min_{z \in p(j\omega, Q)} \|z\|_\infty.$$

That is, at a given frequency $\omega \geq 0$, $H(\omega)$ can be viewed as the distance in max norm from the origin to the closest point (in max norm) to the Kharitonov rectangle $p(j\omega, Q)$.

6.6   Robustness Margin Geometry

For a robustly stable interval polynomial family, there is a temptation to associate the distance of the Kharitonov rectangle to the origin $z = 0$ with the robustness margin. Said another way, when $p(j\omega, Q)$ remains "far away" from $z = 0$ for all $\omega \geq 0$, there is a temptation to conclude that there is a "significant" robustness margin; when $p(j\omega, Q)$ is "close" to zero for some frequencies, the temptation is to conclude that the robustness margin is small. The main objective of this section is to show that such reasoning can be fallacious. This provides motivation for the Tsypkin–Polyak analysis in the next section.

To quantify the idea above, we study the behavior of the distance between the origin $z = 0$ and the Kharitonov rectangle $p(j\omega, Q)$. We want to find the minimum of this distance with respect to frequency; i.e., we want to calculate the closest distance between all possible Kharitonov rectangles and the origin. For example, two natural measures for minimum distance are

$$d_{min} = \min\{\|z\|_\infty : z \in p(j\omega, Q);\ \omega \geq 0\}$$

and

$$d'_{min} = \min\{|z| : z \in p(j\omega, Q);\ \omega \geq 0\}.$$

In the example below, we work with d_{min} noting that the conclusions which we draw are also valid for other minimum distance measures such as d'_{min} above.

EXAMPLE 6.6.1 (Inadequacy of the Distance Measure):   We consider the ϵ-parameterized uncertain polynomial

$$p_\epsilon(s, q) = s^2 + \epsilon s + (100 + q)$$

with uncertainty bounding set $Q_r = [-r, r]$. Since the resulting family of polynomials is monic and second order, stability is equivalent to positivity of coefficients. Hence, robust stability is guaranteed if and only if $r < 100$. In other words, a natural robustness margin based on parameter space considerations leads to $r_{max} = 100$. The important point to note is that this robustness margin is invariant with respect to $\epsilon > 0$.

We now compute the distance measure $d_{min}(\epsilon)$ as a function of the parameter $\epsilon > 0$. Indeed, given any frequency $\omega \geq 0$, we obtain the Kharitonov rectangle by first noting that the value set

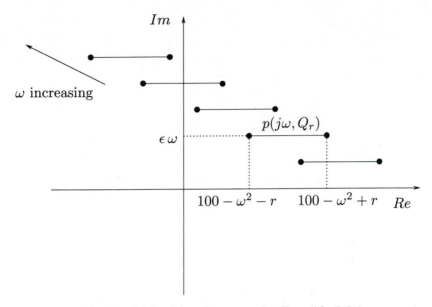

FIGURE 6.6.1   Line Segments for Example 6.6.1

is a straight line segment with constant imaginary part; i.e., with $Re\ p(j\omega, q) = 100 + q - \omega^2$ and $Im\ p(j\omega, q) = \epsilon\omega$, we obtain

$$p_\epsilon(j\omega, Q_r) = \{z \in \mathbf{C} : 100 - \omega^2 - r \le Re\ z \le 100 - \omega^2 + r;\ Im\ z = \epsilon\omega\}$$

which is depicted in Figure 6.6.1. Note that $p_\epsilon(j\omega, Q_r)$ is generated using the two Kharitonov polynomials

$$K_1(s) = K_4(s) = s^2 + \epsilon s + (100 - r)$$

to describe the left endpoint and the two Kharitonov polynomials

$$K_2(s) = K_3(s) = s^2 + \epsilon s + (100 + r)$$

to describe the right endpoint.

Now, with the help of the Figure 6.6.1, for $r < 100$ and fixed frequency $\omega \ge 0$, it is easy to see that

$$\min_{z \in p(j\omega, Q)} \|z\|_\infty = \max\{100 - \omega^2 - r, -100 + \omega^2 - r, \epsilon\omega\}.$$

Next, we compute the frequency at which $p(j\omega, Q)$ is closest to $z = 0$ by setting $\epsilon\omega = 100 - \omega^2 - r_1$. This leads to

$$d_{min}(\epsilon) = -\frac{\epsilon^2}{2} + \frac{\epsilon\sqrt{\epsilon^2 + 4(100 - r)}}{2}.$$

Hence, we conclude that $d_{min}(\epsilon) \to 0$ as $\epsilon \to 0$. Notice that this conclusion holds even if the uncertainty bound r is small. The main point to note is that smallness of $d_{min}(\epsilon)$ for small r is inconsistent with parameter space considerations; i.e., even if ϵ is small, the robustness margin is $r_{max} = 100$. For small $r \geq 0$, the fact that $d_{min}(\epsilon)$ can get arbitrarily small as $\epsilon \to 0$ tells us that d_{min} is not a good robustness indicator.

REMARKS 6.6.2 (Understanding the Example Above):   To enhance our understanding of the example above, we study the roots of $p_\epsilon(s, q)$ as a function of $\epsilon > 0$. By setting $p_\epsilon(s, q) = 0$, we find the pair of roots

$$s_{1,2}(q) = -\frac{\epsilon}{2} \pm j\sqrt{(100 + q) - \left(\frac{\epsilon}{2}\right)^2}.$$

Hence, for $|q| < 100$ and small $\epsilon > 0$, we obtain a lightly damped root pair which approaches the imaginary axis as $\epsilon \to 0$. This manifests itself via smallness of $d_{min}(\epsilon)$ and closeness of the Kharitonov rectangle to the origin. In other words, smallness of d_{min} goes hand in hand with closeness of the roots of $p(s, q)$ to the imaginary axis. As $\epsilon \to 0$, the roots approach the imaginary axis even when the robustness margin for q is large.

6.7   The Tsypkin–Polyak Function

Motivated by the comparison between d_{min} and r_{max} above, the main objective of this section is to describe a technique for graphical visualization of the robustness margin. We want to generate a plot which, upon inspection by eye, provides easily understood information about the robustness margin for stability. Recognizing that a plot of the Kharitonov rectangle does not explicitly provide such information, we now proceed to construct the robust stability testing function described by Tsypkin and Polyak (1991).

In the analysis to follow, we use the same notational convention as in Section 5.9. That is, we emphasize the dependence on the uncertainty bound $r \geq 0$ by writing

$$p_r(s, q) = p_0(s) + r \sum_{i=0}^{n-1} [-\epsilon_i, \epsilon_i] s^i$$

and interpret the $\epsilon_i \geq 0$ above as scale factors which determine the aspect ratios of the uncertainty bounding set Q_r. We call $p_0(s)$ the

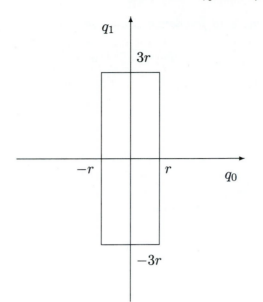

FIGURE 6.7.1   Uncertainty Bounding Set for Example 6.7.1

nominal polynomial and study the interval polynomial family

$$\mathcal{P}_r = \{p_r(\cdot, q) : q \in Q_r\}.$$

To avoid obfuscating technicalities involving degree dropping, notice that uncertain parameters only enter coefficients up to order $n - 1$ above. In Theorem 6.7.2 to follow, we make use of the max norm on $z \in \mathbf{C}$. We recall that $\|z\|_\infty = \max\{|Re\ z|, |Im\ z|\}$.

EXAMPLE 6.7.1 (Uncertainty Bounds): For the interval polynomial family described by $p_r(s, q) = (s^2 + 10s + 5) + r([-3, 3]s + [-1, 1])$, the uncertainty bounding set Q_r is shown in Figure 6.7.1. Notice how the weights $\epsilon_0 = 1$ and $\epsilon_1 = 3$ are used for shaping Q_r and the scalar $r \geq 0$ is used for magnification.

THEOREM 6.7.2 (Tsypkin and Polyak (1991)):   *For fixed $r \geq 0$, consider the interval polynomial family $\mathcal{P}_r$ with order $n \geq 2$, positive weights $\epsilon_0, \epsilon_1, \ldots, \epsilon_{n-1}$ and stable nominal $p_0(s)$. Let*

$$G_{TP}(\omega) = \frac{Re\ p_0(j\omega)}{\displaystyle\sum_{i\ even} \epsilon_i \omega^i} + j\frac{Im\ p_0(j\omega)}{\displaystyle\sum_{i\ odd} \epsilon_i \omega^i}.$$

Then, with max norm on $z \in \mathbf{C}$, it follows that $\mathcal{P}_r$ is robustly stable *if and only if the zero frequency condition*

$$|p_0(j0)| > r\epsilon_0$$

is satisfied and

$$\|G_{TP}(\omega)\|_\infty > r$$

for all frequencies $\omega \geq 0$.

PROOF: Since $p_0(s)$ is assumed stable, application of the Zero Exclusion Condition (Lemma 5.7.9) indicates that $\mathcal{P}_r$ is robustly stable if and only if $0 \notin p_r(j\omega, Q)$ for all $\omega \geq 0$. First, for $\omega = 0$, the necessary and sufficient condition for zero exclusion is $|p_0(j0)| > r\epsilon_0$. Now, using the formula for $p_r(s, q)$, it is easy to see that for fixed $\omega \geq 0$, the Kharitonov rectangle $p_r(j\omega, Q)$ is centered at $p_0(j\omega)$, has width given by

$$d_R(\omega) = 2r \sum_{i \text{ even}} \epsilon_i \omega^i$$

in the real coordinate direction and has height

$$d_I(\omega) = 2r \sum_{i \text{ odd}} \epsilon_i \omega^i$$

in the imaginary coordinate direction. This rectangle is depicted in Figure 6.7.2. From this description of $p_r(j\omega, Q)$, it is obvious that $0 \notin p_r(j\omega, Q)$ if and only if either $|Re\ p_0(j\omega)| > d_R(\omega)/2$ or $|Im\ p_0(j\omega)| > d_I(\omega)/2$. Equivalently, $0 \notin p_r(j\omega, Q)$ if and only if either $|Re\ G_{TP}(\omega)| > r$ or $|Im\ G_{TP}(\omega)| > r$. That is, robust stability is guaranteed if and only if the condition $\|G_{TP}(\omega)\|_\infty > r$ is satisfied for all frequencies $\omega \geq 0$. ∎

REMARKS 6.7.3 (Graphical Visualization):   The theorem of Tsypkin and Polyak suggests a natural procedure for graphical visualization of robustness margin

$$r_{max} = \sup\{r : \mathcal{P}_r \text{ is robustly stable}\}.$$

That is, we generate a Nyquist-like plot of the complex function $G_{TP}(\omega)$ and imagine a square box centered at $z = 0$. The radius r of this box is initially small so that the box fits entirely "inside" the $G_{TP}(\omega)$ plot. Next, we let the radius of the box expand until a first contact is made with the $G_{TP}(\omega)$ plot. Now, in accordance with the theorem, we denote the radius of the box associated with this first contact as r_{max} (see Figure 6.7.3). Now, taking the zero frequency

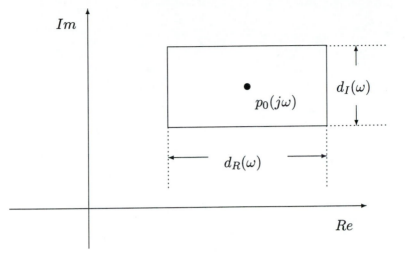

FIGURE 6.7.2   The Kharitonov Rectangle $p_r(j\omega, Q)$

condition into account we let $r_0 = |p_0(j0)|/\epsilon$ and obtain

$$r_{max} = \min\{r_0, r_{max}^+\}.$$

EXAMPLE 6.7.4 (Application of Theorem 6.7.2):   We consider the interval polynomial family $\mathcal{P}_r$ with nominal

$$p_0(s) = s^6 + 15s^5 + 104s^4 + 420s^3 + 1019s^2 + 1365s + 676$$
$$= (s+1)(s+4)(s+2+3j)(s+2-3j)(s+3+2j)(s+3-2j)$$

and scaling factors $\epsilon_0 = 676$, $\epsilon_1 = 682.5$, $\epsilon_2 = 509.5$, $\epsilon_3 = 210$, $\epsilon_4 = 52$, $\epsilon_5 = 15$ and $\epsilon_6 = 1$. Now, the function to be plotted is

$$G_{TP}(\omega) = \frac{-\omega^6 + 104\omega^4 - 1019\omega^2 + 676}{\omega^6 + 52\omega^4 + 509.5\omega^2 + 676} + j\,\frac{15\omega^4 - 420\omega^2 + 1365}{15\omega^4 + 210\omega^2 + 682.5}.$$

By examination of Figure 6.7.4, the radius of the largest inscribed box is $r_{max}^+ \approx 0.2227$. Hence, $r_{max} \approx \min\{1, 0.2227\} = 0.2227$.

EXERCISE 6.7.5 (Pitch Control Loop):   In this exercise, we consider a robust stability problem associated with a pitch control loop for an unmanned free-swimming submersible vehicle. We begin with the

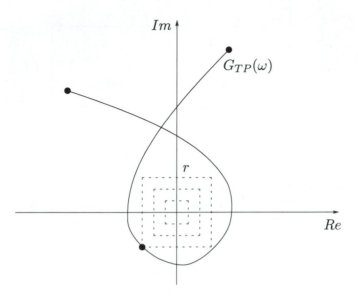

FIGURE 6.7.3   First Contact with the $G_{TP}(\omega)$ Plot

nominal model described in Nise (1992). The plant transfer function, given in Figure 6.7.5, represents the aggregation of elevator actuator, vehicle dynamics and pitch rate sensor.

(a) Verify that the nominal closed loop system is stable for $K = 4$.

(b) With a nominal closed loop polynomial $p_0(s) = \sum_{i=0}^{n} a_i s^i$ obtained in (a), take uncertainty weights ϵ_i based on percentages of the a_i; i.e., with $\epsilon_i = a_i$ for $i = 0, 1, 2, \ldots, n-1$, find the robustness margin r_{max} by plotting the function $G_{TP}(\omega)$.

(c) Using the percentage weighting scheme in (b), study the effect of increasing the loop gain K on r_{max}; note that instability occurs when the gain satisfies $K > 25.9$.

EXERCISE 6.7.6 (Extension): In the setup for Theorem 6.7.2, note that all weights ϵ_i were assumed positive. This exercise is concerned with the case when only a subset of the weights is positive. In other words, only a subset of coefficients is uncertain. Indeed, we now assume that $\epsilon_i \geq 0$ for $i = 0, 1, 2, \ldots, n-1$, allowing for the possibility that $\epsilon_i = 0$.

(a) For the weakened hypotheses that $\epsilon_i = 0$ for all i odd and $\epsilon_i \neq 0$

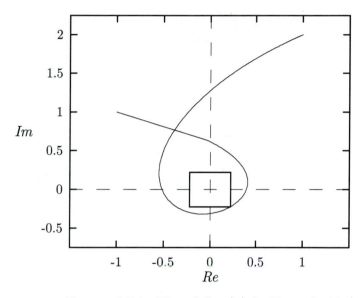

FIGURE 6.7.4   Plot of $G_{TP}(\omega)$ for Example 6.7.4

for at least one i even, argue that Theorem 6.7.2 remains valid using

$$
G_{TP}(\omega) = \begin{cases} \dfrac{Re\ p_0(j\omega)}{\displaystyle\sum_{i\ even} \epsilon_i \omega^i} & \text{if } Im\ p_0(j\omega) = 0; \\[2em] +\infty & \text{if } Im\ p_0(j\omega) \neq 0. \end{cases}
$$

(b) For the case when $\epsilon_i = 0$ for all i even and $\epsilon_i \neq 0$ for all at least one i odd, describe the appropriately modified $G_{TP}(\omega)$ function.

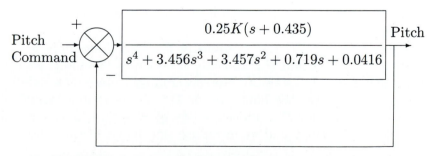

FIGURE 6.7.5   Nominal Pitch Control System for Exercise 6.7.5

EXERCISE 6.7.7 (Reconciliation):   The objective of this exercise is to utilize the Tsypkin–Polyak function to carry out a robust stability analysis for the ϵ-parameterized uncertain polynomial given by $p_\epsilon(s,q) = s^2 + \epsilon s + (100 + q)$ and studied by commonsense analysis in Example 6.6.1. Using the Tsypkin–Polyak framework, show that a formal calculation produces the correct result $r_{max} = 100$ even when the parameter $\epsilon > 0$ is small.

6.8   Complex Coefficients and Transformations

The final set of embellishments which we consider involves uncertain polynomials with complex coefficients. Before proceeding, the obvious question to ask is how complex coefficient polynomials might arise. Although the complex coefficient case can be motivated from modelling considerations, more direct motivation is derived from the fact that the solution of many complex variable problems is often facilitated via transformation. In some cases, a real coefficient polynomial is transformed into a real coefficient polynomial (for example, a bilinear transformation), while in other cases, a transformation of a real coefficient polynomial leads to a complex coefficient polynomial. This is illustrated via the example below.

EXAMPLE 6.8.1 (How Complex Coefficients Arise):   We begin with a family of polynomials $\mathcal{P} = \{p(\cdot, q) : q \in Q\}$ having real coefficients, and suppose that we are dealing with a control problem where damping is of concern. For example, say that the damping cone $\mathcal{D}$ for the roots of $p(s,q)$ is given in Figure 6.8.1. We now argue that we can transform the real coefficient robust $\mathcal{D}$-stability problem at hand to a (strict left half plane) robust stability problem with complex coefficients. Indeed, we first express $\mathcal{D}$ as an intersection of two half planes; i.e., we write $\mathcal{D} = \mathcal{D}^+ \cap \mathcal{D}^-$, where $\mathcal{D}^- = \{z \in \mathbf{C} : \pi - \phi < 4z < 2\pi - \phi\}$ is the lower halfplane and $\mathcal{D}^+ = \{z \in \mathbf{C} : \phi < 4z < \pi + \phi\}$ is the upper halfplane. These two regions are shown in Figure 6.8.2.

We now observe that $\mathcal{P}$ is robustly $\mathcal{D}$-stable if and only if $\mathcal{P}$ is robustly $\mathcal{D}^+$-stable and $\mathcal{P}$ is robustly $\mathcal{D}^-$-stable. Therefore, we can study the robust $\mathcal{D}^+$-stability problem and the robust $\mathcal{D}^-$-stability problem separately. We illustrate for $\mathcal{D}^+$ noting that an identical argument is used for $\mathcal{D}^-$. Indeed, if $p(s,q) = \sum_{i=0}^{n} a_i(q)s^i$ is a real coefficient polynomial and we introduce the change of variables

$$z = se^{-j(\frac{\pi}{2} - \phi)},$$

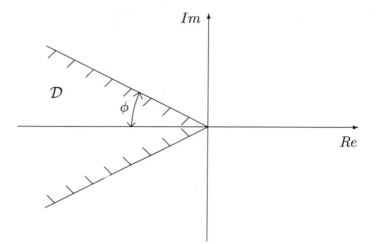

FIGURE 6.8.1   Damping Cone $\mathcal{D}$ for Example 6.8.1

we obtain the *transformed polynomial*

$$\tilde{p}(z, q) = p(ze^{j(\frac{\pi}{2} - \phi)}, q)$$
$$= a_0(q) + a_1(q)ze^{j(\frac{\pi}{2} - \phi)} + \cdots + a_n(q)z^n e^{jn(\frac{\pi}{2} - \phi)}$$
$$= a_0(q) + a_1(q)[\sin \phi + j \cos \phi]z + \cdots$$
$$+ a_n(q) \left[\cos\left(\frac{n\pi}{2} - n\phi\right) + j \sin\left(\frac{n\pi}{2} - n\phi\right) \right] z^n.$$

Notice that the original family of polynomials $\mathcal{P}$ is $\mathcal{D}^+$-stable if and only if the transformed family of polynomials $\tilde{\mathcal{P}} = \{\tilde{p}(\cdot, q) : q \in Q\}$ is robustly stable; i.e., we take $\mathcal{D}$ to be the strict left half plane for the transformed problem. It is also apparent that $\tilde{p}(z, q)$ has complex coefficients but $p(s, q)$ does not. In fact, we can write

$$\tilde{p}(z, q) = \sum_{i=0}^{n} [\alpha_i(q) + j\beta_i(q)]z^i,$$

where

$$\alpha_i(q) = a_i(q) \cos i\left(\frac{\pi}{2} - \phi\right);$$

$$\beta_i(q) = a_i(q) \sin i\left(\frac{\pi}{2} - \phi\right)$$

for $i = 0, 1, \ldots, n$.

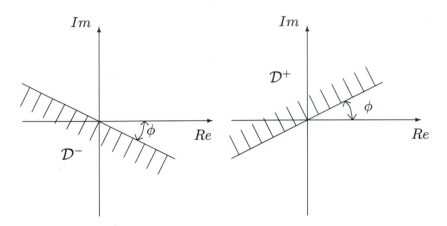

FIGURE 6.8.2   The Halfplanes $\mathcal{D}^-$ and $\mathcal{D}^+$ for Example 6.8.1

REMARKS 6.8.2 (Preservation of Uncertainty Structure):   In the example above, notice that the uncertainty structure for the transformed coefficients $\alpha_i(q)$ and $\beta_i(q)$ looks quite similar to the uncertainty structure for the original coefficients $a_i(q)$. For example, if $a_i(q)$ depends affine linearly on q, then so do $\alpha_i(q)$ and $\beta_i(q)$. In fact, this nice property is not particular to the transformation which we used relating cones and halfplanes. There is a large class of transformations for which affine linearity is preserved.

6.9   Kharitonov's Theorem with Complex Coefficients

Given an uncertain polynomial with complex coefficients, we use real uncertain parameters q_i and r_i to denote uncertainty in the real and imaginary parts of the coefficients of s^i, respectively; i.e., we write

$$p(s, q, r) = \sum_{i=0}^{n}(q_i + jr_i)s^i.$$

Given boxes Q and R as the uncertainty bounding sets for q and r, respectively, we call $\mathcal{P} = \{p(\cdot, q, r) : q \in Q; r \in R\}$ a *complex coefficient interval polynomial family*. Analogous to the real coefficient case, we use bounds $q_i^- \leq q_i \leq q_i^+$ and $r_i^- \leq r_i \leq r_i^+$, and the shorthand notation

$$p(s, q, r) = \sum_{i=0}^{n}([q_i^-, q_i^+] + j[r_i^-, r_i^+])s^i.$$

For brevity, we can simply refer to $p(s, q, r)$ above as a *complex co-efficient interval polynomial*.

In contrast to the real coefficient case, robust stability analysis for the complex coefficient case requires eight Kharitonov polynomials rather than four. To provide some insight as to why four "extra" Kharitonov polynomials are required, we mention a fundamental difference between the real and complex cases. Namely, if $p(s)$ is a real coefficient polynomial, we can restrict frequency sweeps to $\omega \geq 0$; that is, for $\omega \geq 0$ and $p^*(j\omega)$ denoting the complex conjugate of $p(j\omega)$, we have

$$p(-j\omega) = p^*(j\omega).$$

In contrast, if $p(s)$ is a complex coefficient polynomial, the equality above typically does not hold. In the proof of Kharitonov's Theorem for the real coefficient case in Section 5.8, we exploited this *conjugacy property* in restricting the analysis to frequencies $\omega \geq 0$; i.e., if $p(j\omega, Q)$ denotes the Kharitonov rectangle at a given frequency $\omega \in \mathbf{R}$, the conjugacy property tells us that $z \in p(j\omega, Q)$ if and only if $z^* \in p(-j\omega, Q)$.

The major difference in the complex coefficient case is that we need to consider both $\omega \geq 0$ and $\omega < 0$ separately. Although exploitation of the conjugacy property is no longer valid, we can still work with a Kharitonov rectangle taking care to discriminate between vertices for $\omega < 0$ versus $\omega \geq 0$. In other words, four Kharitonov polynomials is used for $\omega \geq 0$ and a "different" set of four Kharitonov polynomials is used for $\omega < 0$.

EXERCISE 6.9.1 (Lack of Conjugacy Property):   For the complex coefficient interval polynomial family $\mathcal{P} = \{p(\cdot, q) : q \in Q\}$ described by $p(s, q) = s^3 + (5 + jq)s^2 + 4s + 5$ and $Q = [-1, 1]$, show that for $\omega = 1$, $p(-j\omega, Q) \neq p^*(j\omega, Q)$.

DEFINITION 6.9.2 (Complex Coefficient Kharitonov Polynomials): Associated with the complex coefficient interval polynomial given as $p(s, q, r) = \sum_{i=0}^{n}([q_i^-, q_i^+] + j[r_i^-, r_i^+])s^i$ are eight fixed *Kharitonov polynomials*. The first four polynomials

$$K_1^+(s) = (q_0^- + jr_0^-) + (q_1^- + jr_1^+)s + (q_2^+ + jr_2^+)s^2 + (q_3^+ + jr_3^-)s^3 + \cdots;$$

$$K_2^+(s) = (q_0^+ + jr_0^+) + (q_1^+ + jr_1^-)s + (q_2^- + jr_2^-)s^2 + (q_3^- + jr_3^+)s^3 + \cdots;$$

$$K_3^+(s) = (q_0^+ + jr_0^-) + (q_1^- + jr_1^-)s + (q_2^- + jr_2^+)s^2 + (q_3^+ + jr_3^+)s^3 + \cdots;$$

$$K_4^+(s) = (q_0^- + jr_0^+) + (q_1^+ + jr_1^+)s + (q_2^+ + jr_2^-)s^2 + (q_3^- + jr_3^-)s^3 + \cdots$$

are associated with $\omega \geq 0$, and the second four polynomials

$$K_1^-(s) = (q_0^- + jr_0^-) + (q_1^+ + jr_1^-)s + (q_2^+ + jr_2^+)s^2 + (q_3^- + jr_3^+)s^3 + \cdots;$$

$$K_2^-(s) = (q_0^+ + jr_0^+) + (q_1^- + jr_1^+)s + (q_2^- + jr_2^-)s^2 + (q_3^+ + jr_3^-)s^3 + \cdots;$$

$$K_3^-(s) = (q_0^+ + jr_0^-) + (q_1^+ + jr_1^+)s + (q_2^- + jr_2^+)s^2 + (q_3^- + jr_3^-)s^3 + \cdots;$$

$$K_4^-(s) = (q_0^- + jr_0^+) + (q_1^- + jr_1^-)s + (q_2^+ + jr_2^-)s^2 + (q_3^+ + jr_3^+)s^3 + \cdots$$

are associated with $\omega < 0$.

REMARKS 6.9.3 (Coefficient Pattern):   Analogous to the real coefficient case, we see a basic pattern in the real and imaginary parts of the coefficients—two lower bounds followed by two lower bounds followed by two upper bounds, etc. For the sake of completeness, it is also important to mention that for complex coefficient polynomials, we use the same definition of invariant degree as in the real coefficient case; i.e., $\mathcal{P}$ has invariant degree if $\deg\ p(s, q^1, r^1) = \deg p(s, q^2, r^2)$ for all pairs (q^1, r^1) and (q^2, r^2) in the bounding set $Q \times R$.

THEOREM 6.9.4 (Kharitonov (1978b)):   *A complex coefficient interval polynomial family* $\mathcal{P} = \{p(\cdot, q, r) : q \in Q; r \in R\}$ *with invariant degree is robustly stable if and only if its eight Kharitonov polynomials are stable.*

PROOF: We only sketch the proof because it is conceptually identical to the one used for the real coefficient case; see Section 5.8. We begin by noting that both the Zero Exclusion Condition (Lemma 5.7.9) for robust stability and the Monotonic Angle Property (Lemma 5.7.6) for stable polynomials remain valid in the complex coefficient case. Now, to study the Kharitonov rectangle, we fix a frequency $\omega = \omega_0$ and seek a description of the set

$$p(j\omega_0, Q, R) = \{p(j\omega_0, q, r) : q \in Q; r \in R\}.$$

For $\omega_0 \geq 0$, arguing as in Subsection 5.7.1, the set $p(j\omega_0, Q)$ is seen to be a rectangle with southwest vertex $K_1^+(j\omega_0)$, northeast vertex $K_2^+(j\omega_0)$, southeast vertex $K_3^+(j\omega_0)$ and northwest vertex $K_4^+(j\omega_0)$; recall Figure 5.7.2. Similarly, for $\omega_0 < 0$, the description of the set remains the same except $K_i^-(j\omega_0)$ replaces $K_i^+(j\omega_0)$.

Analogous to Section 5.8, the proof of necessity is immediate and the proof of sufficiency involves two cases. In Case 1, we assume $0 \in p(j0, Q)$ and contradict the stability of at least one Kharitonov polynomial. In Case 2, we assume that $0 \notin p(j0, Q)$ and proceed by contradiction; i.e., if $\mathcal{P}$ is not robustly stable, the Zero Exclusion Condition implies that $0 \in p(j\omega^*, Q)$ for some $\omega^* \in \mathbf{R}$.

The completion of the proof involves two subcases. In Subcase 2A, we take $\omega^* \geq 0$, and, arguing as in Section 5.8, we arrive at a contradiction using $K_1^+(s)$, $K_2^+(s)$, $K_3^+(s)$ and $K_4^+(s)$. In Subcase 2B, we take $\omega^* < 0$ and use the $K_i^-(s)$ in lieu of the $K_i^+(s)$ to arrive at a contradiction in a similar manner. ∎

6.10   Conclusion

Although the formally stated objective of this chapter was to provide extensions and refinements of Kharitonov's Theorem, there was a second "secret" objective—to further demonstrate the power of the Kharitonov rectangle as a technical device. We are now well prepared to proceed with some generalizations. To this end, the next chapter deals with more general uncertainty structures; we see that the *value set* plays exactly the same role as played by the Kharitonov rectangle. More specifically, instead of working with the zero exclusion from the Kharitonov rectangle, we work with zero exclusion from the value set.

Another important point to note is that the value set formulation is well suited for the more general robust $\mathcal{D}$-stability framework. Instead of the sweeping frequency ω, we sweep a *generalized* frequency variable δ; the sweeping of δ is seen to correspond with a sweep of the boundary of $\mathcal{D}$. For each δ, we obtain a value set, call it $V(\delta)$, and derive a general zero exclusion condition

$$0 \notin V(\delta)$$

for robust $\mathcal{D}$-stability.

Notes and Related Literature

NRL 6.1 The comparison between r_{max} and d_{min} in Section 6.6 raises concern about the robustness margin definition in parameter space. In addition to exposing the fact that r_{max} is uninformative about root locations, the literature also describes ill-conditioning problems associated with numerical computation of r_{max}; e.g., see Barmish, Khargonekar, Shi and Tempo (1990).

NRL 6.2 As pointed out in Tsypkin and Polyak (1991), the function $G_{TP}(\omega)$ in Theorem 6.7.2 has a number of important properties which are consequences of being a Mikhailov-type function; see Mikhailov (1938) and Lemma 5.7.6. For instance, with $p_0(s)$ assumed stable, $G_{TP}(\omega)$ has a monotonically increasing phase which moves through n quadrants in turn with total phase increase of $n\pi/2$.

NRL 6.3 In Tsypkin and Polyak (1991), a more general setting is considered than that provided in Theorem 6.7.2; the uncertainty bounding set Q_r is a ball in any prescribed ℓ^p norm. For example, for the case $p = 2$, if one begins with nominal

$$p_r(s, q) = p_0(s) + r \sum_{i=0}^{n-1} \epsilon_i s^i$$

and the testing function to be plotted turns out to be

$$G_{TP}(\omega) = \frac{Re \; p_0(j\omega)}{\sqrt{\sum_i \epsilon_i \omega^{2i}}} + j \frac{Im \; p_0(j\omega)}{\sqrt{\sum_i \epsilon_i \omega^{2i}}}.$$

By specializing to the case when all weights are equal, the formula above leads immediately to the solution given in Soh, Berger and Dabke (1985) and its control theoretic version given by Biernacki, Huang and Bhattacharyya (1987). However, in both of these papers, the graphical visualization of robustness margin is missing. The distinguishing feature of the Tsypkin–Polyak function is that it is complex-valued, whereas, in the papers cited above, a real-valued testing function is used.

NRL 6.4 Further motivation for studying the complex coefficient case is given in the paper by Bose and Shi (1987). Whirling shafts, vibrational systems and filters are mentioned as examples of systems whose models involve complex coefficient polynomials.

NRL 6.5 The transformation of the damping cone problem into a strict left half plane problem in Example 6.8.1 is only one of many possible ways by which conformal mappings induce complex coefficient polynomials. In the robustness analysis of Sondergeld (1983), a table of other useful transformations is provided.

Chapter 7

The Value Set Concept

Synopsis

In this chapter, the value set is defined and seen to be a generalization of the Kharitonov rectangle. Sweeping the imaginary axis is replaced by sweeping the boundary of a desired root location region $\mathcal{D}$. Subsequently, a link is established between robust $\mathcal{D}$-stability and the value set; i.e., a more general zero exclusion condition is established.

7.1 Introduction

When we exposed Kharitonov's Theorem and its embellishments in Chapters 5 and 6, the Zero Exclusion Condition (Lemma 5.7.9) played a critical role. That is, for an interval polynomial family with invariant degree and at least one stable member, robust stability is equivalent to zero exclusion from the Kharitonov rectangle $p(j\omega, Q)$ at all frequencies $\omega \geq 0$.

In this chapter, we no longer restrict our attention to interval polynomials. For rather general uncertainty structures, we define the value set which is seen to be a generalization of the Kharitonov rectangle. Subsequently, we see that the Zero Exclusion Condition is still meaningful in this more general setting. In fact, as the chapter progresses, we also dispense with left half plane stability and consider the value set concept in the more general framework of robust $\mathcal{D}$-stability as defined in Section 4.13.

It is not our intention to solve the robust $\mathcal{D}$-stability problem in this chapter. Our primary objective is to demonstrate that the robust $\mathcal{D}$-stability problem is equivalent to a zero exclusion from an appropriately constructed value set. This sets the stage for later chapters which address the problem of value set characterization and subsequent solution of the robust $\mathcal{D}$-stability problem. In this chapter, only relatively simple value sets are characterized.

In the remainder of this text, there are two contexts within which value sets arise. First, in many situations we use the value set as a technical stepping stone within a proof. For such cases, understanding the final result does not require any knowledge about value sets—the value set can be hidden from the user. Kharitonov's Theorem exemplifies this situation; i.e., from a user's point of view, one does not need to know about the Kharitonov rectangle to use the theorem in testing for robust stability.

The second context within which the value set arises: There are many results which are communicated in a computer-aided graphics framework. For such cases, we provide a recipe for construction of the value set, and, the robustness test amounts to displaying this set as a function of a generalized frequency variable. In this context, the value set is on "center stage" in the sense that the user must understand the meaning of the value set in order to generate the appropriate graphical display.

7.2   The Value Set

Roughly speaking, given an uncertain polynomial $p(s, q)$ and an uncertainty bounding set Q, then, at a fixed frequency $\omega \in \mathbf{R}$, the value set is the subset of the complex plane consisting of all values which can be assumed by $p(j\omega, q)$ as q ranges over Q. Said another way, $p(j\omega, Q)$ is the range of $p(j\omega, \cdot)$. This idea is stated formally in the definition below.

DEFINITION 7.2.1 (The Value Set): Given a family of polynomials $\mathcal{P} = \{p(\cdot, q) : q \in Q\}$, the *value set at frequency* $\omega \in \mathbf{R}$ is given by

$$p(j\omega, Q) = \{p(j\omega, q) : q \in Q\}.$$

That is, $p(j\omega, Q)$ is the image of Q under $p(j\omega, \cdot)$.

EXAMPLE 7.2.2 (Value Set as a Straight Line Segment): We consider the uncertain polynomial $p(s, q) = s^2 + (2 - q)s + (3 - q)$ and uncertainty bounding set $Q = [0, 4]$. Notice that for fixed $\omega \in \mathbf{R}$,

$Re\, p(j\omega, q) = 3 - \omega^2 - q$ and $Im\, p(j\omega) = (2 - q)\omega$. Hence, for each $\omega \in \mathbf{R}$, the value set $p(j\omega, Q)$ is a straight line segment joining $p(j\omega, 0) = (3 - \omega^2) + 2j\omega$ and $p(j\omega, 4) = -(1 + \omega^2) - 2j\omega$. This is illustrated in Figure 7.2.1 for $0 \leq \omega \leq 4$. Notice that $p(s, q)$ fits ex-

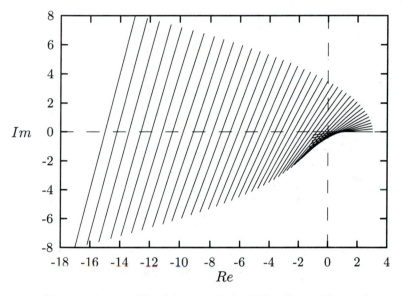

FIGURE 7.2.1   The Motion of the Value Set in Example 7.2.2

actly into the one-parameter framework of Chapter 4. For example, with $f(s) = s^2 + 2s + 3$ and $g(s) = -4s - 4$, the family of polynomials described by $p(s, \lambda) = f(s) + \lambda g(s)$ and $\lambda = [0, 1]$ is the same as the original family $\mathcal{P} = \{p(\cdot, q) : q \in Q\}$. These ideas are generalized in the lemma below; the nearly trivial proof is omitted.

LEMMA 7.2.3 (Value Set with One Parameter Entering Affinely): *Let $f(s)$ and $g(s)$ be fixed polynomials. Then, for the family of polynomials described by $p(s, \lambda) = f(s) + \lambda g(s)$ and $\lambda \in \Lambda = [0, 1]$, with fixed $\omega \in \mathbf{R}$, the value set $p(j\omega, \Lambda)$ is the straight line segment joining the points $p(j\omega, 0) = f(j\omega)$ and $p(j\omega, 1) = f(j\omega) + g(j\omega)$.*

EXERCISE 7.2.4 (Value Set Parallelogram):   Given the family of polynomials described by

$$p(s, q) = s^4 + (3q_1 + 4q_2 + 12)s^3 + (q_1 - 2q_2 + 6)s^2 \\ + (2q_1 - 3q_2 + 8)s + (6 + q_1),$$

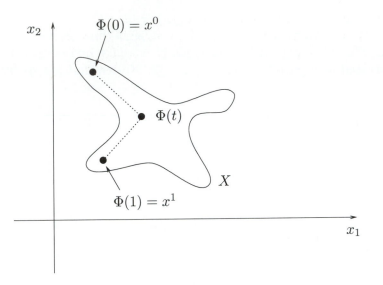

FIGURE 7.3.1   Pathwise Connected Set in $\mathbf{R}^2$

$|q_1| \leq 1$ and $|q_2| \leq 1$, argue that the value set $p(j\omega, Q)$ is a parallel-ogram and plot it for ten frequencies evenly spaced between $\omega = 0$ and $\omega = 2$. This problem serves as a preview of the polygonal value set theory in the chapter to follow.

7.3   The Zero Exclusion Condition

When we introduced the Zero Exclusion Condition for interval poly-nomials in Section 5.7.8, it was noted that this condition actually applies to much more general uncertainty structures. In this section, we establish a Zero Exclusion Condition for a family of polynomials $\mathcal{P}$ having continuous coefficient functions $a_i(q)$ and pathwise con-nected uncertainty bounding set Q. For completeness, we now define pathwise connectedness.

DEFINITION 7.3.1 (Pathwise Connectedness):   A set $X \subseteq \mathbf{R}^k$ is said to be *pathwise connected* if the following condition holds: Given any two points $x^0, x^1 \in X$, there is a continuous function $\Phi : [0, 1] \to X$ such that $\Phi(0) = x^0$ and $\Phi(1) = x^1$.

REMARKS 7.3.2 (Pathwise Connectedness):   The notion of path-wise connectedness is illustrated in Figure 7.3.1. From an applica-tions point of view, assuming pathwise connectedness of the uncertainty bounding set Q is typically quite reasonable. In this regard, note that every convex set Q (such as a sphere or a box) is pathwise connected. This is easily established by noting that if $q^0, q^1 \in Q$, the function $\Phi : [0, 1] \to Q$ given by

$$\Phi(t) = (1 - t)q^0 + tq^1$$

is continuous and satisfies the conditions $\Phi(0) = q^0$, $\Phi(1) = q^1$ and $\Phi(t) \in Q$ for all $t \in [0, 1]$.

THEOREM 7.3.3 (Zero Exclusion Condition): *Suppose that a family of polynomials $\mathcal{P} = \{p(\cdot, q) : q \in Q\}$ has invariant degree with associated uncertainty bounding set Q which is pathwise connected, continuous coefficient functions $a_i(q)$ for $i = 0, 1, 2, \ldots, n$ and at least one stable member $p(s, q^0)$. Then $\mathcal{P}$ is robustly stable if and only if the origin, $z = 0$, is excluded from the value set $p(j\omega, Q)$ at all frequencies $\omega \geq 0$; i.e., $\mathcal{P}$ is robustly stable if and only if*

$$0 \notin p(j\omega, Q)$$

for all frequencies $\omega \geq 0$.

PROOF: To establish necessity, we assume that $\mathcal{P}$ is robustly stable and must show that $0 \notin p(j\omega, Q)$ for all $\omega \geq 0$. Proceeding by contradiction, suppose $0 \in p(j\omega^*, Q)$ for some $\omega^* \geq 0$. Then, $p(j\omega^*, q^*) = 0$ for some $q^* \in Q$. This contradicts the assumed robust stability of $\mathcal{P}$.

   To establish sufficiency, we assume that $0 \notin p(j\omega, Q)$ for all $\omega \geq 0$ and must show that $\mathcal{P}$ is robustly stable. Proceeding by contradiction, suppose that $p(s, q^1)$ is unstable for some $q^1 \in Q$. Then, by pathwise connectedness of Q, there exists a continuous function $\Phi : [0, 1] \to Q$ such that $\Phi(0) = q^0$ and $\Phi(1) = q^1$. Now, in accordance with Lemma 4.8.2, let $s_1(q), s_2(q), \ldots, s_n(q)$ denote root functions for $p(s, q)$ which vary continuously with respect to $q \in Q$. Since $p(s, q^1)$ is unstable, it follows that for some $i^* \in \{1, 2, \ldots, n\}$, $s_{i^*}(q^1)$ is in the right half plane. On the other hand, we know that $s_{i^*}(q^0)$ lies in the strict left half plane. Next, for $t \in [0, 1]$, notice that $s_{i^*}(\Phi(t))$ describes a continuously varying root of $p(s, \Phi(t))$ which begins in the strict left half plane at $s_{i^*}(\Phi(0))$ and terminates in the right half plane at $s_{i^*}(\Phi(1))$. In view of continuity of $s_{i^*}(\Phi(t))$ with respect to t, there must exist some $t^* \in (0, 1]$ such that $s_{i^*}(\Phi(t^*))$ lies on the imaginary axis. Taking $q^* = \Phi(t^*)$ and noting that $p(s, q^*)$ has an imaginary root, it follows that $p(j\omega^*, q^*) = 0$ for some $\omega^* \geq 0$. Hence, $0 \in p(j\omega^*, Q)$, which is the contradiction we seek. ∎

EXERCISE 7.3.4 (Finite Union of Pathwise Connected Sets):   Give an example showing that Theorem 7.3.3 no longer remains valid when Q is a finite union of pathwise connected sets.

EXERCISE 7.3.5 (Cutoff Frequency):   In addition to the hypotheses associated with the Zero Exclusion Condition, assume that Q is bounded. Prove that there exists some frequency $\omega_c \geq 0$ having the property that $0 \notin p(j\omega, Q)$ for $\omega > \omega_c$. *Hint*: Consider the function

$$f(\omega) = (\min_{q \in Q} |a_n(q)|)\omega^n - \sum_{i=0}^{n-1}(\max_{q \in Q} |a_i(q)|)\omega^i.$$

7.4   Zero Exclusion Condition for Robust $\mathcal{D}$-Stability

The objective of this section is to show that the zero exclusion concept can easily be extended to the more general robust $\mathcal{D}$-stability framework. To this end, we first provide a more general definition of the value set. Instead of evaluating an uncertain polynomial along the imaginary axis, we consider an arbitrary evaluation point $z \in \mathbf{C}$. Furthermore, instead of sweeping the imaginary axis, we sweep the boundary of $\mathcal{D}$. The final result, Theorem 7.4.2, provides the most general version of the Zero Exclusion Condition, which is given in this book. In later chapters, this theorem is frequently invoked.

DEFINITION 7.4.1 (The Value Set at $z \in \mathbf{C}$):   Given a family of polynomials $\mathcal{P} = \{p(\cdot, q) : q \in Q\}$, the *value set* at $z \in \mathbf{C}$ is given by

$$p(z, Q) = \{p(z, q) : q \in Q\}.$$

That is, $p(z, Q)$ is the image of Q under $p(z, \cdot)$.

THEOREM 7.4.2 (Zero Exclusion Condition):   *Let $\mathcal{D}$ be an open subset of the complex plane and suppose that $\mathcal{P} = \{p(\cdot, q) : q \in Q\}$ is a family of polynomials with invariant degree, uncertainty bounding set Q which is pathwise connected. Furthermore, assume that the coefficient functions $a_i(q)$ are continuous and that $\mathcal{P}$ has at least one $\mathcal{D}$-stable member $p(s, q^0)$. Then $\mathcal{P}$ is robustly $\mathcal{D}$-stable if and only if*

$$0 \notin p(z, Q)$$

for all $z \in \partial\mathcal{D}$, where $\partial\mathcal{D}$ denotes the boundary of $\mathcal{D}$.

PROOF: We omit the proof since it is nearly identical to that given for Theorem 7.3.3. ∎

EXERCISE 7.4.3 (Bounded $\mathcal{D}$ and Invariant Degree):   Under the strengthened hypothesis that $\mathcal{D}$ is bounded, argue that the invariant degree assumption is no longer needed in Theorem 7.4.2.

EXAMPLE 7.4.4 (Robust Schur Stability):   We take $\mathcal{D}$ to be the interior of the unit disc and consider the family of polynomials $\mathcal{P}$ described by $p(s, q) = s^3 + (0.5 + q)s^2 - (0.25 + q)$ and $Q = [-0.1, 0.1]$. To apply the Zero Exclusion Condition, we first identify one stable member of $\mathcal{P}$. Indeed, by taking $q = q^0 = 0$ and computing roots $s_{1,2} = -0.5 \pm j0.5$ and $s_3 = 0.5$, we see that $p(s, q^0)$ is Schur stable. Next, we sweep z around the unit circle while plotting the value set $p(z, Q)$. Since q enters affine linearly into the coefficients, a minor extension of Lemma 7.2.3 (with z replacing $j\omega$) enables us to assert that $p(z, Q)$ is the straight line segment joining $p(z, -0.1)$ and $p(z, 0.1)$. In Figure 7.4.1, the value set $p(z, Q)$ is shown using 500

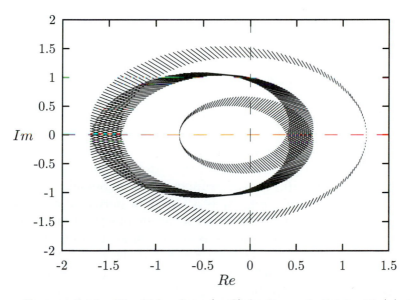

FIGURE 7.4.1   The Value Set $p(z, Q)$ for Example 7.4.4 with $|q| \leq 0.1$

evaluation points $z = z_i$ evenly spaced around the unit circle. It is clear by inspection that $0 \notin p(z, Q)$ for all z. Hence, we conclude that $\mathcal{P}$ is robustly $\mathcal{D}$-stable.

We now entertain an increase in the uncertainty bound. Taking $Q = [-0.3, 0.3]$, the value set is plotted again using 500 evaluation points and $\mathcal{P}$ is still seen to be robustly stable; see Figure 7.4.2. By further increasing the uncertainty bound, we can determine the

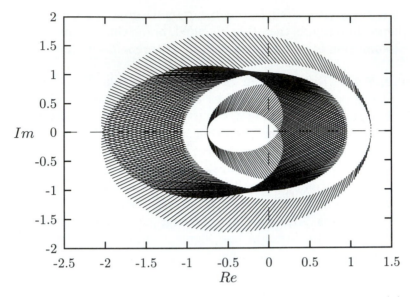

FIGURE 7.4.2   The Value Set $p(z, Q)$ for Example 7.4.4 with $|q| \leq 0.3$

robustness margin; i.e., if we take $Q = [-r, r]$, we seek the supremal value of r, call it r_{max}, for which robust $\mathcal{D}$-stability is guaranteed. By gradually increasing r while checking the Zero Exclusion Condition, we obtain robustness margin $r_{max} \approx 0.38$.

EXERCISE 7.4.5 (Cutoff for Boundary Sweep):   In this exercise, we consider the $\mathcal{D}$-stability analogue of the cutoff frequency concept discussed in Section 5.10.   Indeed, suppose $\mathcal{D}$ is unbounded, Q is bounded and $\mathcal{P} = \{p(\cdot, q) : q \in Q\}$ is a family of polynomials with invariant degree and with continuous coefficient functions $a_i(q)$ for $i = 0, 1, 2, \ldots, n$. Now, prove there exists a bounded subset $\mathcal{D}_0$ of $\mathcal{D}$ such that if $z \in \mathcal{D}/\mathcal{D}_0 = \{z \in \mathcal{D} : z \notin \mathcal{D}_0\}$, then $0 \notin p(z, Q)$. *Hint*: Construct a function $f(z)$ such that $\lim_{|z| \to \infty} f(z) = +\infty$ and $0 \notin p(z, Q)$ whenever $f(z) > 0$.

7.5   Boundary Sweeping Functions

This section introduces the notion of a boundary sweeping function. We see that such a function facilitates calculations associated with value set generation and zero exclusion testing. Indeed, since testing for satisfaction of the Zero Exclusion Condition involves sweeping the boundary $\partial \mathcal{D}$ of $\mathcal{D}$, it is convenient to have a scalar parameter

δ which can be used to parameterize motion along $\partial\mathcal{D}$. For robust $\mathcal{D}$-stability analysis, this scalar δ plays a role which is analogous to the frequency ω in ordinary robust stability analysis.

DEFINITION 7.5.1 (Boundary Sweeping Function):   Suppose that $\mathcal{D}$ is an open subset of the complex plane with boundary $\partial\mathcal{D}$. Then, given an interval (perhaps semi-infinite or infinite) $I \subseteq \mathbf{R}$, a mapping $\Phi_{\mathcal{D}} : I \to \partial\mathcal{D}$ is said to be a *boundary sweeping function* for $\mathcal{D}$ if $\Phi_{\mathcal{D}}$ is continuous and onto; i.e., $\Phi_{\mathcal{D}}$ is continuous and for each point $z \in \partial\mathcal{D}$, there exists some $\delta \in I$ such that

$$\Phi_{\mathcal{D}}(\delta) = z.$$

The scalar δ is called a *generalized frequency* variable for $\mathcal{D}$.

EXAMPLE 7.5.2 (Halfplanes):   When $\mathcal{D}$ is the strict left half plane,

$$\Phi_{\mathcal{D}}(\delta) = j\delta$$

corresponds to setting $s = j\omega$. Notice that there are infinitely many other possibilities for a boundary sweeping function for the strict left half plane; e.g., take $I = (-\infty, \infty)$ and $\Phi_{\mathcal{D}}(\delta) = j\delta \sin\,\delta$. Now, suppose $\mathcal{D}$ is a halfplane reflecting our concern about the degree of stability; that say $\mathcal{D} = \{z \in \mathbf{C} : Re\,z < -\sigma\}$, where $\sigma > 0$ is given. Then we can take $I = (-\infty, \infty)$ and

$$\Phi_{\mathcal{D}}(\delta) = -\sigma + j\delta$$

as a boundary sweeping function.

EXAMPLE 7.5.3 (Unit Disc):   When $\mathcal{D}$ is the interior of the unit disc, we obtain a boundary sweeping function with $I = [0, 1]$ and

$$\Phi_{\mathcal{D}}(\delta) = \cos\,2\pi\delta + j\sin\,2\pi\delta.$$

EXAMPLE 7.5.4 (Damping):   For a robust $\mathcal{D}$-stability problem with the damping cone $\mathcal{D} = \{z \in \mathbf{C} : \pi - \theta < \measuredangle z < \pi + \theta\}$ and $0 < \theta < \pi/2$, a suitable boundary sweeping function is described by $I = (-\infty, \infty)$ and

$$\Phi_{\mathcal{D}}(\delta) = \begin{cases} \delta\cos\,\theta + j\delta\sin\,\theta & \text{if } \delta \leq 0; \\ -\delta\cos\,\theta + j\delta\sin\,\theta & \text{if } \delta > 0. \end{cases}$$

EXERCISE 7.5.5 (Zero Exclusion via Boundary Sweeping Functions): In addition to assumptions associated with the Zero Exclusion Condition in Theorem 7.4.2, suppose that $\mathcal{D}$ has boundary sweeping function $\Phi_{\mathcal{D}} : I \mapsto \partial\mathcal{D}$. Prove that $\mathcal{P}$ is robustly stable if and only if

$$0 \notin p(\Phi_{\mathcal{D}}(\delta), Q)$$

for all generalized frequencies $\delta \in I$.

EXERCISE 7.5.6 (More General Boundary Sweeping Functions): Consider the desired root location region given by $\mathcal{D} = \mathcal{D}_1 \bigcup \mathcal{D}_2$, where $\mathcal{D}_1$ and $\mathcal{D}_2$ are the strips $\mathcal{D}_1 = \{z \in \mathbf{C} : -1 < Re\ z < 0\}$ and $\mathcal{D}_2 = \{z \in \mathbf{C} : Re\ z \leq -1; -1 < Im\ z < 1\}$. The Zero Exclusion Condition, as given in Exercise 7.5.5, cannot be applied in the obvious way because the domain I of $\Phi_{\mathcal{D}}$ must be an interval while $\Phi_{\mathcal{D}}(\delta)$ is continuous. Generalize the statement of the Zero Exclusion Condition to allow for $\mathcal{D}$ regions of the sort described above.

EXERCISE 7.5.7 (More General Interpretation of Zero Exclusion): Let $\mathcal{D}_1, \mathcal{D}_2, \ldots, \mathcal{D}_m$ be disjoint open subsets of the complex plane and suppose $\mathcal{P} = \{p(\cdot, q) : q \in Q\}$ is a family of polynomials with invariant degree, uncertainty bounding set Q which is pathwise connected and continuous coefficient functions $a_i(q)$ for $i = 0, 1, 2, \ldots, n$. For each $q \in Q$ and $i \in \{1, 2, \ldots, m\}$, let $n_i(q)$ denote the number of roots of $p(s, q)$ in $\mathcal{D}_i$. Finally, assume that for some $q^0 \in Q$, $p(s, q^0)$ has no roots in the boundary of $\mathcal{D} = \mathcal{D}_1 \cup \mathcal{D}_2 \cup \cdots \cup \mathcal{D}_m$. Now, prove that each of the root indices $n_i(q)$ remains invariant over Q if and only if the Zero Exclusion Condition $0 \notin p(z, Q)$ is satisfied for all points $z \in \partial\mathcal{D}$.

7.6   More General Value Sets

In this section, we provide further testimony to the power of the value set point of view. To this end, we begin by noting that the definition of the value set is trivially generalized to uncertain functions which are not necessarily polynomials. After all, the value set is nothing more than the range of a function. However, the value set is special in the sense that the range of concern is only two-dimensional. This fact drives much of the theory in this book; i.e., we replace a multidimensional robustness problem over the uncertainty bounding set Q with a two-dimensional geometry problem over the value set. With these ideas in mind, the definition below is natural for func-

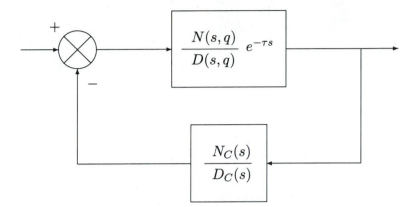

FIGURE 7.6.1   Uncertain Delay System for Example 7.6.2

tions which are not necessarily polynomic. Using the more general value set definition below, we are equipped to enlarge the class of robustness problems which we can address.

DEFINITION 7.6.1 (Value Set at $z \in \mathbf{C}$):   Given an uncertainty bounding set Q and an uncertain function defined by a mapping $F : \mathbf{C} \times Q \to \mathbf{C}$, the *value set* at $z \in \mathbf{C}$ is given by

$$F(z, Q) = \{F(z, q) : q \in Q\}.$$

That is, $F(z, Q)$ is the image of Q under $F(z, \cdot)$.

EXAMPLE 7.6.2 (Delay Systems):   For the uncertain delay system depicted in Figure 7.6.1, the closed loop uncertain *quasipolynomial* is given by

$$p_\tau(s, q) = N(s, q)N_C(s)e^{-\tau s} + D(s, q)D_C(s).$$

Now, given an uncertainty bounding set Q and a fixed delay $\tau > 0$, the associated value set at $s = j\omega$ is

$$p_\tau(j\omega, Q) = \{p_\tau(j\omega, q) : q \in Q\}.$$

Using this definition, there are large classes of robust stability problems for delay systems which can be attacked using the Zero Exclusion Condition $0 \notin p_\tau(j\omega, Q)$; see the notes at the end of the chapter for further discussion.

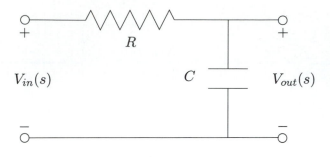

FIGURE 7.6.2 RC Filter for Example 7.6.4

EXERCISE 7.6.3 (Line Segment as Value Set for Delay Systems): With the delay system setup as in the example above, suppose that $N(s,q)$ and $D(s,q)$ have coefficients depending affine linearly on a single parameter $q \in [0,1]$; i.e., there exist fixed polynomials $N_0(s)$, $N_1(s)$, $D_0(s)$ and $D_1(s)$ such that

$$N(s,q) = N_0(s) + qN_1(s)$$

and

$$D(s,q) = D_0(s) + qD_1(s).$$

Argue that for $s = j\omega$, the value set $p_\tau(j\omega, Q)$ for the closed loop polynomial is a straight line segment with endpoints $p_\tau(j\omega, 0)$ and $p_\tau(j\omega, 1)$. In other words, the linear nature of the value set segments (established in Lemma 7.2.3) is preserved in the presence of delay.

EXERCISE 7.6.4 (RC Filter): Consider the simple RC filter with resistance R and capacitance C given in Figure 7.6.2. With uncertainty ranges described by $\pm 20\%$ about the nominal values $R = 1000\Omega$ and $C = 100\mu F$, take $q = RC$ and derive the uncertain transfer function

$$P(s,q) = \frac{1}{1+qs}$$

and associated uncertainty bounding set $Q = [0.064, 0.144]$. For this rational function, characterize the value set $P(j\omega, Q)$ associated with the frequency response and plot for representative frequencies in the range $0 \leq \omega \leq 10$.

EXERCISE 7.6.5 (Uncertainty in a Simple Zero): In the exercise above, note that the value set associated with uncertainty in a pole was nonlinear (curved) at each fixed frequency. For the case of a simple zero, however, the situation is much simpler. Indeed, consider the uncertain transfer function

$$P(s,q) = \frac{K(s-q)\prod\limits_{i=1}^{m}(s-z_i)}{\prod\limits_{i=1}^{n}(s-p_i)}$$

with fixed gain K, all poles $p_i \neq 0$ fixed and all zeros z_i fixed except for the zero at $s = q$. Now, using the uncertainty bounding set which is the interval $Q = [q^-, q^+]$, argue that at frequencies $\omega \geq 0$ not corresponding to a pole, the value set $P(j\omega, Q)$ is the straight line segment joining $P(j\omega, q^-)$ and $P(j\omega, q^+)$. Generalize this result for uncertain transfer functions of the form

$$P(s,q) = \frac{N_0(s) + qN_1(s)}{D_0(s)},$$

where $N_0(s)$, $N_1(s)$ and $D_0(s)$ are fixed polynomials.

EXERCISE 7.6.6 (Conjugacy Property for Rational Functions): Given an uncertain family of plants described by

$$P(s,q) = \frac{N(s,q)}{D(s,q)}$$

and $q \in Q$, let frequency $\omega \geq 0$ be given such that $0 \notin D(j\omega, Q)$. Taking z^* to be the complex conjugate of z, prove that $z \in P(j\omega, Q)$ if and only if

$$z^* \in P(-j\omega, Q).$$

EXERCISE 7.6.7 (Value Set for Rational Function): For the family of plants described by $P(s,q) = (s+q)/(s+3q)$ and $Q = [1,2]$, generate a plot of the value set $p(j\omega, Q)$ for ten evenly spaced frequency points in the range $0 \leq \omega \leq 10$.

7.7   Conclusion

In this chapter, we reformulated a number of robustness problems in terms of zero exclusion from an appropriate value set. This leads us to the following question, which we begin to address in the next chapter: Can we delineate important robustness problems for which value set construction is computationally tractable?

Notes and Related Literature

NRL 7.1 If an uncertain polynomial $p(s, q)$ has continuous coefficient functions and the uncertainty bounding set Q is "nice" (for example, say Q is closed, bounded and convex), there is a temptation to conclude that the value set is simply connected. However, as pointed out in Barmish, Ackermann and Hu (1992), such a conclusion is erroneous. For example, with

$$p(s, q) = (s + q_1)(s + q_2)(s + q_3)$$

and uncertainty bound $|q_i| \leq \sqrt{3}$ for $i = 1, 2, 3$, the value set $p(j0.5, Q)$ has a hole.

NRL 7.2 Rather than working with the value set, another takeoff point for robust stability analysis involves working with the stability boundary in parameter space. Such work begins with the so-called $\mathcal{D}$-partition technique of Neimark (1947) and is further pursued in Ackermann (1980). To illustrate, by setting real and imaginary parts to zero, we see that given the family of polynomials described by $p(s, q) = \sum_{i=0}^{n} a_i(q) s^i$ and $q \in Q$, the condition $0 \in p(j\omega, Q)$ is equivalent to

$$0 = a_0(q) - a_2(q)\omega^2 + a_4(q)\omega^4 - a_6(q)\omega^6 + \cdots$$

and

$$0 = a_1(q)\omega - a_3(q)\omega^3 + a_5(q)\omega^5 - a_7(q)\omega^7 + \cdots$$

for some $q \in Q$. Using the fact that the upper left $(n-1) \times (n-1)$ block $H_{n-1}(p(s, q))$ of the Hurwitz matrix $H(p(s, q))$ is the Sylvester resultant corresponding to the two equations above, we obtain a description for the stability boundary. Namely,

$$a_0(q) = 0, \ldots, a_{n-1}(q) = 0$$

and

$$\det H_{n-1}(p(s, q)) = 0,$$

where $\det H_{n-1}(p(s, q))$ denotes the upper left $(n-1) \times (n-1)$ block of $H(p(s, q))$.

NRL 7.3 For further elaboration on robust stability in a delay systems context, two basic references are Barmish and Shi (1989) and Fu, Olbrot and Polis (1989).

Part III

The Polyhedral Theory

Chapter 8

Polytopes of Polynomials

Synopsis

Polytopes of polynomials are natural objects to study when dealing with affine linear uncertainty structures. From a technical point of view, the most important point to note about polytopes of polynomials is that they have value sets which are convex polygons in the complex plane. This fact facilitates solution of many types of robustness problems.

8.1 Introduction

Primary motivation for this chapter is derived from the fact that a robustness theory for independent uncertainty structures leads to conservative results when applied to more general uncertainty structures. To reinforce this point, we recall the discussion following Kharitonov's Theorem (see Section 5.11): By replacing a family of polynomials $\mathcal{P}$ with an overbounding interval polynomial family $\overline{\mathcal{P}}$, it may turn out that $\mathcal{P}$ is robustly stable but $\overline{\mathcal{P}}$ is not. In other words, overbounding via independent uncertainty structures is conservative in the sense that only sufficient conditions for robustness are obtained.

In this chapter, we attack dependent uncertainty structures in a more direct manner. To this end, we consider the case when the system coefficients depend affine linearly on the vector of uncertain parameters q. In this affine linear framework, polytopes of polynomials are the natural objects to study when the uncertainty bounding set Q is a box. The fact that such families of polynomials turn out to have value sets which are convex polygons paves the way for many results in the chapters to follow.

8.2   Affine Linear Uncertainty Structures

The main objective of this section is to indicate a number of ways by which affine linear uncertainty structures arise. Of foremost importance is the following fact: Affine linear uncertainty structures are preserved under large classes of feedback interconnections. We now proceed to make this statement more precise. Since affine linear uncertainty structures have only been casually mentioned thus far, (for example, see Section 4.4), we include a formal definition for the sake of completeness.

DEFINITION 8.2.1 (Affine Linear Uncertainty Structure):   An uncertain polynomial $p(s, q) = \sum_{i=0}^{n} a_i(q)s^i$ is said to have an *affine linear uncertainty structure* if each coefficient function $a_i(q)$ is an affine linear function of q; i.e., for each $i \in \{0, 1, 2, \ldots, n\}$, there exists a column vector α_i and a scalar β_i such that

$$a_i(q) = \alpha_i^T q + \beta_i,$$

where $\alpha_i^T(q)$ is the transpose of $\alpha_i(q)$. More generally, an uncertain rational function, which we write as $P(s, q) = N(s, q)/D(s, q)$, is said to have an *affine linear uncertainty structure* if both polynomials $N(s, q)$ and $D(s, q)$ have affine linear uncertainty structures.

REMARKS 8.2.2 (Feedback Interconnection):   We now consider an uncertain plant $P(s, q) = N(s, q)/D(s, q)$ connected in the feedback configuration of Figure 8.2.1 with a compensator

$$C(s) = \frac{N_C(s)}{D_C(s)}.$$

A simple calculation leads to the closed loop transfer function

$$P_{CL}(s, q) = \frac{N(s, q)D_C(s)}{N(s, q)N_C(s) + D(s, q)D_C(s)}.$$

In the lemma below, we see that if $P(s, q)$ has an affine linear uncertainty structure, then so does $P_{CL}(s, q)$. In other words, the affine

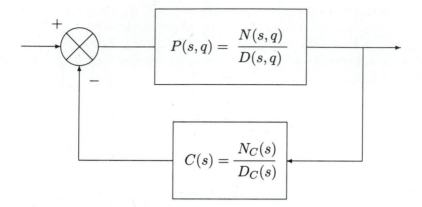

FIGURE 8.2.1   Feedback Connection Preserving Uncertainty Structure

linear uncertainty structure is preserved in going from the open loop to the closed loop.

LEMMA 8.2.3 (Affine Linear Uncertainty Preservation):   *Consider an uncertain plant connected in a feedback configuration as in Figure 8.2.1 and assume that $P(s,q)$ has an affine linear uncertainty structure. Then it follows that the uncertain closed loop transfer function $P_{CL}(s,q)$ also has an affine linear uncertainty structure.*

PROOF: With $q \in \mathbf{R}^\ell$, the affine linear uncertainty structure for $P(s,q)$ enables us to express the numerator and denominator of $P(s,q)$ in the form

$$N(s,q) = N_0(s) + \sum_{i=1}^{\ell} q_i N_i(s)$$

and

$$D(s,q) = D_0(s) + \sum_{i=1}^{\ell} q_i D_i(s)$$

with the $N_i(s)$ and $D_i(s)$ being fixed polynomials for $i \in \{0, 1, \ldots, \ell\}$. Now, it is easy to verify that the closed loop system transfer function has numerator

$$N_{CL}(s,q) = N_0(s)D_C(s) + \sum_{i=0}^{\ell} q_i N_i(s)D_C(s)$$

and denominator

$$D_{CL}(s,q) = N_0(s)N_C(s) + D_0(s)D_C(s)$$
$$+ \sum_{i=1}^{\ell} q_i[N_i(s)N_C(s) + D_i(s)D_C(s)].$$

By inspection, it is obvious that $N_{CL}(s,q)$ and $D_{CL}(s,q)$ also have affine linear uncertainty structures. ∎

REMARKS 8.2.4 (Variations on the Same Theme):   There are many variations of Lemma 8.2.3; i.e., if $P(s,q)$ has an affine linear uncertainty structure, then every transfer function of practical interest has the same structure as well. This is illustrated in the exercise below using the sensitivity function and the complementary sensitivity function.

EXERCISE 8.2.5 (Sensitivity and Complementary Sensitivity):   For the feedback interconnection in Figure 8.2.1, prove that if $P(s,q)$ has an affine linear uncertainty structure, then the *sensitivity function*

$$S(s,q) = \frac{1}{1 + P(s,q)C(s)}$$

and the *complementary sensitivity function*

$$T(s,q) = \frac{P(s,q)C(s)}{1 + P(s,q)C(s)}$$

also have affine linear uncertainty structures.

EXERCISE 8.2.6 (Linear Fractional Transformations):   Take $p(s,q)$ to be an uncertain polynomial of order n having an affine linear uncertainty structure and let γ_1, γ_2, γ_3 and γ_4 be real with either $\gamma_3 \neq 0$ or $\gamma_4 \neq 0$. Now, prove that the transformed polynomial

$$\tilde{p}(s,q) = p\left(\frac{\gamma_1 s + \gamma_2}{\gamma_3 s + \gamma_4}, q\right)(\gamma_3 s + \gamma_4)^n$$

also has an affine linear uncertainty structure. As a special case, associated with the robust stability problem for an interval polynomial family is a robust Schur stability problem with uncertain polynomial having argument z (to emphasize discrete-time) and given by

$$\tilde{p}(z,q) = \sum_{i=0}^{n} q_i(z-1)^{n-i}(z+1)^i.$$

Notice that $\tilde{p}(z, q)$ has an affine linear uncertainty structure.

EXERCISE 8.2.7 (Special Case):   Suppose that $A(q)$ is an uncertain matrix with uncertain parameter vector q entering affine linearly into only one row or one column. Letting

$$p(s, q) = \det(sI - A(q)),$$

prove that the family of polynomials $\mathcal{P} = \{p(\cdot, q) : q \in Q\}$ has an affine linear uncertainty structure. *Hint*: Expand the determinant via the row or column which contains the uncertain parameters.

EXAMPLE 8.2.8 (Overbounding Nonlinear Uncertainty Structures): In order to deal with uncertain parameters entering nonlinearly into a system, it often suffices to generate an overbounding family which has an affine linear uncertainty structure. To illustrate, consider the uncertain polynomial

$$p(s, q) = s^3 + (4q_1^2 + 2q_1 + 3q_2 + 3)s^2$$
$$+ (q_1^2 + 2q_2^2 + 2)s + (6q_1 + q_2 + 4)$$

with uncertainty bounds $|q_i| \leq 1$ for $i = 1, 2$. By defining new variables $\bar{q}_1 = q_1$, $\bar{q}_2 = q_2$, $\bar{q}_3 = q_1^2$ and $\bar{q}_4 = q_2^2$, we can define the *overbounding polynomial*

$$\bar{p}(s, \bar{q}) = s^3 + (2\bar{q}_1 + 3\bar{q}_2 + 4\bar{q}_3 + 3)s^2$$
$$+ (\bar{q}_3 + 2\bar{q}_4 + 2)s + (6\bar{q}_1 + \bar{q}_2 + 4)$$

which has an affine linear uncertainty structure and associated uncertainty bounds $|q_i| \leq 1$ for $i = 1, 2, 3, 4$.

EXERCISE 8.2.9 (Polytope Arising from Rank One μ Problem): We consider the feedback interconnection which is used extensively in the μ theory; see Figure 8.2.2. Furthermore, we assume that the matrix $M(s)$ has entries which are stable rational functions and the system matrix $M(s)$ is of the form

$$M(s) = a(s)b^T(s)$$

with $a(s)$ and $b(s)$ being n-dimensional vectors of rational functions with i-th component $a_i(s)$ and $b_i(s)$, respectively. Prove that

$$\det(I + M(s)\Delta(q)) = 1 + \sum_{i=1}^{n} a_i(s)b_i(s)q_i$$

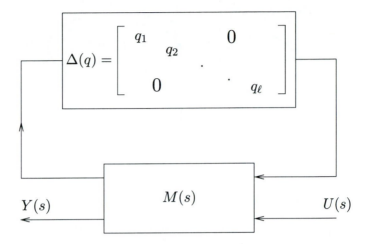

FIGURE 8.2.2   Configuration for Exercise 8.2.9

and argue that robust stability can be studied by using an uncertain polynomial $p(s, q)$ which has an affine linear uncertainty structure. *Hint*: Clear the denominators from the entries of $a_i(s)$ and $b_i(s)$ in the expression for the determinant above.

EXERCISE 8.2.10 (Converse):   Suppose that $p(s, q)$ is an uncertain polynomial having affine linear uncertainty structure. Take $\Delta(q)$ diagonal (as in the exercise above) and show that there is a matrix of the form $M(s) = a(s)b^T(s)$ with both $a(s)$ and $b(s)$ being rational and $p(s, q) = \det(I + M(s)\Delta(q))$.

8.3   A Primer on Polytopes and Polygons

In order to create a foundation for the robustness analysis to follow, we now review some elementary material from the theory of convex analysis. Some readers may opt to skip this section.

8.3.1   Convex Set and Convex Hull

A set $C \subseteq \mathbf{R}^k$ is said to be *convex* if the line joining any two points c^1 and c^2 in C remains entirely within C; i.e., given any $c^1, c^2 \in C$ and $\lambda \in [0, 1]$, it follows that $\lambda c^1 + (1 - \lambda)c^2 \in C$; we call $\lambda c^1 + (1 - \lambda)c^2$ a *convex combination* of c^1 and c^2. In Figure 8.3.1, a convex set and a nonconvex set in $\mathbf{R}^2$ are depicted. The reader can easily verify that common multidimensional sets such as rectangles, spheres and diamonds are convex.

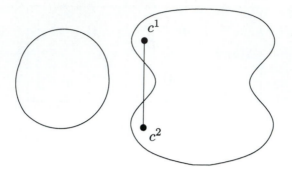

Convex Set   Nonconvex Set

FIGURE 8.3.1   Examples of Convex and Nonconvex Sets

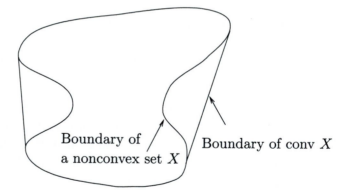

Boundary of / Boundary of conv X
a nonconvex set X

FIGURE 8.3.2   A Nonconvex Set and Its Convex Hull

Given a set $C \subset \mathbf{R}^k$ (not necessarily convex), its *convex hull*, conv C, is the "smallest" convex set which contains C. More precisely, if C^+ denotes the collection of all convex sets which contains the set C, then we have

$$\text{conv } C = \bigcap_{C^+ \in \mathcal{C}^+} C^+.$$

If the given set C is already convex, it follows that conv $C = C$. In Figure 8.3.2, an example of a nonconvex set with its convex hull is given. Notice the set inclusion conv $C \supseteq C$.

8.3.2 Polytopes and Polygons

A *polytope* **P** in $\mathbf{R}^k$ is the convex hull of a finite set of points $\{p^1, p^2, \ldots, p^m\}$. We write

$$\mathbf{P} = \mathrm{conv}\{p^i\}$$

and call $\{p^1, p^2, \ldots, p^m\}$ the *set of generators*. Note that the set of generators can be highly nonunique. For example, in Figure 8.3.3, the points p^3, p^5 and p^7 are optional for inclusion in a generating set

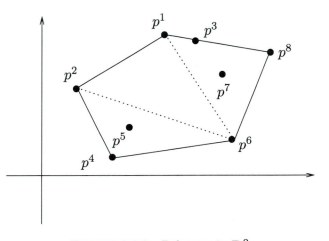

FIGURE 8.3.3   Polytope in $\mathbf{R}^2$

for **P**. The extreme point concept, covered in the next subsection, enables us to identify a unique set of generators.

In the sequel, it is important to make a distinction between polytopes in $\mathbf{R}^2$ and polytopes in $\mathbf{R}^k$ with $k > 2$. When manipulating value sets, we work with polytopes in the two-dimensional complex plane **C**, which we identify with $\mathbf{R}^2$ whenever convenient. Henceforth, we refer to a polytope in $\mathbf{R}^2$ as a *polygon*. According to this convention, both polytopes and polygons are automatically convex. We make note of this point because many authors make a distinction between a polygon and a convex polygon. For example, according to some authors, a star-shaped figure can be a polygon without necessarily being a convex polygon.

8.3.3   Extreme Points

Suppose $\mathbf{P} = \mathrm{conv}\{p^i\}$ is a polytope in $\mathbf{R}^k$. Then a point $p \in \mathbf{P}$ is said to be an *extreme point* of **P** if it cannot be expressed as a convex combination of two distinct points in **P**. That is, there does not exist $p^a, p^b \in \mathbf{P}$ with $p^a \neq p^b$ and $\lambda \in (0,1)$ such that $\lambda p^a + (1-\lambda)p^b = p$. For example, in Figure 8.3.4, the extreme points are p^1, p^2, p^3, p^4 and

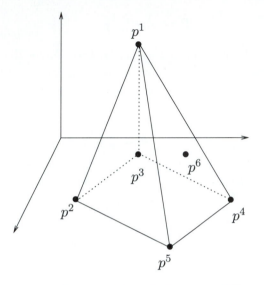

FIGURE 8.3.4   Polytope in $\mathbf{R}^3$

p^5. Although the interior point p^6 might be included in a generating set, it is not an extreme point. Given a finite set of generators $\{p^i\}$ for a polytope **P**, the set of extreme points is a subset of the set of generators. Furthermore, the set of extreme points can be called a *minimal generating set* in the sense that any other generating set contains the set of extremes.

In many applications, generators or extreme points of a polytope are specified implicitly rather than explicitly. A prime example occurs in the theory of linear programming where polytopes are described by a set of linear inequalities in the matrix form $Ax \leq b$.

8.3.4   Convex Combination Property

Given a polytope $\mathbf{P} = \text{conv}\{p^1, p^2, \ldots, p^m\}$, every point $p \in \mathbf{P}$ can be expressed as a *convex combination* of the p^i; that is, there exist real scalars $\lambda_1, \lambda_2, \ldots, \lambda_m \geq 0$ such that

$$p = \sum_{i=1}^{m} \lambda_i p^i$$

and

$$\sum_{i=1}^{m} \lambda_i = 1.$$

In the sequel, it is sometimes convenient to describe the constraint set for λ using the notation

$$\Lambda = \{\lambda \in \mathbf{R}^m : \lambda_i \geq 0 \text{ for } i = 1, 2, \ldots, m \text{ and } \sum_{i=1}^{m} \lambda_i = 1\}.$$

For such cases, Λ is called a *unit simplex*.

To illustrate the notion of convex combinations, consider the polygon $\mathbf{P}$ in Figure 8.3.3. Observe that $\mathbf{P}$ is the union of three triangles given by $\mathbf{P}_1 = \text{conv}\{p^1, p^6, p^8\}$, $\mathbf{P}_2 = \text{conv}\{p^1, p^2, p^6\}$ and $\mathbf{P}_3 = \text{conv}\{p^2, p^4, p^6\}$. Now, any point $p \in \mathbf{P}_1$ can be expressed as a convex combination $\lambda_1 p^1 + \lambda_6 p^6 + \lambda_8 p^8$. For example, a point such as p^7 might be obtained with $\lambda_1 = \lambda_6 = \lambda_8 = 1/3$, a point such as p^3 is generated with $\lambda_1 \neq 0$, $\lambda_8 \neq 0$ and $\lambda_6 = 0$ and finally, an extreme point such as p^6 is obtained with $\lambda_6 = 1$ and $\lambda_1 = \lambda_8 = 0$. To conclude, we observe that the description of a point $p \in \mathbf{P}$ as a convex combination of extreme points is nonunique. For example, a point such as p^7 can be expressed as a convex combination of p^1, p^6 and p^8 or p^2, p^6 and p^8.

The fact that we can describe every point $p \in \mathbf{P}$ in Figure 8.3.3 as a convex combination of three or less extreme points is not particular to the example at hand. In fact, Cartheodory's Theorem tells us: Every point in a polytope $\mathbf{P} \subset \mathbf{R}^k$ is expressible as a convex combination of $k+1$ extreme points at most; e.g., see Rockafellar (1970).

8.3.5   Edges of a Polytope

Given any two points x^a and x^b in $\mathbf{R}^k$, we denote the straight line segment joining these points by $[x^a, x^b]$. Notice that every point $x \in [x^a, x^b]$ can be expressed uniquely as a convex combination of x^a and x^b; that is,

$$x = \lambda x^a + (1 - \lambda)x^b$$

for some unique $\lambda \in [0, 1]$. Furthermore, if $x^a = x^b$, $[x^a, x^b]$ degenerates to a *point* which is viewed as a special case of a line.

We now consider lines of the form $[p^{i_1}, p^{i_2}]$, where p^{i_1} and p^{i_2} are extremes of a given polytope $\mathbf{P}$ and $p^{i_1} \neq p^{i_2}$. We say that $[p^{i_1}, p^{i_2}]$ is an *edge* of $\mathbf{P}$ if the following condition holds: Given any $p^a, p^b \in \mathbf{P}$ with $p^a, p^b \notin [p^{i_1}, p^{i_2}]$, it follows that $[p^a, p^b] \cap [p^{i_1}, p^{i_2}] = \phi$.

In two or three dimensions, the edges of a polytope are apparent by inspection. For example, in Figure 8.3.3, edges of the polytope $\mathbf{P}$ are $[p^1, p^2]$, $[p^2, p^4]$, $[p^4, p^6]$, $[p^6, p^8]$ and $[p^8, p^1]$ and in Figure 8.3.4, the edges of the polygon $\mathbf{P}$ are $[p^1, p^2]$, $[p^1, p^3]$, $[p^1, p^4]$, $[p^1, p^5]$, $[p^2, p^3]$, $[p^2, p^5]$, $[p^3, p^4]$ and $[p^4, p^5]$.

8.3.6 Operations on Polytopes

In this subsection, we provide a number of basic facts about operations on polytopes.

LEMMA 8.3.7 (Direct Sum for Two Polytopes):   *Given two polytopes* $\mathbf{P}_1 = conv\{p^{1,i_1}\}$ *and* $\mathbf{P}_2 = conv\{p^{2,i_2}\}$ *in* $\mathbf{R}^k$, *the direct sum*

$$\mathbf{P}_1 + \mathbf{P}_2 = \{p^1 + p^2 : p^1 \in \mathbf{P}_1; p^2 \in \mathbf{P}_2\}$$

is a polytope. Moreover,

$$\mathbf{P}_1 + \mathbf{P}_2 = conv\{p^{1,i_1} + p^{2,i_2}\}.$$

REMARKS 8.3.8 (Direct Sum for Polytope and Point):   For the special case when $\mathbf{P}_2$ consists of a single point, $\mathbf{P}_1 + \mathbf{P}_2$ corresponds to a translation of $\mathbf{P}$. In the lemma below, we provide another useful characterization of $\mathbf{P}_1 + \mathbf{P}_2$.

LEMMA 8.3.9 (Another Direct Sum Description):   *Given two polytopes* $\mathbf{P}_1 = conv\{p^{1,i}\}$ *and* $\mathbf{P}_2$ *in* $\mathbf{R}^k$, *it follows that*

$$\mathbf{P}_1 + \mathbf{P}_2 = conv \bigcup_i (p^{1,i} + \mathbf{P}_2).$$

EXAMPLE 8.3.10 (Illustration of Direct Sum):   To illustrate formation of the direct sum via application of the lemma above, suppose that $\mathbf{P}_1 = conv\{2 + 2j, 4 - 2j, 6 + 6j\}$ and $\mathbf{P}_2$ is the unit square in the complex plane. Then, the lemma leads to the direct sum, which is shown in Figure 8.3.5.

EXERCISE 8.3.11 (Less Restriction on $\mathbf{P}_2$):   Argue that Lemma 8.3.9 remains valid when $\mathbf{P}_2$ is an arbitrary convex set which is not necessarily polytopic. Illustrate by considering Example 8.3.10 with $\mathbf{P}_2$ being the unit disc rather than the unit square. Sketch the resulting direct sum $\mathbf{P}_1 + \mathbf{P}_2$.

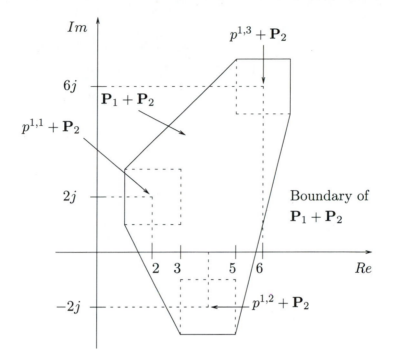

FIGURE 8.3.5   Formation of Direct Sum in Example 8.3.10

LEMMA 8.3.12 (Intersection of Two Polytopes):   *Let* $\mathbf{P}_1$ *and* $\mathbf{P}_2$ *be two polytopes in* $\mathbf{R}^k$. *Then it follows that* $\mathbf{P}_1 \cap \mathbf{P}_2$ *is a polytope.*

REMARKS 8.3.13 (Intersection):   If $\mathbf{P}_1$ and $\mathbf{P}_2$ are polytopes, $\mathbf{P}_1 \cap \mathbf{P}_2$ may have extreme points which are not extreme points of either $\mathbf{P}_1$ or $\mathbf{P}_2$; e.g., in Figure 8.3.6, the points $\hat{p}^a$ and $\hat{p}^b$ are extreme points of $\mathbf{P}_1 \cap \mathbf{P}_2$ but are not extreme points of $\mathbf{P}_1$ or $\mathbf{P}_2$.

LEMMA 8.3.14 (Multiplication of a Scalar and a Polytope):   *Given a polytope* $\mathbf{P} = conv\{p^i\}$ *and a real scalar* α, *it follows that the set*

$$\alpha\mathbf{P} = \{\alpha p : p \in \mathbf{P}\}$$

is a polytope. Moreover,

$$\alpha\mathbf{P} = conv\{\alpha p^i\}.$$

LEMMA 8.3.15 (Convex Hull of a Union):   *Given any two polytopes* $\mathbf{P}_1 = conv\{p^{1,i_1}\}$ *and* $\mathbf{P}_2 = conv\{p^{2,i_2}\}$ *in* $\mathbf{R}^k$, *it follows that* $conv(\mathbf{P}_1 \cup \mathbf{P}_2)$ *is a polytope with generating set* $\{p^{1,i_1}\} \cup \{p^{2,i_2}\}$.

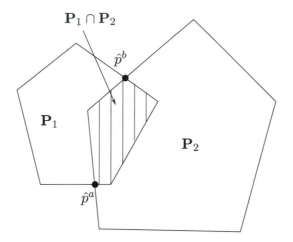

FIGURE 8.3.6   New Extreme Points Created by Intersection

REMARKS 8.3.16 (Loss of Extreme Points):   Given two polytopes $\mathbf{P}_1$ and $\mathbf{P}_2$, the geometry associated with formation of $\mathrm{conv}(\mathbf{P}_1 \bigcup \mathbf{P}_2)$ is depicted in Figure 8.3.7. Note that some of the extreme points of $\mathbf{P}_1$ and $\mathbf{P}_2$ are no longer extremes of $\mathrm{conv}(\mathbf{P}_1 \bigcup \mathbf{P}_2)$.

LEMMA 8.3.17 (Affine Linear Transformation of a Polytope):   *Suppose that* $\mathbf{P} = conv\{p^i\}$ *is a polytope in* $\mathbf{R}^{k_1}$ *and* $T : \mathbf{R}^{k_1} \to \mathbf{R}^{k_2}$ *is an affine linear transformation. Then the set* $T\mathbf{P} = \{Tp : p \in \mathbf{P}\}$ *is a polytope in* $\mathbf{R}^{k_2}$. *Moreover,*

$$TP = conv\{Tp^i\}$$

and every edge point of $T\mathbf{P}$ *is the image of some edge point of* $\mathbf{P}$. *That is, if* x *is an edge point of* $T\mathbf{P}$, *then* $x = Tp$ *for some edge point* $p \in \mathbf{P}$.

REMARKS 8.3.18 (Edge Mapping):   In the lemma above, note that not all edge points of $\mathbf{P}$ map into edge points of $T\mathbf{P}$. Roughly speaking, some of the edges of $\mathbf{P}$ can map into the "inside" of $T\mathbf{P}$. Also, the lemma does not rule out the possibility that points which are not on the edge of $\mathbf{P}$ are mapped onto an edge of $T\mathbf{P}$. As a simple illustration, suppose $\mathbf{P}$ is the unit square in $\mathbf{R}^2$ and take $T : \mathbf{R}^2 \to \mathbf{R}^2$ to be the mapping which projects a point $p \in \mathbf{R}^2$ onto its first component p_1; i.e., the coordinates of Tp are $T_1 p = p_1$ and $T_2 p = 0$. Observe that every point in $\mathbf{P}$ is mapped into an edge point of $T\mathbf{P}$.

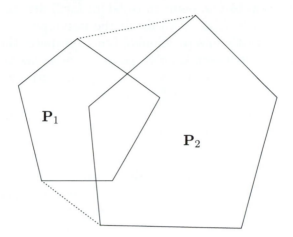

FIGURE 8.3.7   Formation of conv($\mathbf{P}_1 \bigcup \mathbf{P}_2$)

EXAMPLE 8.3.19 (Two-Dimensional Domain and Range): An important special case of Lemma 8.3.17 is obtained when $\mathbf{P}$ is a polygon in the complex plane and we seek a description of the set

$$z\mathbf{P} = \{zp : p \in \mathbf{P}\}$$

with $z \in \mathbf{C}$ being a given complex number. Note that if $p \in \mathbf{P}$, we can write

$$\begin{bmatrix} Re\ zp \\ Im\ zp \end{bmatrix} = \begin{bmatrix} Re\ z & -Im\ z \\ Im\ z & Re\ z \end{bmatrix} \begin{bmatrix} Re\ p \\ Im\ p \end{bmatrix}$$

and view the formation of $z\mathbf{P}$ as a linear transformation from $\mathbf{R}^2$ to $\mathbf{R}^2$. Now, Lemma 8.3.17 provides us with a description of the generators of $T\mathbf{P}$. In fact, if $\mathbf{P} = \text{conv}\{p^i\}$ and

$$z = Re^{j\theta},$$

we obtain

$$T\mathbf{P} = \text{conv}\{Re^{j\theta}p^i\}.$$

In other words, the i-th generator for $T\mathbf{P}$ is simply obtained from p^i via a scaling and a rotation.

8.4   Introduction to Polytopes of Polynomials

We are now prepared to consider polytopes in the context of polynomials. However, before proceeding, it is important to draw the reader's attention to one point in order to facilitate understanding of the exposition to follow: Although the polytopes in this section are defined abstractly in a polynomial function space, there is a natural isomorphism between a polytope of polynomials and its set of coefficients. Hence, once things are set up correctly, we can apply all the machinery in Section 8.3. When we perform operations on a polytope of polynomials in a convex analysis context, there is always an appropriate interpretation in coefficient space.

DEFINITION 8.4.1 (Polytope of Polynomials): A family of polynomials $\mathcal{P} = \{p(\cdot, q) : q \in Q\}$ is said to be a *polytope of polynomials* if $p(s, q)$ has an affine linear uncertainty structure and Q is a polytope. If $Q = \text{conv}\{q^i\}$, then we call $p(s, q^i)$ the *i-th generator* for $\mathcal{P}$.

EXAMPLE 8.4.2 (Polytope of Polynomials): If a polytope of polynomials $\mathcal{P}$ is described by $p(s, q) = s^2 + (4q_1 + 3q_2 + 2)s + (2q_1 - q_2 + 5)$, $|q_1| \leq 1$ and $|q_2| \leq 1$, the uncertainty bounding set Q has four extremes $q^1 = (-1, -1)$, $q^2 = (-1, 1)$, $q^3 = (1, -1)$ and $q^4 = (1, 1)$. The four associated generators are given by $p(s, q^1) = s^2 - 5s + 4$, $p(s, q^2) = s^2 + s + 2$, $p(s, q^3) = s^2 + 3s + 8$ and $p(s, q^4) = s^2 + 9s + 6$.

EXERCISE 8.4.3 (Generators in Coefficient Space): Consider a polytope of polynomials $\mathcal{P} = \{p(\cdot, q) : q \in Q\}$ with *coefficient vector* $a(q)$ for $p(s, q)$. Prove that the *coefficient set*

$$a(Q) = \{a(q) : q \in Q\}$$

is a polytope and, moreover, argue that if $Q = \text{conv}\{q^i\}$, then

$$a(Q) = \text{conv}\{a(q^i)\}.$$

8.5 Generators for a Polytope of Polynomials

Our objective in this section is to show that the $p(s, q^i)$ in Definition 8.4.1 rightly deserve to be called generators. To this end, notice that since Q is a polytope with i-th generator q^i, any $q \in Q$ can be expressed as a convex combination

$$q = \sum_i \lambda_i q^i$$

for appropriate λ in the unit simplex. Now, using the fact that the coefficients of $p(s, q)$ depend affine linearly on q, we obtain

$$p(s, q) = \sum_{i=0}^{n} a_i (\sum_k \lambda_k q^k) s^i = \sum_{i=0}^{n} \sum_k \lambda_k a_i(q^k) s^i = \sum_k \lambda_k p(s, q^k).$$

Hence, for each $q \in Q$, $p(s, q)$ is a convex combination of the $p(s, q^i)$. This justifies calling $p(s, q^i)$ the i-th generator for $\mathcal{P}$ and writing

$$\mathcal{P} = \text{conv}\{p(\cdot, q^i)\}$$

with the understanding that operations (such as taking the convex hull) in the space of polynomials can be associated with operations on q or the coefficients. For example, if q^{i_1} and q^{i_2} are generators of the uncertainty bounding set Q, then the convex combination

$$\tilde{p}(s, \lambda) = \lambda p(s, q^{i_1}) + (1 - \lambda) p(s, q^{i_2})$$

is associated with the polynomial $p(s, q^\lambda)$ with $q^\lambda = \lambda q^{i_1} + (1 - \lambda) q^{i_2}$. Equivalently, we can associate $\tilde{p}(s, \lambda)$ with the coefficient vector

$$a^\lambda = \lambda a(q^{i_1}) + (1 - \lambda) a(q^{i_2}).$$

EXAMPLE 8.5.1 (Interval Polynomial as a Special Case):   In this example, we view an interval polynomial

$$p(s, q) = \sum_{i=0}^{n} [q_i^-, q_i^+] s^i$$

in the polytopic framework. Indeed, if q^k denotes the k-*th extreme point* of the associated uncertainty bounding set Q, the i-th component q_i^k of q^k is q_i^- or q_i^+. Hence, this interval polynomial family can be described using at most 2^{n+1} generators. Associated with the k-th extreme point q^k of Q is the k-*th extreme*

$$p(s, q^k) = \sum_{i=0}^{n} q_i^k s^i.$$

EXERCISE 8.5.2 (Enumeration of Extremes):   Enumerate the eight extreme polynomials for the polytope associated with the interval polynomial $p(s, q) = 2s^4 + [1, 2]s^3 + 5s^2 + [3, 4]s + [5, 6]$.

REMARKS 8.5.3 (Generators Using the Unit Simplex):   In view
of the discussion above, it is often convenient to describe a family
of polynomials over the unit simplex rather than the uncertainty
bounding set Q. In other words, if $p_1(s)$, $p_2(s)$, ..., $p_m(s)$ are fixed
polynomials and Λ is the unit simplex, we can define the family
$\mathcal{P} = \{p(\cdot, \lambda) : \lambda \in \Lambda\}$ consisting of all convex combinations of the
$p_i(s)$; that is, a polynomial $p(s)$ is in $\mathcal{P}$ if there exists some vector
$\lambda \in \Lambda$ such that

$$p(s) \equiv \sum_{i=1}^{m} \lambda_i p_i(s).$$

We write

$$\mathcal{P} = \text{conv}\{p_i(\cdot)\}$$

and view the simplex parameter vector λ as a surrogate for the vector of uncertain parameters q. Note that if $\mathcal{P} = \text{conv}\{p(\cdot, q^i)\}$ is a
polytope of polynomials defined over q space, then, by taking gen-
erator $p_i(s) = p(s, q^i)$, we obtain a polytope over a unit simplex. In
view of the change of variables from q to λ, we refer to $p(s, \lambda)$ as an
uncertain polynomial.

EXERCISE 8.5.4 (Polytope over the Unit Simplex):   Starting with
fixed polynomials $p_0(s), p_1(s), \ldots, p_\ell(s)$, consider the polytope of
polynomials $\mathcal{P}$ described by

$$p(s, q) = p_0(s) + \sum_{i=1}^{\ell} q_i p_i(s),$$

$q \in Q$ and $Q = \text{conv}\{q^i\}$. Describe this family of polynomials $\mathcal{P}$ in
terms of the unit simplex.

EXERCISE 8.5.5 (Diamond Family of Polynomials):   Repeat Exercise 8.5.4 with the uncertainty bounding set Q described by

$$\sum_{i=1}^{\ell} w_i |q_i| \leq 1,$$

where $w_i > 0$ for $i = 1, 2, \ldots, \ell$. We refer to the w_i as *weights* for
the uncertain parameters. The resulting set $\mathcal{P} = \{p(\cdot, q) : q \in Q\}$ is
called a *diamond family* of polynomials.

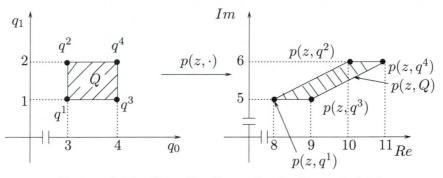

FIGURE 8.6.1   Value Set Generation for Example 8.6.1

8.6   Polygonal Value Sets

In this section, we concentrate on characterization of value sets for a
polytope of polynomials. Since the interval polynomial framework is
a special case of the polytopic framework, the value set which we obtain is a generalization of the Kharitonov rectangle of Section 5.7.1.
For polytopes of polynomials, we now argue that the relevant value
sets are polygons in the complex plane. To see one of the many possible ways by which polygonal value sets arise, we provide a simple
motivating example.

EXAMPLE 8.6.1 (Polygonal Value Set):   We consider the interval
polynomial $p(s, q) = s^2 + [1, 2]s + [3, 4]$ and take $z = 2 + j$ for
construction of the value set $p(z, Q)$. Indeed, by substitution, we
find $p(2 + j, q) = (3 + q_0 + 2q_1) + j(4 + q_1)$ and then seek to map the
box Q into $p(2 + j, Q)$. This is accomplished by first mapping the
extreme points of Q; i.e., $q^1 = (3, 1) \mapsto 8 + j5$, $q^2 = (3, 2) \mapsto 10 + j6$,
$q^3 = (4, 1) \mapsto 9 + j5$ and $q^4 = (4, 2) \mapsto 11 + j6$. Using the four points
$p(z, q^i)$ and noting that the real and imaginary parts of $p(2+j, q)$ are
affine linear in q, we obtain the complete value set $p(z, Q)$ by taking
the convex hull as shown in Figure 8.6.1. We see that the value set
$p(z, Q)$ is a parallelogram in the complex plane with edges which are
obtained by mapping the edges of Q through $p(z, \cdot)$. For example,
the edge of Q obtained by joining $p(z, q^1)$ and $p(z, q^2)$ comes from
the edge joining q^1 and q^2.

REMARKS 8.6.2 (Generalization):   In view of the motivating example above, our goal is to characterize the value set $p(z, Q)$ for a
more general polytope of polynomials. An informal statement of the

technical result given below is as follows: For a polytope of polynomials, the value set $p(z, Q)$ is a polygon with generating set $\{p(z, q^i)\}$ and edges which come from the edges of Q. The reader should be forewarned, however, that not all edges of Q necessarily map into edges of $p(z, Q)$. To illustrate, notice that in Figure 8.6.2, the edge

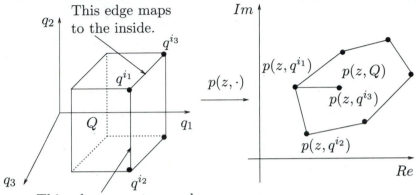

FIGURE 8.6.2   Mapping an Edge of Q into the Interior of $p(z, Q)$

$[q^{i_1}, q^{i_3}]$ is mapped to the interior of the set $p(z, Q)$.

8.7   Value Set for a Polytope of Polynomials

Given that the discussion of polygonal value sets above is in the context of numerical examples, the objective in this section is to provide a general result. We now provide a polygonal characterization of value sets for a polytope of polynomials.

PROPOSITION 8.7.1 (Value Set for a Polytope of Polynomials):   *Let* $\mathcal{P} = \{p(\cdot, q) : q \in Q\}$ *be a polytope of polynomials with uncertainty bounding set* $Q = conv\{q^i\}$. *Then, for fixed* $z \in \mathbf{C}$, *the value set* $p(z, Q)$ *is a polygon with generating set* $\{p(z, q^i)\}$. *That is,*

$$p(z, Q) = conv\{p(z, q^i)\}.$$

Furthermore, all edges of the polygon $p(z, Q)$ *are obtained from the edges of* Q *in the following sense: If* z_0 *is a point on an edge of* $p(z, Q)$, *then* $z_0 = p(z, q^0)$ *for some* q^0 *on an edge of* Q.

PROOF: For fixed $z \in \mathbf{C}$, we note that the mapping $T : Q \to \mathbf{C}$ defined by

$$Tq = p(z, q)$$

is affine linear. Since Q is a polytope and $TQ = p(z, Q)$, the proposition follows immediately from Lemma 8.3.17. That is, the value set $p(z, Q)$ is the image of Q under T. ∎

EXAMPLE 8.7.2 (Polygonal Value Sets): Given the uncertain polynomial

$$p(s, q) = (2q_1 - q_2 + q_3 + 1)s^3 + (3q_1 - 3q_2 + q_3 + 3)s^2$$
$$+ (3q_1 + q_2 + q_3 + 3)s + (q_1 - q_2 + 2q_3 + 3)$$

and uncertainty bounding set Q which is described by $|q_i| \leq 0.245$ for $i = 1, 2, 3$, Proposition 8.7.1 indicates that the value set $p(z, Q)$ is a polygon with eight generators. For example, corresponding to the particular generator $q^5 = (0.245, -0.245, 0.245)$, we obtain the polynomial $p(s, q^5) = 1.98s^3 + 4.715s^2 + 3.735s + 3.98$. In Figure 8.7.2, the value set $p(j\omega, Q)$ is shown for thirty frequencies evenly spaced

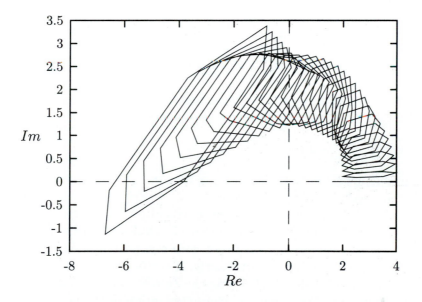

FIGURE 8.7.1   Polygonal Value Sets for Example 8.7.2

between $\omega = 0$ and $\omega = 1.5$. Notice that even though there are eight generators, the value set has only six extreme points; i.e., two of the extreme points of the box Q are mapped into the interior of $p(j\omega, Q)$. Now, we test for robust stability using the value set plot in conjunction with the Zero Exclusion Condition (Theorem 7.4.2). To this end, notice that $p(s, q^0) = p(s, 0)$ serves as a stable member and it can be easily verified that $\omega_c = 1.5$ serves as a cutoff frequency for the plot; i.e., $0 \notin p(j\omega, Q)$ for $\omega > \omega_c$. Finally, for $0 \leq \omega \leq 1.5$, we verify from the plot that $0 \notin p(j\omega, Q)$. Hence, we deem the family $\mathcal{P}$ to be robustly stable.

EXERCISE 8.7.3 (Satellite Attitude Control):   The transfer function for a satellite attitude control problem, as given by Franklin, Powell and Emani-Naeini (1986), is

$$\frac{\theta(s)}{T(s)} = \frac{10ds + 10k}{s^2(s^2 + 11ds + 11k)},$$

where $\theta(s)$ is the Laplace transform of the angle, $T(s)$ is the Laplace transform of the applied torque and k and d are uncertain parameters associated with spring and viscous damping constants, respectively. With the uncertainty bounds $0.09 \leq k \leq 0.4$, $0.004 \leq d \leq 0.04$ and PD Controller $C(s) = 0.001(30s + 1)$ connected in feedback configuration as in Figure 8.2.1, determine if the closed loop system is robustly stable. Over the relevant range of frequencies, display the value set associated with the family of closed loop polynomials.

EXERCISE 8.7.4 (Dominant Pole Specification):   A feedback system gives rise to the polytope of polynomials $\mathcal{P}_r$ described by

$$p(s, q) = s^3 + (10 + q_2)s^2 + (29 + q_1)s + (30 + q_1 + q_2)$$

with variable uncertainty bound $|q_i| \leq r$ for $i = 1, 2$. Notice that for $q = 0$, the *nominal* polynomial $p(s, 0) = s^3 + 10s^2 + 29s + 30$ has roots $s_1 = -6$ and $s_{2,3} = -2 \pm j$. The specification for the system is to have two "dominant" roots remain within circles of radius $\epsilon = 1$ centered at $s = -2 \pm j$ and a third root with real part ≤ 5. We want this specification to be satisfied robustly. That is, this specification should be satisfied for all $q \in Q$. Using the Zero Exclusion Condition in combination with the polygonal characterization of the value set, verify that the robustness margin

$$r_{max} = \sup\{r : \mathcal{P}_r \text{ has the specified root distribution}\}$$

is given by $r_{max} \approx 0.35$. Use the more general interpretation of zero exclusion in Exercise 7.5.7 and generate a value set plot for an appropriate range of frequencies.

EXERCISE 8.7.5 (DC Motor with Resonant Load):   The goal in this exercise is to apply polygonal value set theory using the model of a DC motor in Example 2.8.1. In contrast to the analysis in Chapter 2, consider uncertain parameters $0.5 \times 10^{-2} \le J_L \le 1.5 \times 10^{-2}$ and $2 \times 10^{-3} \le B_L \le 4 \times 10^{-1}$. The remaining parameters are fixed at $L = 5 \times 10^{-3}$, $R = 1$, $J_m = 2 \times 10^{-3}$, $B_m = 2 \times 10^{-3}$, $K = 0.5$ and $K_s = 2 \times 10^3$. In order to study pole locations as a function of the uncertain parameters, take $q_1 = J_L$ and $q_2 = B_L$ and concentrate on the transfer function from armature voltage to shaft speed. This transfer function is given by

$$P(s) = \frac{K J_L s^2 + K B_L s + K K_s}{\Delta(s)},$$

where

$$
\begin{aligned}
\Delta(s) = {} & J_m J_L L s^4 + (B_m J_L L + B_L J_m L + J_m J_L R)s^3 \\
& + (B_m B_L L + J_m K_s L + K_s J_L L \\
& \quad + B_m J_L R + B_L J_m R + K^2 J_L)s^2 \\
& + (B_m K_s L + B_L K_s L + B_m B_L R \\
& \quad + J_m K_s R + K_s J_L R + K^2 B_L)s \\
& + (B_m K_s R + B_L K_s R + K^2 K_s).
\end{aligned}
$$

(a) Verify that the uncertain denominator polynomial is given by

$$
\begin{aligned}
p(s,q) \approx {} & 10^{-5} q_1 s^4 + (2 \times 10^{-3} q_1 + 10^{-5} q_2)s^3 \\
& + (10.252 q_1 + 2 \times 10^{-3} q_2 + 2 \times 10^{-2})s^2 \\
& + (4.02 + 2 \times 10^3 q_1 + 10.252 q_2)s \\
& + 5.04 \times 10^2 + 2 \times 10^3 q_2.
\end{aligned}
$$

(b) With uncertainty bounding set

$$Q = \{q \in \mathbf{R}^2 : 0.005 \le q_1 \le 0.015; \; 0.002 \le q_2 \le 0.4\}$$

obtained from the data above, take desired pole location region $\mathcal{D}$ to be a damping cone with angle $\phi = 45^\circ$ as in Figure 6.8.1. By sweeping the boundary of $\mathcal{D}$, generate appropriate value sets $p(z, Q)$ and use the Zero Exclusion Condition (Theorem 7.4.2) to determine whether $\mathcal{P}$ is robustly $\mathcal{D}$-stable; i.e., determine whether the damping specification is robustly satisfied.

8.8   Improvement over Rectangular Bounds

To demonstrate that a polygonal value set plot leads to better results than overbounding via a Kharitonov rectangle (see Section 5.11), we compare results using both techniques in the example below.

EXAMPLE 8.8.1 (Conservatism of Overbounding):   For the polytope of polynomials $\mathcal{P}$ described by

$$p(s,q) = s^4 + (2q_2 + 1)s^3 + (2q_1 - q_2 + 4)s^2$$
$$+ (q_2 + 1)s + (q_1 - 2q_2 + 2)$$

and $Q = \{q \in \mathbf{R}^2 : -0.5 \le q_1 \le 2; -0.3 \le q_2 \le 0.3\}$, we carry out two robust stability analyses. First, we replace $\mathcal{P}$ by the overbounding interval polynomial family $\overline{\mathcal{P}}$ described by

$$\overline{p}(s,\overline{q}) = s^4 + [0.4, 1.6]s^3 + [2.7, 8.3]s^2 + [0.7, 1.3]s + [0.9, 4.6]$$

and reach the conclusion that $\overline{\mathcal{P}}$ is not robustly stable. That is, by a straightforward calculation, it is easy to verify that the Kharitonov polynomial

$$K_3(s) = s^4 + 1.6s^3 + 2.7s^2 + 0.7s + 4.6$$

has an unstable root pair $s_{1,2} \approx 0.4099 \pm j1.1106$.

To begin the second analysis, we verify the critical precondition for application of the Zero Exclusion Condition (Theorem 7.4.2). Indeed, with $q^0 = 0$, $p(s, q^0) = s^4 + s^3 + 4s^2 + s + 2$ is a stable member of $\mathcal{P}$. Next, a preliminary computation (see Section 7.3) indicates that a suitable cutoff frequency is $\omega_c = 1$; i.e., $0 \notin p(j\omega, Q)$ for frequency $\omega > \omega_c$. Subsequently, in accordance with Proposition 8.7.1, we generate 40 polygonal value sets corresponding to frequencies evenly spaced between $\omega = 0$ and $\omega = 1$; see Figure 8.8.1. Within computational limits, we conclude that $0 \notin p(j\omega, Q)$ for all $\omega \ge 0$. Hence, by the Zero Exclusion Condition, we say that $\mathcal{P}$ is robustly stable.

The conclusion to be drawn is that working with the overbounding family $\overline{\mathcal{P}}$ is inconclusive but working with polygonal value sets leads to the conclusion that $\mathcal{P}$ is robustly stable.

EXERCISE 8.8.2 (Value Set Comparison):   With the same setup as in the example above, generate value sets $\overline{p}(j\omega, \overline{Q})$ corresponding to the overbounding interval polynomial $\overline{p}(j\omega, \overline{Q})$. For purposes of comparison with Figure 8.8.1, use 40 frequencies evenly spaced between $\omega = 0$ and $\omega = 1$. Finally, observe that the Zero Exclusion

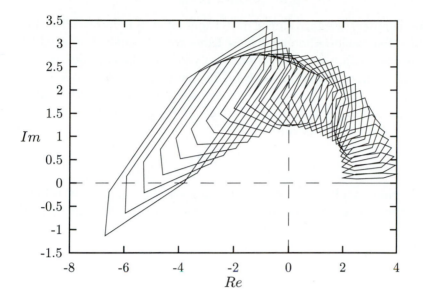

FIGURE 8.8.1   Value Set for Example 8.8.1

Condition is violated for the Kharitonov rectangles but holds for the polygons $p(j\omega, Q)$.

EXERCISE 8.8.3 (Total Subtended Angle):   There are many approaches for numerical implementation of a zero exclusion test. To illustrate, suppose that $\mathcal{P} = \{p(\cdot, q) : q \in Q\}$ is a polytope of polynomials with Q having set of extremes $\{q^i\}$. For a given frequency $\omega \geq 0$, let

$$\theta_i(\omega) = \tan^{-1}\left(\frac{Im\ p(j\omega, q^i)}{Re\ p(j\omega, q^i)}\right)$$

and define the *total subtended angle* by

$$\theta(\omega) = \max_i \theta_i(\delta) - \min_i \theta_i(\delta).$$

Now, at frequency $\omega \geq 0$, argue that the Zero Exclusion Condition $0 \notin p(j\omega, Q)$ is satisfied if and only if

$$\theta(\omega) < \pi.$$

Generalize this result to the robust $\mathcal{D}$-stability problem.

8.9 Conclusion

In this chapter, we introduced polytopes of polynomials. To this end, we first reviewed some basics from convex analysis and then characterized value sets as (convex) polygons in the complex plane. This characterization is quite useful in many chapters to follow. In particular, in the next chapter, we see that the edges of the value set polygon are very important. Indeed, if $\mathcal{P} = \{p(\cdot, q) : q \in Q\}$ is a polytope of polynomials and

$$\mathcal{E} = \{\mathcal{E}_i : i \in I\}$$

denotes the set of edges of Q, then we see that a loss of robust $\mathcal{D}$-stability is synonymous with satisfaction of an *edge penetration condition*

$$0 \in p(z, \mathcal{E}_i)$$

for some $z \in \partial \mathcal{D}$ and some $i \in I$. Subsequently, it becomes possible to reduce the robust $\mathcal{D}$-stability problem, formulated over the multidimensional set Q, to a finite set of single-parameter problems.

Notes and Related Literature

NRL 8.1 With regard to Exercises 8.2.9 and 8.2.10, a more detailed discussion on the relationship between μ theory and the theory in this text is given in Chen, Fan and Nett (1992).

NRL 8.2 For a more detailed exposition of the material on convex analysis in Section 8.3, see references such as Rockafellar (1970), Stöer and Witzgall (1970) and Aubin and Vinter (1980).

NRL 8.3 For the DC motor with resonant load in Exercise 8.7.5, Bailey, Panzer and Gu (1988) concentrate on generation of the value set of the transfer function rather than the denominator polynomial. When these value sets are displayed on the Nichol's chart, we obtain the Horowitz templates; see Horowitz and Sidi (1972).

Chapter 9

The Edge Theorem

Synopsis

The focal point of this chapter is the celebrated Edge Theorem of Bartlett, Hollot and Huang: Under mild conditions, a polytope of polynomials $\mathcal{P}$ is robustly $\mathcal{D}$-stable if and only if every polynomial on an edge of $\mathcal{P}$ is $\mathcal{D}$-stable. This has strong ramifications throughout the remainder of this text. The fact that the number of edges can be prohibitively large paves the way for results in later chapters.

9.1 Introduction

Having established the fact that every polytope of polynomials has polygonal value sets, we are now prepared to study the Edge Theorem and its ramifications. In this chapter, we see that the interior of these polygonal value sets are unimportant as far as robust stability is concerned. With the machinery we have in place, it is easy to show that the problem of robust stability is solved by working with a set of "edge polynomials" which can be identified with edges of the value set. Since our description of these edge polynomials involves only one-parameter, we end up with a set of one-parameter robust stability problems in lieu of the original problem. Hence, the one-parameter formulation of Chapter 4, which may have seemed quite specialized at the time it was introduced, is actually quite important.

The key idea is that we can reduce a problem with a multidimensional uncertainty set to a finite number of one-parameter problems whose solutions are readily available.

9.2   Lack of Extreme Point Results for Polytopes

Perhaps the most important motivation for the analysis to follow is derived from the fact that Kharitonov-like extreme point results do not hold for a general polytope of polynomials. For example, recalling the pair of polynomials $f(s) = 10s^3 + s^2 + 6s + 0.57$ and $g(s) = s^2 + 2s + 1$ analyzed in Exercise 4.15.1, we found that with

$$p(s, \lambda) = f(s) + \lambda g(s),$$

both $p(s, 0)$ and $p(s, 1)$ are stable but $p(s, 0.5)$ is unstable.

In view of such examples, we proceed as follows: In this chapter, we concentrate on results which apply to *all polytopes*; these are called edge results. In later chapters, we identify rich classes of $(f(s), g(s))$ pairs for which extreme point results hold. That is, stability of $p(s, 0)$ and $p(s, 1)$ implies stability of $p(s, \lambda)$ for all $\lambda \in [0, 1]$. In other words, under strengthened hypotheses, we can do "better" than edges. We see that such hypotheses have an interpretation in a feedback control context; i.e., when the plant has an affine linear uncertainty structure.

9.3   Heuristics Underlying the Edge Theorem

Before providing the main results of this chapter, we give a heuristic argument which motivates the Edge Theorem. Indeed, suppose that $\mathcal{P} = \{p(\cdot, q) : q \in Q\}$ is a polytope of polynomials with at least one stable member and the desired root location region $\mathcal{D}$ is an open subset of the complex plane. Now, with the Zero Exclusion Condition (Theorem 7.4.2) in mind, we sweep z over the boundary $\partial \mathcal{D}$ of $\mathcal{D}$ and obtain the value set $p(z, Q)$. We are interested to see if $z = 0$ ever enters this set.

Suppose that initially, $0 \notin p(z_0, Q)$ with $z_0 \in \partial \mathcal{D}$. Now, as we move away from z_0 along $\partial \mathcal{D}$, we know from Proposition 8.7.1 that the value set $p(z, Q)$ is a polygon. Furthermore, with a z moving along $\partial \mathcal{D}$, either $z = 0$ remains outside of $p(z, Q)$ or we arrive at some critical $z_1 \in \partial \mathcal{D}$ for which the origin, $z = 0$, lies on the edge of the polygon $p(z, Q)$. In other words, the situation which we associate with a loss of robust $\mathcal{D}$-stability occurs when the origin, $z = 0$, lies on an edge of $p(z_1, Q)$ for some $z_1 \in \partial \mathcal{D}$; this situation is depicted in Figure 9.3.1. Recalling that edges of the value set are obtained

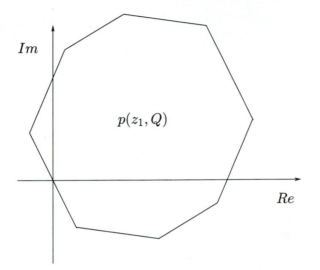

FIGURE 9.3.1   Loss of Robust $\mathcal{D}$-Stability

from edges of Q (see Lemma 8.3.17), this heuristic argument leads us to conjecture that under suitable regularity conditions, a necessary and sufficient condition for robust $\mathcal{D}$-stability of $\mathcal{P}$ is $\mathcal{D}$-stability of all *edge polynomials* of the form

$$p_{i_1,i_2}(s, \lambda) = \lambda p_{i_1}(s) + (1 - \lambda)p_{i_2}(s),$$

where $p_{i_1}(s) = p(s, q^{i_1})$, $p_{i_2}(s) = p(s, q^{i_2})$ and q^{i_1} and q^{i_2} are extreme points of Q having the property that the straight line joining q^{i_1} and q^{i_2} is an edge of Q. For example, if Q is the three-dimensional box shown in Figure 9.3.2, then this heuristic argument indicates that we need only check for robust $\mathcal{D}$-stability with q restricted to one of the twelve edges e_1 through e_{12}.

To further illustrate the ideas above, suppose that each component of q has bounds $q_i^- \leq q_i \leq q_i^+$. Then edge polynomials of $\mathcal{P}$ associated with edge e_5 of Q in Figure 9.3.2 are of the form

$$p_{i_1,i_2}(s, \lambda) = \lambda p(s, q_1^-, q_2^-, q_3^-) + (1 - \lambda)p(s, q_1^-, q_2^-, q_3^+),$$

and the edge polynomials associated with e_9 are of the form

$$p_{i_3,i_4}(s, \lambda) = \lambda p(s, q_1^+, q_2^-, q_3^-) + (1 - \lambda)p(s, q_1^+, q_2^+, q_3^-).$$

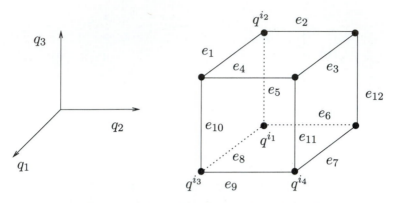

FIGURE 9.3.2   The Edges of the Box Q

REMARKS 9.3.1 (Checking the Edges):   Recalling the analysis for the one-parameter case in Chapter 4, we note that checking robust $\mathcal{D}$-stability of an edge polynomial is readily accomplished in a number of ways ranging from root locus to Nyquist methods. Another alternative for edge testing involves value set generation. With $p_{i_1,i_2}(s,\lambda) = \lambda p_{i_1}(s) + (1-\lambda)p_{i_2}(s)$ and $\lambda \in \Lambda = [0,1]$, we recall that the value set $p(z,\Lambda)$ is a straight line segment with endpoints $p(z,0)$ and $p(z,1)$; see Lemma 7.2.3. Hence, for each such edge, we verify $\mathcal{D}$-stability of one endpoint and check for satisfaction of the Zero Exclusion Condition

$$0 \notin p(z,\Lambda)$$

for all $z \in \partial \mathcal{D}$.

9.4   The Edge Theorem

We now consolidate the ideas above by formally stating the celebrated Edge Theorem. Note that in the theorem statement below, a boundary sweeping function is assumed for the desired root location region $\mathcal{D}$. Although the theorem holds under weaker assumption that $\mathcal{D}$ is simply connected, we include a boundary sweeping function because the proof becomes much simpler to understand without seriously compromising the domain of applicability. In the chapters to follow, we see numerous applications of the theorem below; see also the notes at the end of this chapter for additional discussion of various extensions and refinements of the theorem.

THEOREM 9.4.1 (Bartlett, Hollot and Huang (1988)): *Suppose that* $\mathcal{D}$ *is an open subset of the complex plane with boundary sweeping function* $\Phi_{\mathcal{D}} : I \to \mathbf{C}$, *and let* $\mathcal{P} = \{p(\cdot, q) : q \in Q\}$ *be a polytope of polynomials with invariant degree. Then* $\mathcal{P}$ *is robustly* $\mathcal{D}$*-stable if and only if for each pair of extreme points* q^{i_1} *and* q^{i_2} *corresponding to an edge of the set* Q, *the polynomial*

$$p_{i_1, i_2}(s, \lambda) = \lambda p(s, q^{i_1}) + (1 - \lambda)p(s, q^{i_2})$$

is $\mathcal{D}$*-stable for all* $\lambda \in [0, 1]$.

PROOF: The proof of necessity is trivial; i.e., for all extreme points q^{i_1} and q^{i_2} of Q and all $\lambda \in [0, 1]$, the polynomial $p_{i_1, i_2}(s, \lambda) \in \mathcal{P}$ is a member of $\mathcal{P}$. Hence, robust $\mathcal{D}$-stability of $\mathcal{P}$ implies $\mathcal{D}$-stability of the polynomial $p_{i_1, i_2}(s, \lambda)$ for all $\lambda \in [0, 1]$.

To establish sufficiency, we assume $\mathcal{D}$-stability of each of the edge polynomials $p_{i_1, i_2}(s, \lambda)$ and must prove that the family $\mathcal{P}$ is robustly $\mathcal{D}$-stable. Proceeding by contradiction, we assume $\mathcal{P}$ is not robustly $\mathcal{D}$-stable. Then, by the Zero Exclusion Condition (Theorem 7.4.2), there exists some $\delta_0 \in I$ such that $0 \in p(\Phi_{\mathcal{D}}(\delta_0), Q)$. The proof now breaks down into two cases.

Case 1: There exists some $\delta_1 \subset I$ such that $0 \notin p(\Phi_{\mathcal{D}}(\delta_1), Q)$. Without loss of generality, say $\delta_1 > \delta_0$. Corresponding to extreme point q^i of Q, we let

$$p_i(s) = p(s, q^i)$$

denote the i-th generator of $\mathcal{P}$. Since $\Phi_{\mathcal{D}}(\delta)$ varies continuously, it follows that

$$p(\Phi_{\mathcal{D}}(\delta), Q) = \text{conv}\{p_i(\Phi_{\mathcal{D}}(\delta))\}$$

varies continuously as a set. Now, using the continuity of $p(\Phi_{\mathcal{D}}(\delta), Q)$, there must exist some $\delta^* \in I$ such that $z = 0$ lies on an edge of

$$p(\Phi_{\mathcal{D}}(\delta^*), Q) = \text{conv}\{p_i(\Phi_{\mathcal{D}}(\delta^*)\}.$$

Since the edges of $p(\Phi_{\mathcal{D}}(\delta^*), Q)$ come from the edges of Q (see Lemma 8.3.17), it follows that there are two extreme points q^{i_1} and q^{i_2} corresponding to an edge of Q and some $\lambda^* \in [0, 1]$ such that

$$\lambda^* p_{i_1}(\Phi_{\mathcal{D}}(\delta^*)) + (1 - \lambda^*)p_{i_2}(\Phi_{\mathcal{D}}(\delta^*)) = 0.$$

However, this contradicts the assumed $\mathcal{D}$-stability of $p_{i_1, i_2}(s, \lambda^*)$; i.e., $\Phi_{\mathcal{D}}(\delta^*)$ is a root of $p_{i_1, i_2}(s, \lambda^*)$ which is not in $\mathcal{D}$.

Case 2: For all $\delta \in I$, $0 \in p(\Phi_{\mathcal{D}}(\delta), Q)$. We first pick $s_0 \in \mathbf{C}$ such that $p(s_0, q) \neq 0$ for all $q \in Q$. Note that the existence of s_0 is guaranteed because we can construct a bound for the roots of $p(s,q)$ for $q \in Q$; see Section 5.10. Hence, $0 \notin p(s_0, Q) = \text{conv}\{p_i(s_0)\}$. Now, we fix any $\delta_0 \in I$ and observe that our standing hypotheses guarantee that $0 \in p(\Phi_{\mathcal{D}}(\delta_0), Q)$.

Next, we consider a path from s_0 to $\Phi_{\mathcal{D}}(\delta_0)$. For $\alpha \in [0,1]$, let $s(\alpha) = (1 - \alpha)s_0 + \alpha \Phi_{\mathcal{D}}(\delta_0)$. Now, since $p(s_0, Q) = p(s(0), Q)$, $0 \notin p(s(0), Q)$ and $0 \in p(s(1), Q)$, continuous variation of the set $p(s(\alpha), Q) = \text{conv}\{p(s(\alpha), q^i)\}$ implies that there must exist some "first" $\alpha^* \in [0,1]$ such that $0 \in p(s(\alpha^*), Q)$. Letting

$$\alpha^* = \sup\{\alpha : 0 \notin p(s(\alpha), Q)\},$$

we claim that $s(\alpha^*) \notin \mathcal{D}$. To prove this claim, we proceed by contradiction and assume that $s(\alpha^*) \in \mathcal{D}$. Hence, there exists some $\hat{\alpha} < \alpha^*$ such that $s(\hat{\alpha}) \in \partial\mathcal{D}$. However, in view of our standing assumption that $0 \in p(\Phi_{\mathcal{D}}(\delta), Q)$ for all $\delta \in I$, it follows that $0 \in p(s(\hat{\alpha}), Q)$. This, however, contradicts the maximality of α^*.

Our next claim is that $z = 0$ lies on the boundary of $p(s(\alpha^*), Q)$; i.e., $0 \in \partial p(s(\alpha^*), Q)$. Proceeding again by contradiction, if $z = 0$ lies interior to $p(s(\alpha^*), Q)$, then, by continuity of $p(s(\alpha), Q)$ with respect to α, there exists some some $\hat{\alpha} < \alpha^*$ such that $0 \in p(s(\hat{\alpha}), Q)$. Again, we have contradicted the maximality of α^*.

The proof of Case 2 is now completed as in Case 1; i.e., $s(\alpha^*)$ lies in the boundary of the polygon $p(s(\alpha^*), Q)$ if and only if there exists some pair of extreme points q^{i_1} and q^{i_2} corresponding to an edge of Q and some $\lambda^* \in [0,1]$ such that $p_{i_1, i_2}(s(\alpha^*), \lambda^*) = 0$. This contradicts the $\mathcal{D}$-stability of $p_{ij}(s, \lambda^*)$. ∎

EXERCISE 9.4.2 (Stronger Version):   Modify the proof above to show that the Edge Theorem remains valid under the weaker hypothesis that q^{i_1} and q^{i_2} are extreme points of Q corresponding to the edges of the coefficient set $a(Q)$. Although the number of edges of $a(Q)$ is typically less than the number of edges of Q, the identification of these edges is generally nontrivial. Hence, from an applications point of view, it can be argued that this stronger version of the Edge Theorem is "less useful" than Theorem 9.4.1.

9.5   Fiat Dedra Engine Revisited

In this section, we illustrate the application of the Edge Theorem using the Fiat Dedra engine model derived in Section 3.2. The starting point for the analysis is the seventh order closed loop polynomial $p(s,q)$ given in Appendix A. In accordance with Fiat specifications, we consider three operating conditions. Using our standard q notation, the three operating conditions of interest are the vectors $q = q^A$, $q = q^B$ and $q = q^C$ associated with the rows given in the table below.

	q_1	q_2	q_3	q_4	q_5	q_6	q_7
A	2.1608	0.1027	0.0357	0.5607	0.0100	4.4962	1.0000
B	3.4329	0.1627	0.1139	0.2539	0.0208	2.0247	1.0000
C	2.1608	0.1027	0.0357	0.5607	0.0208	4.4962	10.000

Three Operating Conditions of Interest in q–Space

The three operating points are interpreted as follows: Operating point q^A represents the completely unloaded engine at idle speed. Operating point q^B is the most common of the three and represents a slightly loaded engine at idle speed. It can be viewed as the *nominal*. Finally, operating point q^C represents deviations from the nominal parameter set as in A, but with a larger inertia and different values for the motor gains K_5 and K_7.

9.5.1   The Polytope of Polynomials

Corresponding to each of the three operating points in the table, we compute a closed loop polynomial by evaluating the uncertain polynomial $p(s, q)$ in Appendix A. After a lengthy but straightforward computation, we obtain

$$p(s, q^A) \approx s^7 + 1.444309 s^6 + 0.7361252 s^5 + 0.1772927 s^4 + 0.02648999 s^3$$
$$+ 2.442136 \times 10^{-3} s^2 + 1.13555 \times 10^{-4} s + 1.73903 \times 10^{-6};$$

$$p(s, q^B) \approx s^7 + 1.336368 s^6 + 0.6779808 s^5 + 0.1802489 s^4 + 0.02924289 s^3$$
$$+ 0.002765453 s^2 + 1.270495 \times 10^{-4} s + 1.920464 \times 10^{-6};$$

$$p(s, q^C) \approx 100 s^7 + 68.95635 s^6 + 16.31662 s^5 + 2.120266 s^4 + 0.1708726 s^3$$
$$+ 0.008123757 s^2 + 1.965509 \times 10^{-4} s + 1.830692 \times 10^{-6}.$$

There are many possibilities for studying robust stability with respect to transitions between operating points. In this illustrative application of the Edge Theorem, we consider transitions associated with the polytope of polynomials

$$\mathcal{P} = \text{conv}\{p(\cdot, q^A), p(\cdot, q^B), p(\cdot, q^C)\}.$$

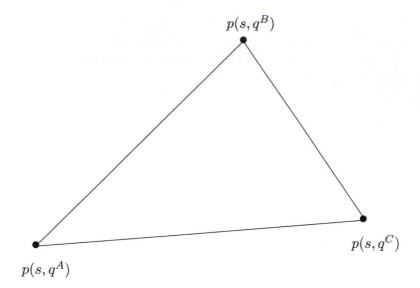

$$p(s, q^B)$$

$$p(s, q^C)$$

$$p(s, q^A)$$

FIGURE 9.5.1   A Triangle of Polynomials for the Fiat Dedra

Note that this family can be viewed as a triangle in the space of polynomials or the space of coefficients; see Figure 9.5.1.

9.5.2   Application of the Edge Theorem

By inspection, we observe that the polytope of polynomials $\mathcal{P}$ satisfies the preconditions for application of the Edge Theorem: The theorem indicates that we do not need to check stability of all polynomials in the triangle—we need only examine the three edges described by

$$p_{AB}(s, \lambda) = \lambda p(s, q^A) + (1 - \lambda)p(s, q^B);$$
$$p_{AC}(s, \lambda) = \lambda p(s, q^A) + (1 - \lambda)p(s, q^C);$$
$$p_{BC}(s, \lambda) = \lambda p(s, q^B) + (1 - \lambda)p(s, q^C)$$

and $\lambda \in \Lambda = [0, 1]$.

We now summarize the results of numerical computations: We first calculate the roots of $p(s, q^A)$, $p(s, q^B)$ and $p(s, q^C)$ and find that each of these three polynomials is stable. For example, the roots of $p(s, q^A)$ are found to be $s_1 \approx -0.0276$, $s_{2,3} \approx -0.1047 \pm j0.0041$, $s_{4,5} \approx -0.0540 \pm j0.1332$, $s_6 \approx -0.4810$ and $s_7 \approx -0.5785$.

The next step in the computation is to identify the "interesting" frequency ranges associated with zero exclusion testing for each edge. By carrying out a preliminary frequency sweep for each edge, it is easily verified that a suitable cutoff frequency is $\omega_c = 0.4$. In other words, for $\omega > \omega_c$, the three Zero Exclusion Conditions $0 \notin p_{AB}(j\omega, \Lambda)$, $0 \notin p_{AC}(j\omega, \Lambda)$ and $0 \notin p_{BC}(j\omega, \Lambda)$ all hold.

To compute the robust stability test, we now generate value sets for each edge. In accordance with Remarks 9.3.1, each of these value sets is a straight line segment; e.g., for $p_{AB}(s, \lambda)$, the value set $p(j\omega, \Lambda)$ is the line segment joining $p_{AB}(j\omega, 0)$ and $p_{AB}(j\omega, 1)$. We provide three representative value set plots in Figures 9.5.2, 9.5.3 and 9.5.4 using the critical frequency range in each case. Since the

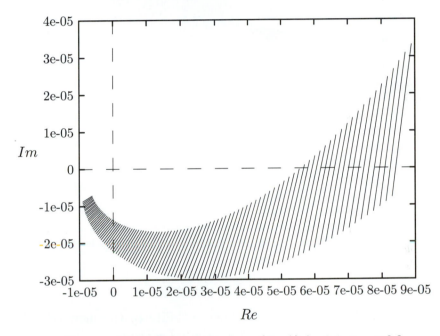

FIGURE 9.5.2   Value Sets of $p_{AB}(j\omega, \Lambda)$ for $0.1 \leq \omega \leq 0.2$

Zero Exclusion Condition is satisfied in all cases, we conclude from Theorem 7.4.2 that each of the edges is stable. Hence, by the Edge Theorem, we conclude that $\mathcal{P}$ is robustly stable.

EXERCISE 9.5.3 (Numerical Conditioning):   Using the triangle of polynomials $\mathcal{P}$ for the Fiat Dedra engine above, verification of the Zero Exclusion Condition for edge AC required a computation at rather high resolution. Hence, the following question arises: Given our judgment call in reaching the conclusion $0 \notin p_{AC}(j\omega, \Lambda)$ for all $\omega \geq 0$, what can be done to boost our confidence that the computed solution is correct? This issue is addressed below.

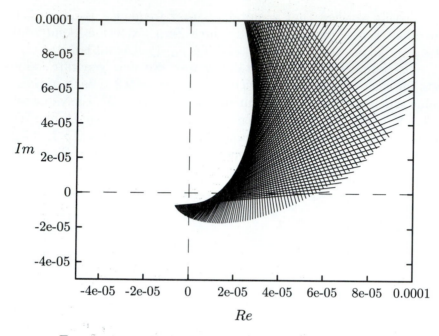

FIGURE 9.5.3   Value Sets of $p_{AC}(j\omega, \Lambda)$ for $0.1 \leq \omega \leq 0.2$

(a) Verify that $p(s, q^A)$ and $p(s, q^C)$ have a common root $s \approx -0.0276$.
(b) Argue that $s \approx -0.0276$ is a root of $p_{AC}(s, \lambda)$ for all $\lambda \in \Lambda$.
(c) Motivated by (b), define a new family of polynomials by

$$\tilde{p}_{AC}(s, \lambda) = \frac{p_{AC}(s, \lambda)}{s + 0.0276}$$

and $\lambda \in \Lambda$. Now, argue that robust stability of this new edge is equivalent to robust stability of the original edge.
(d) Generate representative value sets $\tilde{p}_{AC}(j\omega, \Lambda)$ and check for satisfaction of the Zero Exclusion Condition. Discuss your findings in the context of the judgment call associated with the original edge polynomial $p_{AC}(s, \lambda)$.

EXERCISE 9.5.4 (Application of the Edge Theorem): Use the Edge Theorem to investigate robust stability of the polytope of polynomials considered in Bartlett, Hollot and Huang (1988); i.e., take

$$p_1(s) = s^3 + 9.77s^2 + 30.6s + 18.27;$$
$$p_2(s) = s^3 + 15s^2 + 75s + 25;$$
$$p_3(s) = s^3 + 8.96s^2 + 21.9s + 15.61;$$

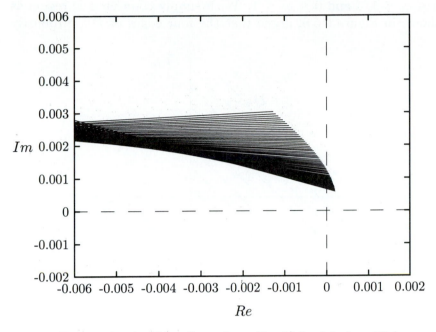

FIGURE 9.5.4   Value Sets of $p_{BC}(j\omega, \Lambda)$ for $0.3 \leq \omega \leq 0.4$

$$p_4(s) = s^3 + 11.43s^2 + 20.2s + 82.5$$

and study robust stability of $\mathcal{P} = \mathrm{conv}\{p_1(\cdot), p_2(\cdot), p_3(\cdot), p_4(\cdot)\}$.

EXERCISE 9.5.5 (The Example of Soh and Foo (1990)):   Consider the polytope of polynomials with generators $p_1(s) = s^2 + 2s + 2$, $p_2(s) = s^2 + 4s + 5$ and $p_3(s) = s^2 + 2s + 5$. Next, with $z_0 = -1.5 + j1.5$ and $z_0^* = -1.5 - j1.5$, let $\mathcal{D} = \mathcal{D}_1 \cap \mathcal{D}_2$, where

$$\mathcal{D}_1 = \{z \in \mathbf{C} : |z - z_0| > \epsilon\};$$

$$\mathcal{D}_2 = \{z \in \mathbf{C} : |z - z_0^*| > \epsilon\}.$$

With $\epsilon > 0$ suitably small, show that the edges of $\mathcal{P}$ are robustly $\mathcal{D}$-stable but $\mathcal{P}$ is not robustly $\mathcal{D}$-stable. Explain this pathology in light of the Edge Theorem.

EXERCISE 9.5.6 (The Role of Invariant Degree):   To demonstrate the importance of the invariant degree requirement in the Edge Theorem, consider the polytope of polynomials $\mathcal{P}$ of Sideris and Barmish (1989) described by $p(s, q) = q_1 s^2 + (2q_1 + q_2)s + (q_1 + q_2 + 1)$,

$0 \leq q_1 \leq 1/4$ and $0 \leq q_2 \leq 1$. With simply connected $\mathcal{D}$ region as shown in Figure 9.5.5, argue that the four edges of $\mathcal{P}$ are robustly

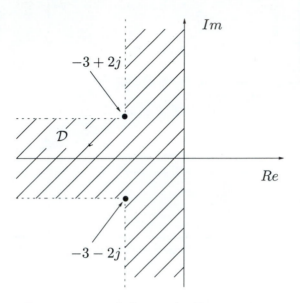

FIGURE 9.5.5 $\mathcal{D}$ Region for Exercise 9.5.6

$\mathcal{D}$-stable but with $q_1 = 1/128$ and $q_2 = 1/8$, $p(s, q)$ has a root outside $\mathcal{D}$.

9.6   Root Version of the Edge Theorem

To motivate the development below, note that the Edge Theorem provides a yes or no solution to the robust $\mathcal{D}$-stability problem for a polytope of polynomials. We now present a lesser known version of the Edge Theorem which provides more detailed information about the actual root locations for a polytope of polynomials. Using the roots associated with the edge polynomials, we can obtain the boundary of the so-called spectral set of $\mathcal{P}$.

DEFINITION 9.6.1 (The Spectral Set):   Given a family of polynomials $\mathcal{P} = \{p(\cdot, q) : q \in Q\}$, we call

$$\sigma[\mathcal{P}] = \{z \in \mathbf{C} : p(z, q) = 0 \text{ for some } q \in Q\}$$

the *spectral set* (or *root set*) of $\mathcal{P}$.

REMARKS 9.6.2 (Spectral Set and Robust $\mathcal{D}$-Stability):   The spectral set generation problem generalizes the robust $\mathcal{D}$-stability problem in the following sense: Once we have the spectral set $\sigma[\mathcal{P}]$, we have a complete characterization of all possible $\mathcal{D}$-regions for which robust $\mathcal{D}$-stability is guaranteed; i.e., if $\sigma[\mathcal{P}]$ is in hand and $\mathcal{D} \subseteq \mathbf{C}$ is given, robust $\mathcal{D}$-stability is guaranteed if and only if the condition

$$\sigma[\mathcal{P}] \subseteq \mathcal{D}$$

is satisfied.

EXERCISE 9.6.3 (Basic Properties of $\sigma[\mathcal{P}]$):   Given a polytope of polynomials $\mathcal{P}$ with invariant degree, show that $\sigma[\mathcal{P}]$ is both closed and bounded.

NOTATION 9.6.4 (Edges and Boundary):   Given a polytope $\mathbf{P}$, we take $\mathcal{E}(\mathbf{P})$ to be its set of edges. Now, if $\mathcal{P} = \{p(\cdot, q) : q \in Q\}$ is a polytope of polynomials, we can use the natural isomorphism between polynomials and coefficients to define edges of $\mathcal{P}$. Namely, if $\mathcal{E}(a(Q))$ denotes the set of edges of $a(Q)$ in $\mathbf{R}^{n+1}$, we take

$$\mathcal{E}(\mathcal{P}) = \{p(\cdot, q) : a(q) \in \mathcal{E}(a(Q))\}$$

for the set of edges of $\mathcal{P}$.

REMARKS 9.6.5 (Edges of Q):   From an applications point of view, however, it is more convenient to work with the edges of Q rather than the edges of $a(Q)$; recall the discussion in Exercise 9.4.2. Hence, if $\mathcal{E}(Q)$ denotes the set of edges of Q, we work with the set

$$\mathcal{E}_Q(\mathcal{P}) = \{p(\cdot, q) : q \in \mathcal{E}(Q)\}$$

in lieu of $\mathcal{E}(\mathcal{P})$. In view of the fact that $a(Q)$ is the image of Q under an affine linear transformation, it follows from Lemma 8.3.17 that

$$\mathcal{E}(\mathcal{P}) \subseteq \mathcal{E}_Q(\mathcal{P}).$$

Finally, note that in the theorem to follow, we use the notation $\partial\sigma[\mathcal{P}]$ to denote the boundary of the spectral set $\sigma[\mathcal{P}]$.

THEOREM 9.6.6 (Root Version of the Edge Theorem):   *Given a polytope of polynomials $\mathcal{P} = \{p(\cdot, q) : q \in Q\}$ with invariant degree, it follows that*

$$\partial\sigma[\mathcal{P}] \subseteq \sigma[\mathcal{E}_Q(\mathcal{P})].$$

PROOF: We fix $z \in \partial\sigma[\mathcal{P}]$ and must show that $z \in \sigma[\mathcal{E}(\mathcal{P})]$. Proceeding by contradiction, suppose $z \notin \sigma[\mathcal{E}(\mathcal{P})]$. For each extreme point pair (q^{i_1}, q^{i_2}) defining an edge of Q, we obtain an edge polynomial $p_{i_1,i_2}(s, \lambda) = \lambda p(s, q^{i_1}) + (1-\lambda)p(s, q^{i_2})$ for $\mathcal{P}$. We know that $0 \notin p_{i_1,i_2}(z, \Lambda)$ and by closedness of $\sigma[\mathcal{P}]$, $z \in \sigma[\mathcal{P}]$. Now, for each (q^{i_1}, q^{i_2}) pair defining an edge polynomial of $\mathcal{P}$, continuity of $p_{i_1,i_2}(\cdot, \lambda)$ implies that there exists some $\epsilon_{i_1,i_2} > 0$ such that if $z' \in \mathbf{C}$ with $|z' - z| \le \epsilon_{i_1,i_2}$, then $0 \notin p_{i_1,i_2}(z', \Lambda)$ and $z' \in \sigma[\mathcal{P}]$.

Now, with

$$\epsilon = \min_{i_1,i_2} \epsilon_{i_1,i_2},$$

it follows that the disc $B_\epsilon(z) = \{z' : |z' - z| < \epsilon\}$ is wholly contained in the spectral set of $\mathcal{P}$; i.e., $B_\epsilon(z) \subseteq \sigma[\mathcal{P}]$. However, this contradicts the standing assumption that $z \in \partial\sigma[\mathcal{P}]$. ∎

EXERCISE 9.6.7 (Stronger Version):   Modify the proof above to show that

$$\partial\sigma[\mathcal{P}] \subseteq \sigma[\mathcal{E}(\mathcal{P})].$$

EXERCISE 9.6.8 (Spectral Set Generation):   Use the root version of the Edge Theorem to generate the spectral set for the triangle of polynomials $\mathcal{P}$ associated with the Fiat Dedra engine; see Section 9.5.

9.7   Conclusion

The Edge Theorem is the takeoff point for the generation of many results in the sequel. For example, when dealing with interval plants and first order compensators in Chapter 11, the robust stability problem for the closed loop is first reduced to an edge problem and subsequently, new machinery makes it possible to reduce the checking of edges to the checking of extreme points. Hence, the statement of the final result makes no mention of edges—the reliance on the Edge Theorem occurs at the level of proof rather than at the level of application. As far as more direct application of the Edge Theorem is concerned, a fundamental limitation is encountered because the number of edges of a polytope can grow exponentially fast with respect to the number of variables describing it. This issue is the focal point of the next chapter. For polytopes generated from feedback control systems, this exponential growth problem can be overcome.

Notes and Related Literature

NRL 9.1 The proof of the Edge Theorem by Bartlett, Hollot and Huang (1988) is carried out in coefficient space without recourse to the value set.

NRL 9.2 In the papers by Fu and Barmish (1989), Tits (1990) and Soh and Foo (1990), refinements of the Edge Theorem are provided for more general classes of $\mathcal{D}$ regions. For example, in Fu and Barmish (1989), it is assumed that $\mathcal{D}$ has the following property: Through every point in $\mathcal{D}^c$, there is an unbounded path that remains within $\mathcal{D}^c$. In Tits (1990), an even less restrictive condition is used: Either the condition of Fu and Barmish (1989) holds or $0 \in \mathcal{D}^c$, and through every point in $\mathcal{D}^c$ there is a continuous path to the origin which remains in $\mathcal{D}^c$.

NRL 9.3 At a more general level, an edge theorem can be provided for delay systems; see Fu, Olbrot and Polis (1989). In fact, it is even possible to provide a similar result for a polytope of analytic functions; see Soh and Foo (1989) and Dasgupta, Parker, Anderson, Kraus and Mansour (1991).

NRL 9.4 The discovery of the λ-invariant root $s \approx -0.0276$ found in Exercise 9.5.3 motivates a number of interesting research problems associated with numerical conditioning of robust stability computations. For example, given a polytope of polynomials $\mathcal{P}$, provide classes of isomorphisms $\mathcal{P} \to \tilde{\mathcal{P}}$ such that robust stability is invariant and computations for $\tilde{\mathcal{P}}$ are "simpler" in some quantifiable sense.

NRL 9.5 The connection between value set theory and the root version of the Edge Theorem is recognized in the Ph.D. dissertation of Bartlett (1990a); our proof is an adaptation of the one given in this reference.

NRL 9.6 The topic of transitional models for robust stability analysis seems ripe for future research. To briefly illustrate what is meant, consider the following alternative to the polytope of polynomials used in the Fiat Dedra analysis: Given the three operating points q^A, q^B and q^C, study robust stability of the family of polynomials described by $p(s, \lambda) = p(s, \lambda_1 q^A + \lambda_2 q^B + \lambda_3 q^C)$ with λ restricted to the unit simplex Λ. Note that this family of polynomials has coefficients depending nonlinearly on q. There are many interesting modeling issues motivating new research. For example, suppose that transitions between operating points may be restricted; e.g., one must pass through q^B in going from q^A to q^C. Such restrictions can be incorporated into the formulation of new robustness problems.

Chapter 10

Distinguished Edges

Synopsis

The main obstacle associated with application of the Edge Theorem is computational complexity; for arbitrary polytopes, the number of edges can be excessively large. In this chapter, we see that a "small" distinguished subset of the edges can often be identified. These edges are distinguished in the sense that the remaining edges need not be checked in a robust $\mathcal{D}$-stability analysis. Of particular interest is the Thirty-Two Edge Theorem of Chapellat and Bhattacharyya. This result applies to feedback systems involving interval plants.

10.1   Introduction

From an applications point of view, there is one difficulty associated with the Edge Theorem—a *combinatoric explosion* in the number of edges of Q as the number of uncertain parameters increases. For example, if Q is an ℓ-dimensional box, then the number of edges of Q is given by

$$N_{edges} = \ell 2^{\ell-1}.$$

This exponential growth with respect to ℓ can lead to serious computational difficulties. For ℓ suitably large, the required robustness computation can easily exceed the capability of modern computers.

In order to overcome the combinatoric problem above, we consider the following question: From the $\ell 2^{\ell-1}$ possible edges of Q, can we identify a "small" distinguished subset of *critical* edges which are important as far as robust $\mathcal{D}$-stability analysis is concerned? In other words, once we have found these distinguished edges, we can ignore all remaining edges in a robust stability analysis. When such a subset is readily identifiable, then the Edge Theorem becomes more useful as an application tool for problems with ℓ being large.

10.2   Parallelotopes

Strong motivation for the technical discussion to follow is derived from an important observation: Given a polytope of polynomials $\mathcal{P} = \{p(\cdot, q) : q \in Q\}$, even though the uncertainty bounding set $Q \subseteq \mathbf{R}^\ell$ has a total number of edges which increases exponentially with respect to ℓ, at fixed frequency $\omega \in \mathbf{R}$, the value set $p(j\omega, Q)$ may have a number of edges which increases at a much slower rate with respect to ℓ. In fact, we see below that for the case when Q is a box, we obtain a value set which is a "parallelotope" having at most 2ℓ edges. This name is derived from the following fact: Whenever the value set has a nonempty interior, for each edge e_{i_1}, there is a second edge $e_{i_2} \neq e_{i_1}$ such that e_{i_1} and e_{i_2} are parallel. Before proceeding, the reader should be forewarned that care must be exercised in exploiting the parallelotope property in a frequency sweeping context. That is, the 2ℓ distinguished edges may change as a function of frequency; further discussion of this issue is provided in the notes at the end of this chapter.

DEFINITION 10.2.1 (Parallelotopes):   Let $\mathcal{P} = \{p(\cdot, q) : q \in Q\}$ be a polytope of polynomials with $Q \subseteq \mathbf{R}^\ell$ being a box. Then $\mathcal{P}$ is called a *parallelotope of polynomials*.

REMARKS 10.2.2 (Boxes Versus Polytopes):   From an applications point of view, working with parallelotopes rather than general polytopes can hardly be viewed as restrictive. Since Q represents uncertainty bounds for underlying physical parameters, in most applications, the box model is quite appropriate.

EXERCISE 10.2.3 (Number of Edges of a Polygon):   Take $\mathbf{P}$ to be a polygon in the complex plane and let $M = \{m_1, m_2, \ldots, m_N\}$ denote the finite set of numbers representing the "slopes" of its edges. Argue that the total number of edges of $\mathbf{P}$ is at most $2N$.

LEMMA 10.2.4 (Value Set):   *Let $\mathcal{P} = \{p(\cdot, q) : q \in Q\}$ be a parallelotope of polynomials with $Q \subseteq \mathbf{R}^\ell$. Then, given any $z \in \mathbf{C}$, the value set $p(z, Q)$ is a polygon with 2ℓ sides at most.*

PROOF: Since $p(s, q)$ has an affine linear uncertainty structure, we can write

$$p(s, q) = p_0(s) + \sum_{i=0}^{\ell} q_i p_i(s),$$

where $p_0(s), p_1(s), \ldots, p_\ell(s)$ are fixed polynomials. Now, in accordance with Proposition 8.7.1, the value set $p(z, Q)$ is a convex polygon with edges which are obtained from the edges of Q. Since Q is a box, each edge of Q is obtained by setting all but one q_i to one of its extreme values and varying the remaining uncertain parameter between its bounds; i.e., given uncertainty bounds $q_i^- \le q_i \le q_i^+$, all edges of $p(z, Q)$ are generated using the expression

$$\bar{p}(z, q_k) = p_0(z) + q_k p_k(z) + \sum_{i \neq k} q_i^{\pm} p_i(z),$$

where $q_i^{\pm} = q_i^+$ or q_i^- for $i \neq k$, $q_k^- \le q_k \le q_k^+$ and $p_k(z) \neq 0$.

   To complete the proof, we observe that the edge of $p(z, Q)$ associated with q_k has slope (perhaps infinite) given by

$$m_k = \frac{Im\ p_k(z)}{Re\ p_k(z)}.$$

Since the slope m_k does not depend on the fixed choices of $q_i^{\pm}$ above, every edge of $p(z, Q)$ has a slope which assumes at most one of ℓ possible values. In view of Exercise 10.2.3, we conclude that $p(z, Q)$ has at most 2ℓ edges. ∎

REMARKS 10.2.5 (Subtlety):   Consistent with the discussion at the beginning of this section, we make note of one subtlety associated with the practical exploitation of Lemma 10.2.4. Namely, the 2ℓ distinguished edges in the lemma above may vary with z. For example, in robust stability analysis with $z = j\omega$, as ω is swept from 0 to $+\infty$, the following situation can occur: As one increases the frequency from $\omega = \omega_1$ to $\omega = \omega_2$, some of the distinguished edges of $p(j\omega, Q)$ can move into the interior of $p(j\omega, Q)$ and new edges can emerge as members of the distinguished set.

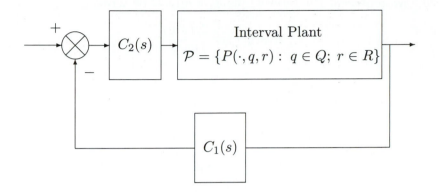

FIGURE 10.4.1   Basic Setup for this Chapter

10.3   Parpolygons

In this section, we provide a minor embellishment on Lemma 10.2.4. We see that the value set not only has at most 2ℓ sides but also has opposite sides which are parallel.

DEFINITION 10.3.1 (Parpolygon):   Let **P** be a polygon with distinct edges $e_1, e_2, \ldots, e_N$. We call **P** a *parpolygon* if either $N = 1$ (the degenerate case) or the following condition is satisfied: For each $i_1 \in \{1, 2, \ldots, N\}$, there exits some $i_2 \in \{1, 2, \ldots, N\}$ such that $i_1 \neq i_2$ and e_{i_1} and e_{i_2} are parallel.

EXERCISE 10.3.2 (Value Set as a Parpolygon):   Under the hypotheses of Lemma 10.2.4, prove that the value set $p(z, Q)$ is a parpolygon. *Hint*: Identify parallel edges of $p(z, Q)$ using parallel edges of Q.

10.4   Setup with an Interval Plant

Throughout the remainder of this chapter, the focal point is avoidance of the combinatoric explosion problem associated with robust stability testing for a polytope of polynomials; recall the discussion in Section 10.1. We now describe an important class of polytopes for which this problem can be totally eliminated. Namely, we consider the stability problem which arises upon interconnection of an interval plant and a fixed compensator as shown in Figure 10.4.1.

We represent the interval plant family $\mathcal{P}$ by writing

$$P(s, q, r) = \frac{N(s, q)}{D(s, r)}$$

with

$$N(s, q) = \sum_{i=0}^{m} [q_i^-, q_i^+] s^i; \, D(s, r) = \sum_{i=0}^{n} [r_i^-, r_i^+] s^i.$$

Subsequently, we express

$$C(s) = C_1(s) C_2(s)$$

as a quotient of polynomials

$$C(s) = \frac{N_C(s)}{D_C(s)}$$

and obtain the family of closed loop polynomials $\mathcal{P}_{CL}$ described by

$$p(s, q, r) = N(s, q) N_C(s) + D(s, r) D_C(s),$$

$q \in Q$ and $r \in R$; i.e., $\mathcal{P}_{CL} = \{p(\cdot, q, r) : q \in Q; r \in R\}$. Since the uncertainty bounding set $Q \times R$ can have as many as

$$N_{edges} = (n + m + 2) \times 2^{n+m+1}$$

edges, there is a temptation to conclude that robust stability computations quickly become intractable as m and n increase. The objective of the next few sections is to demonstrate that reaching such a conclusion is entirely erroneous. To this end, we demonstrate a remarkable property enjoyed by the value set $p(j\omega, Q, R)$. Namely, at each frequency $\omega \in \mathbf{R}$, $p(j\omega, Q, R)$ has at most eight edges. Furthermore, as a function of the frequency ω, there are only four possible groups of eight edges which can arise. This leads to a total of thirty-two distinguished edges which are instrumental to robust stability analysis. In other words, for the feedback system under consideration, the special nature of the polytope which arises enables us to restrict attention to just a small number of "critical edges."

10.5   The Thirty-Two Edge Theorem

The proof of the theorem below is relegated to the next two sections.

THEOREM 10.5.1 (Chapellat and Bhattacharyya (1989)):   *Consider an interval plant family $\mathcal{P}$ having Kharitonov polynomials $N_1(s)$,*

$N_2(s)$, $N_3(s)$, $N_4(s)$ and $D_1(s)$, $D_2(s)$, $D_3(s)$, $D_4(s)$ *for the numer-ator and denominator, respectively, and compensation blocks* $C_1(s)$ *and* $C_2(s)$ *as indicated in Figure 10.4.1. Assuming the family of closed loop polynomials* $\mathcal{P}_{CL}$ *has invariant degree, robust stability of* $\mathcal{P}_{CL}$ *is guaranteed if and only if all edge polynomials of the form*

$$e(s, \lambda) = N_{i_1}(s)N_C(s) + D_{i_2,i_3}(s, \lambda)D_C(s)$$

with $i_1 \in \{1, 2, 3, 4\}$ *and* $(i_2, i_3) \in \{(1,3), (1,4), (2,3), (2,4)\}$ *or*

$$e(s, \lambda) = N_{i_1,i_2}(s, \lambda)N_C(s) + D_{i_3}(s)D_C(s)$$

with $(i_1, i_2) \in \{(1,3), (1,4), (2,3), (2,4)\}$ *and* $i_3 \in \{1, 2, 3, 4\}$ *are sta-ble for all* $\lambda \in [0, 1]$.

10.6   Octagonality of the Value Set

Our proof of Theorem 10.5.1 exploits a characterization of the value set $p(j\omega, Q, R)$. In addition to facilitating the proof of the theorem, this characterization is useful in its own right; it can be used to generate a graphics display of $p(j\omega, Q, R)$. Hence, satisfaction or failure of the Zero Exclusion Condition

$$0 \notin p(j\omega, Q, R)$$

can be verified by inspection.

Indeed, we begin with the setup in Section 10.4. Now, associated with the interval plant family $\mathcal{P}$ are four Kharitonov polynomials for the numerator

$$N_1(s) = q_0^- + q_1^- s + q_2^+ s^2 + q_3^+ s^3 + q_4^- s^4 + q_5^- s^5 + q_6^+ s^6 + \cdots;$$
$$N_2(s) = q_0^+ + q_1^+ s + q_2^- s^2 + q_3^- s^3 + q_4^+ s^4 + q_5^+ s^5 + q_6^- s^6 + \cdots;$$
$$N_3(s) = q_0^+ + q_1^- s + q_2^- s^2 + q_3^+ s^3 + q_4^+ s^4 + q_5^- s^5 + q_6^- s^6 + \cdots;$$
$$N_4(s) = q_0^- + q_1^+ s + q_2^+ s^2 + q_3^- s^3 + q_4^- s^4 + q_5^+ s^5 + q_6^+ s^6 + \cdots$$

and four Kharitonov polynomials for the denominator

$$D_1(s) = r_0^- + r_1^- s + r_2^+ s^2 + r_3^+ s^3 + r_4^- s^4 + r_5^- s^5 + r_6^+ s^6 + \cdots;$$
$$D_2(s) = r_0^+ + r_1^+ s + r_2^- s^2 + r_3^- s^3 + r_4^+ s^4 + r_5^+ s^5 + r_6^- s^6 + \cdots;$$
$$D_3(s) = r_0^+ + r_1^- s + r_2^- s^2 + r_3^+ s^3 + r_4^+ s^4 + r_5^- s^5 + r_6^- s^6 + \cdots;$$
$$D_4(s) = r_0^- + r_1^+ s + r_2^+ s^2 + r_3^- s^3 + r_4^- s^4 + r_5^+ s^5 + r_6^+ s^6 + \cdots.$$

Recalling the discussion on interval polynomials in Section 5.3, the two value sets $N(j\omega, Q)$ and $D(j\omega, R)$ are rectangles and obtain the value set for the closed loop polynomial as the direct sum

$$p(j\omega, Q, R) = N_C(j\omega)N(j\omega, Q) + D_C(j\omega)D(j\omega, R).$$

To characterize this set, we consider two cases and combine the results at the end.

Case 1 (The Degenerate Case): If $D_C(j\omega) = 0$, then

$$p(j\omega, Q, R) = N_C(j\omega)N(j\omega, Q).$$

Recalling the remarks associated with Example 8.3.19, we can view $N_C(j\omega)$ as a linear transformation on $N(j\omega, Q)$. Hence, we then obtain $p(j\omega, Q, R)$ by rotating the Kharitonov rectangle $N(j\omega, Q)$ through an angle $\theta = \measuredangle N_C(j\omega)$ and scaling all points by $|N_C(j\omega)|$. This situation is depicted in Figure 10.6.1. It is plain to see that the

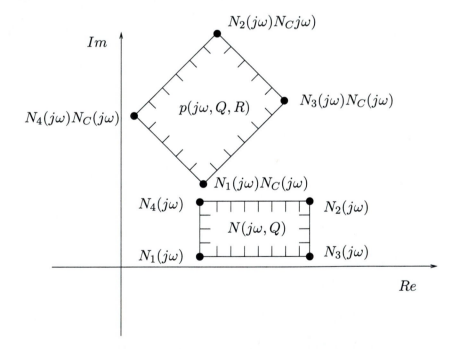

FIGURE 10.6.1   The Value Set for the Degenerate Case

value set $p(j\omega, Q, R)$ is described using the four edge polynomials

$$e_1(s, \lambda) = [\lambda N_2(s) + (1 - \lambda)N_3(s)]N_C(s);$$

$$e_2(s, \lambda) = [\lambda N_1(s) + (1 - \lambda)N_3(s)]N_C(s);$$
$$e_3(s, \lambda) = [\lambda N_1(s) + (1 - \lambda)N_4(s)]N_C(s);$$
$$e_4(s, \lambda) = [\lambda N_2(s) + (1 - \lambda)N_4(s)]N_C(s).$$

This partial result will later be combined with the results obtained for the nondegenerate case below.

Case 2 (The Nondegenerate Case): If $D_C(j\omega) \neq 0$, then division by $D_C(j\omega)$ leads to

$$p(j\omega, Q, R) = D_C(j\omega)[C(j\omega)N(j\omega, Q) + D(j\omega, R)].$$

Now, by viewing $C(j\omega)$ as a linear transformation on $N(j\omega, Q)$, it follows that $C(j\omega)N(j\omega, Q)$ is simply a rectangle which is a rotated and scaled version of $N(j\omega, Q)$; see Example 8.3.19. Applying Lemma 8.3.9, we arrive at the formula

$$p(j\omega, Q, R) = D_C(j\omega) \cdot \text{conv} \bigcup_i \{C(j\omega)N(j\omega, Q) + D_i(j\omega)\}.$$

Now, we can easily identify the edges of $p(j\omega, Q, R)$ by first obtaining the edges of the set

$$\mathbf{P}(\omega) = \text{conv} \bigcup_i \{C(j\omega)N(j\omega, Q) + D_i(j\omega)\}.$$

To this end, we claim that $\mathbf{P}(\omega)$ is an octagon (or a degenerate octagon). To prove this, there are four cases to consider; i.e., the i-th case corresponds to

$$0 \leq \measuredangle C(j\omega) \leq i\frac{\pi}{2}.$$

We now concentrate on the first case, $0 \leq \measuredangle C(j\omega) \leq \pi/2$. By viewing $C(j\omega)$ as a linear transformation on $N(j\omega, Q)$ and treating $D_i(j\omega)$ as a translation, it now follows that each of the four sets $C(j\omega)N(j\omega, Q) + D_i(j\omega)$ is a rotated rectangle with center obtained by "applying" $C(j\omega)$ to the centerpoint of $N(j\omega, Q)$, followed by translation by $D_i(j\omega)$. Now, to complete the construction, we form $\mathbf{P}(\omega)$ as the convex hull of the union of these four rectangles. Subsequently, Lemma 8.3.15 leads us to conclude that $\mathbf{P}(\omega)$ is an octagon with eight edges as shown in Figure 10.6.2. Of course, the figure must be interpreted in the appropriate manner because it does not represent all possible geometries; e.g., if $Im\ D_2(j\omega) = Im\ D_3(j\omega)$, we can obtain a hexagon instead of an octagon. Nevertheless, the

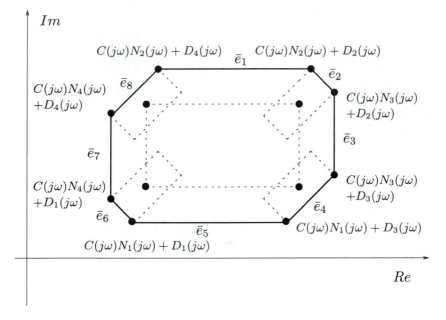

FIGURE 10.6.2   Construction of $\mathbf{P}(\omega)$

figure gives us the complete description of the eight edges of the value set octagon with the understanding that some of the edges in our list may be nonunique. In other words, although we use the expression *value set octagon*, the understanding is that in some cases the formulas to follow can lead to degenerate octagons such as hexagons, quadrilaterals or line segments. Nevertheless, eight is the upper bound on the number of edges.

To summarize, associated with the eight edges $\bar{e}_1, \bar{e}_2, \ldots, \bar{e}_8$ in the figure are eight *edge polynomials* for the value set $p(j\omega, Q, R)$. For example, beginning with edge $\bar{e}_1$, multiplication by $D_C(j\omega)$ leads to the edge polynomial

$$\bar{e}_1(s, \lambda) = N_2(s)N_C(s) + D_{2,4}(s, \lambda)D_C(s),$$

where

$$D_{2,4}(s, \lambda) = \lambda D_2(s) + (1 - \lambda)D_4(s).$$

In a nearly identical manner, we can obtain formulas for the remaining seven edges of the value set octagon. The presentation of these formulas is facilitated with the definition below.

DEFINITION 10.6.1 (Critical Numerator and Denominator Edges): Consider an interval plant family $\mathcal{P}$ with Kharitonov numerator and denominator polynomials $N_1(s)$, $N_2(s)$, $N_3(s)$ and $N_4(s)$ and $D_1(s)$, $D_2(s)$, $D_3(s)$ and $D_4(s)$, respectively. Then, for each pair

$$(i_1, i_2) \in \{(1,3), (1,4), (2,3), (2,4)\},$$

we define a *critical numerator edge polynomial*

$$N_{i_1,i_2}(s, \lambda) = \lambda N_{i_1}(s) + (1 - \lambda)N_{i_2}(s)$$

and a *critical denominator edge polynomial*

$$D_{i_1,i_2}(s, \lambda) = \lambda D_{i_1}(s) + (1 - \lambda)D_{i_2}(s).$$

REMARKS 10.6.2 (Formulae for Edges of the Value Set):   Using Figure 10.6.2 and arguments identical to the one used to obtain $e_1(s, \lambda)$, we still continue to assume that

$$0 \le \measuredangle C(j\omega) \le \frac{\pi}{2}$$

and obtain the complete set of eight edge polynomials characterizing the value set octagon. Namely, with $\Lambda = [0, 1]$ and

$$e_1(s, \lambda) = N_2(s)N_C(s) + D_{2,4}(s, \lambda)D_C(s);$$

$$e_2(s, \lambda) = N_{2,3}(s)N_C(s) + D_2(s, \lambda)D_C(s);$$

$$e_3(s, \lambda) = N_3(s)N_C(s) + D_{2,3}(s, \lambda)D_C(s);$$

$$e_4(s, \lambda) = N_{1,3}(s)N_C(s) + D_3(s, \lambda)D_C(s);$$

$$e_5(s, \lambda) = N_1(s)N_C(s) + D_{1,3}(s, \lambda)D_C(s);$$

$$e_6(s, \lambda) = N_{1,4}(s)N_C(s) + D_1(s, \lambda)D_C(s);$$

$$e_7(s, \lambda) = N_4(s)N_C(s) + D_{1,4}(s, \lambda)D_C(s);$$

$$e_8(s, \lambda) = N_{2,4}(s)N_C(s) + D_4(s, \lambda)D_C(s),$$

the value set octagon $p(j\omega, Q, R)$ has i-th edge

$$e_i(j\omega, \Lambda) = \{e_i(j\omega, \lambda) : \lambda \in \Lambda\}.$$

Note that these formulas hold for both the degenerate and nondegenerate cases.

EXERCISE 10.6.3 (Other Possibilities for $\measuredangle C(j\omega)$):   For the remaining cases characterized by $\pi/2 \le \measuredangle C(j\omega) < \pi$, $\pi \le \measuredangle C(j\omega) < 3\pi/2$

and $3\pi/2 \leq \angle C(j\omega) < 2\pi$, describe the edge polynomials associated with the value set octagon $p(j\omega, Q, R)$. Note that in each case, different (i_1, i_2) combinations are used to describe the critical edges for the plant numerator and denominator.

REMARKS 10.6.4 (Summary): Taking the union of all four cases for $\angle C(j\omega)$, we obtain a maximum of thirty-two (not necessarily distinct) edges which describe the behavior of the value set at all frequencies; these edges are listed in Theorem 10.5.1. At any fixed frequency, however, only eight edges at most are in play. We now provide an example illustrating how the value set octagon is generated in graphics. Subsequently, we visually inspect the graphical plot for satisfaction of the Zero Exclusion Condition (Theorem 7.4.2).

EXAMPLE 10.6.5 (Zero Exclusion Testing): We consider the interval plant

$$P(s, q, r) = \frac{[4, 6]s^3 + [3, 5]s^2 + [2, 4]s + [6, 8]}{s^3 + [4, 6]s^2 + [5, 7]s + [7, 9]}$$

connected in the feedback configuration of Figure 10.4.1 with compensators $C_1(s) = 1$ and $C_2(s) = 1/(s + 1)$. We analyze the robust stability of this system by generating the value set octagon and checking for satisfaction of the Zero Exclusion Condition. Indeed, we first compute the closed loop polynomial

$$p(s, q, r) = s^4 + (11 + q_3 + r_2)s^3 + (15 + q_2 + r_1 + r_2)s^2$$
$$+ (17 + q_1 + r_0 + r_1)s + (15 + q_0 + r_0)$$

and easily verify that with all $q_i = 0$, the nominal $p(s, 0, 0)$ is stable. Next, we generate the value set octagon. Noting the fact that $-\pi/2 \leq \angle C(j\omega) < 0$ for all $\omega \geq 0$, the eight distinguished edges of the value set octagon are invariant over the entire frequency range. This fact is confirmed by the value set plot in Figure 10.6.3. The plot was generated using 50 evenly spaced frequency points over the critical range $0 \leq \omega \leq 1.3$. From the plot, we also observe that the Zero Exclusion Condition (Theorem 7.4.2) is violated. Hence, the family of closed loop polynomials is not robustly stable.

REMARKS 10.6.6 (Edge Reduction): The power of the octagonal value set characterization of this chapter is demonstrated in the example above. Suppose that instead of using the theory in this chapter, one attacks the example above via direct application of the Edge

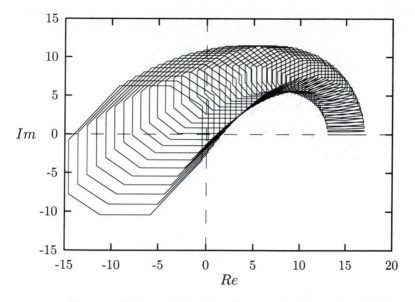

FIGURE 10.6.3   Value Set Octagon for Example 10.6.5

Theorem (see Section 9.3). Since the number of uncertain parameters is $\ell = 7$, the uncertainty bounding box $Q \times R$ has $\ell 2^{\ell-1} = 448$ edges. Without the octagonal value set characterization, direct application of the Edge Theorem dictates that each of these 448 edges must be individually tested.

EXERCISE 10.6.7 (Robustness Margin): For the feedback system described in Example 10.6.5, replace the uncertainty bounding set $Q \times R$ by a uniformly proportioned box with variable radius $r \geq 0$. Now, verify that the robustness margin is $r_{max} \approx 0.75$. Furthermore, when carrying out computations, note that one needs to be careful that the family of closed loop polynomials $\mathcal{P}_{CL}$ continues to have invariant degree as the uncertainty bound is increased.

10.7   Proof of Thirty-Two Edge Theorem

With the help of the octagonal value set description in the preceding section, the proof of Theorem 10.5.1 becomes quite simple. We sketch the key points. First, we observe that necessity is trivial since each edge polynomial is a member of $\mathcal{P}_{CL}$. To establish sufficiency, we assume stability of the $e(s, \lambda)$ and must prove that $\mathcal{P}_{CL}$ is robustly stable. This is accomplished using a simple zero exclusion argument; i.e., we begin by noting that $\mathcal{P}_{CL}$ has at least one stable member and at sufficiently high $\omega > 0$, satisfaction of the condition

$$0 \notin p(j\omega, Q, R)$$

is guaranteed by invariant degree of $\mathcal{P}_{CL}$. Subsequently, if the Zero Exclusion Condition (Theorem 7.4.2) is violated, there must be some value set penetration frequency $\omega^* > 0$ for which $z = 0$ lies on an edge of $p(j\omega^*, Q, R)$. However, in view of the thirty-two edge characterization of $p(j\omega, Q, R)$ in the preceding section, such a penetration would contradict the standing assumption that all of the edge polynomials $e(s, \lambda)$ are stable. ∎

10.8   Conclusion

The results in this chapter were obtained by strengthening the hypotheses of Chapter 9. That is, instead of allowing arbitrary polytopes of polynomials for the numerator and denominator of the plant, we worked with interval polynomials. The main payoff was a dramatic reduction in computational complexity. For problems with ℓ uncertain parameters, the value set is an octagon at each frequency. Furthermore, taking the union over all frequencies, no more than thirty-two edges ever come into play.

In the next chapter, we specialize the result even further. We remain within the realm of interval plants but restrict our attention to first order compensators. The benefit associated with this restriction on the compensator is quite simply explained: Instead of thirty-two edges, we obtain an extreme point result involving sixteen distinguished plants. Since many industrial controllers (such as the classical lead, lag and PI compensators) are first order, there is strong motivation for consideration of this special case.

Notes and Related Literature

NRL 10.1 With the goal of further exploiting the 2ℓ edge property for a parallelotope, the paper by Djaferis and Hollot (1989a) identifies a finite number of fixed frequencies for which "edge switching" can occur. The key idea is described roughly as follows: One creates a frequency partition $(0, \infty) = \cup_{i=1}^{N}(\omega_i, \omega_{i+1})$ with each ω_i obtained by finding the roots of an appropriately constructed fixed polynomial. This partition has the property that on each interval (ω_i, ω_{i+1}), the distinguished edges of $p(j\omega, Q)$ are invariant. The identification of a finite set of distinguished frequencies is also central to the the work of Sideris (1991), where a more complete analysis of computational complexity is considered .

NRL 10.2 To address the issue of computational complexity, some authors dispense entirely with the Edge Theorem and work more directly toward verification of the Zero Exclusion Condition; e.g., see Barmish (1989). In Saridereli and Kern (1987) and Vicino (1989), a frequency parameterized linear program is used. Roughly speaking, it is possible to take the data describing the robust stability problem and create a *frequency parameterized* linear program with objective $c^T(\omega)x$ and constraint $A(\omega)x \leq b(\omega)$. Letting $r_{max}(\omega)$ denote the infimum at frequency $\omega \geq 0$, we can compute the robustness margin r_{max} by minimizing $r_{max}(\omega)$ with respect to $\omega \geq 0$.

NRL 10.3 In all of the papers cited in the note above, the issue of combinatoric explosion prevails. More effective methods for overcoming the combinatoric explosion problem are given in Sideris (1991) and Kraus and Truöl (1991). In the case of Sideris (1991), computational complexity is reduced via special computations between simplex steps, and in the case of Kraus and Truöl (1991), an algorithm for value set construction is given which does not require enumeration of either edges or extreme points.

NRL 10.4 The paper by Rantzer (1992a) also addresses the issue of computational complexity by identifying a "small" number of distinguished edges. This more general framework involves *testing sets* which might also include extreme points.

Chapter 11

The Sixteen Plant Theorem

Synopsis

If the compensator for an interval plant is first order, testing thirty-two edges for robust stability of the closed loop is no longer required. Under this strengthening of hypothesis, it is shown that stability of sixteen distinguished closed loop systems implies robust stability of the entire family. In view of this extreme point result and the fact that the number of parameters entering into the controller is at most three (the pole, the zero and the gain), it becomes possible to use the results in this chapter in a synthesis context.

11.1   Introduction

The takeoff point for this chapter is the following question: Given an interval plant $\mathcal{P}$ with compensator interconnected as in Figure 10.4.1, under what conditions can we establish robust stability of the closed loop by testing a "small" finite subset of systems corresponding to extreme members of $\mathcal{P}$? In other words, when can we dispense with the thirty-two edges of Theorem 10.5.1 and work solely with the set of extreme plants?

We have already encountered one situation which leads to extreme point results of the sort described above. Recalling Exercise 6.4.3 for the special case when $C(s) = K$ is a pure gain compensator, we know that robust stability of the closed loop is guaranteed

178

if and only if four distinguished extreme plants are stabilized; we need only eight plants if the sign of K is not specified. This result is a simple consequence of the fact that with pure gain compensation, the interval polynomial structure is preserved in going from the open loop to the closed loop. In contrast, we saw in Chapters 9 and 10 that systems with more general compensators do not have this property. That is, an interval plant problem gets converted into a polytopic problem under feedback.

In this chapter, we concentrate on a special case motivated by industrial applications. Namely, we restrict our attention to classical first order compensators such as the classical lead, lag and PI controllers. The main result of this chapter is the Sixteen Plant Theorem of Barmish, Hollot, Kraus and Tempo (1992). We see that it is necessary and sufficient to stabilize only *sixteen* extreme plants in order to stabilize the entire family.

11.2   Setup with an Interval Plant

We concentrate on the interval plant

$$P(s,q,r) = \frac{\displaystyle\sum_{i=0}^{m}[q_i^-,q_i^+]s^i}{s^n + \displaystyle\sum_{i=0}^{n-1}[r_i^-,r_i^+]}.$$

As usual, Q and R denote the boxes bounding the uncertain parameter vectors q and r, respectively, and we assume that the resulting family of plants $\mathcal{P} = \{P(\cdot,q,r) : q \in Q; r \in R\}$ is strictly proper; that is, $m < n$. We consider the compensation scheme indicated in Figure 11.2.1 and express $C(s) = C_1(s)C_2(s)$ as a quotient of polynomials by writing

$$C(s) = \frac{N_C(s)}{D_C(s)}.$$

Since our sole concern is first order compensators, we take

$$N_C(s) = K(s - z)$$

and

$$D_C(s) = s - p$$

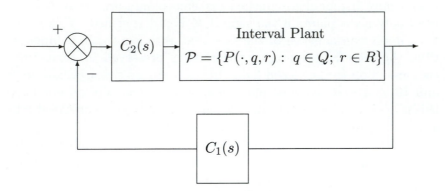

FIGURE 11.2.1   Basic Setup for this Chapter

with $K \neq 0$ and either $p \neq 0$ or $z \neq 0$. Given the compensator $C(s)$ above and expressing the uncertain plant as the quotient of uncertain polynomials

$$P(s, q, r) = \frac{N(s, q)}{D(s, r)},$$

the resulting closed loop polynomial is

$$p(s, q, r) = K(s - z)N(s, q) + (s - p)D(s, r).$$

This leads to the family of closed loop polynomials

$$\mathcal{P}_{CL} = \{p(\cdot, q, r) : q \in Q; r \in R\}.$$

11.3   Sixteen Distinguished Plants

In this section, we make a special selection from the set of extreme plants. For motivation, observe that if Q is $(m+1)$-dimensional and R is n-dimensional, the uncertainty bounding set $Q \times R$ has number of extreme points given by

$$N_{ext} = 2^{m+n+1}.$$

Furthermore, we can associate extreme plants with the extremes of $Q \times R$ in a simple manner: If $\{q^i\}$ denotes the set of extremes for Q and $\{r^i\}$ denotes the set of extremes for R, we generate the N_{ext} extreme plants by considering all plants of the form

$$P_{i_1, i_2}(s) = \frac{N(s, q^{i_1})}{D(s, r^{i_2})}.$$

From the set of extremes above, we now select a distinguished subset of sixteen plants in accordance with the definition below.

DEFINITION 11.3.1 (The Sixteen Kharitonov Plants):   Given an interval plant family $\mathcal{P}$ with Kharitonov polynomials $N_1(s)$, $N_2(s)$, $N_3(s)$ and $N_4(s)$ and $D_1(s)$, $D_2(s)$, $D_3(s)$ and $D_4(s)$ for the numerator and denominator, respectively, (see Section 5.5), we define *sixteen Kharitonov plants* by

$$P_{i_1,i_2}(s) = \frac{N_{i_1}(s)}{D_{i_2}(s)}$$

with $i_1, i_2 \in \{1, 2, 3, 4\}$.

DEFINITION 11.3.2 (Associated Closed Loop Polynomials):   For the feedback system under consideration (see Section 11.2), we associate a closed loop polynomial

$$p_{i_1,i_2}(s) = K(s - z)N_{i_1}(s) + (s - p)D_{i_2}(s)$$

with each Kharitonov plant $P_{i_1,i_2}(s)$.

EXAMPLE 11.3.3 (Sixteen Plants and Polynomials):   We consider the interval plant family $\mathcal{P}$,

$$P(s, q, r) = \frac{[4.5, 5.5]s^3 + [3.5, 4.5]s^2 + [2.5, 3.5]s + [6.5, 7.5]}{s^3 + [4.5, 5.5]s^2 + [5.5, 6.5]s + [7.5, 8.5]},$$

and note that Exercise 10.6.7 indicates that with the first order compensator $C(s) = 1/(s+1)$, robust stability is guaranteed. In contrast to Chapter 10, where an octagonal value set was used, the results in this chapter enable us to reach the same conclusion more simply because $C(s)$ is first order. To this end, note that it is straightforward to calculate the sixteen Kharitonov plants and the associated closed loop polynomials. To illustrate, it is readily verified that

$$P_{2,3}(s) = \frac{4.5s^3 + 3.5s^2 + 3.5s + 7.5}{s^3 + 4.5s^2 + 5.5s + 8.5}$$

and the associated closed loop polynomial is

$$p_{2,3}(s) = s^4 + 10s^3 + 13.5s^2 + 17.5s + 16.$$

After stating the theorem below, the robust stability analysis for this system is completed.

11.4   The Sixteen Plant Theorem

We are now prepared to provide the main result of this chapter. The proof is relegated to Sections 11.6 and 11.7.

THEOREM 11.4.1 (Barmish, Hollot, Kraus and Tempo (1992)):
Consider the strictly proper interval plant family $\mathcal{P}$ with first order compensator $C(s)$ as described in Section 11.2. Then $C(s)$ robustly stabilizes $\mathcal{P}$ if and only if it stabilizes each of the sixteen Kharitonov plants; i.e., the family of closed loop polynomials $\mathcal{P}_{CL}$ is robustly stable if and only if $p_{i_1,i_2}(s)$ is stable for $i_1, i_2 \in \{1, 2, 3, 4\}$.

EXERCISE 11.4.2 (Application of the Theorem):   Consider the interval plant family $\mathcal{P}$ with compensator $C(s)$ in Example 11.3.3. Using the Sixteen Plant Theorem, show that $C(s)$ robustly stabilizes $\mathcal{P}$.

11.5   Controller Synthesis Technique

Although the Sixteen Plant Theorem is stated as an analysis result, a moment's reflection indicates that the theorem is quite useful in a synthesis context as well. To explain this point, consider the following illustration: Suppose that one wants to construct a robustly stabilizing PI controller

$$C(s) = K_1 + \frac{K_2}{s}$$

for an interval plant family $\mathcal{P}$.

Then, to determine if appropriate gains K_1 and K_2 exist, we first set up sixteen Routh tables—one for each Kharitonov plant with compensator $C(s)$. Noting that the first column entries of these tables are functions of K_1 and K_2, the positivity requirement for stability leads to a set of inequalities. Since these inequalities only involve the two parameters K_1 and K_2, a graphical description of the set of stabilizing gains is easily generated. In conclusion, a necessary and sufficient condition for the existence of a robust stabilizing controller with the specified form is nonemptiness of the set of gains satisfying the inequalities associated with the Routh tables. Moreover, any feasible point (K_1^*, K_2^*) in this set of gains is associated with a robustly stabilizing PI controller. Note that the idea is readily extended to a more general class of first order compensators having the form

$$C(s) = \frac{K(s-z)}{s-p}.$$

Using the Sixteen Plant Theorem, one can characterize the set of stabilizing triples (K, z, p).

EXAMPLE 11.5.1 (Synthesis Using the Sixteen Plant Theorem): We consider the model of an experimental oblique wing aircraft given in Dorf (1974). In the absence of uncertainty, the aircraft transfer function is

$$P(s) = \frac{64s + 128}{s^4 + 3.7s^3 + 65.6s^2 + 32s}.$$

Now, to illustrate application of the Sixteen Plant Theorem in a robust synthesis context, we replace $P(s)$ by the interval plant family $\mathcal{P}$ described by

$$P(s, q, r) = \frac{q_1 s + q_0}{s^4 + r_3 s^3 + r_2 s^2 + r_1 s + r_0}$$

and consider uncertainty bounds $90 \leq q_0 \leq 166$, $54 \leq q_1 \leq 74$, $-0.1 \leq r_0 \leq 0.1$, $30.1 \leq r_1 \leq 33.9$, $50.4 \leq r_2 \leq 80.8$ and $2.8 \leq r_3 \leq 4.6$. For this interval plant family $\mathcal{P}$, the objective is to determine if a robustly stabilizing PI compensator

$$C(s) = K_1 + \frac{K_2}{s}$$

exists. If we determine that a robust stabilizer exists, we also want to compute appropriate gains K_1 and K_2. For the sake of brevity, we do not show all numerical calculations below. However, we provide enough detail so that the reader can replicate all calculations.

The first step in the synthesis procedure is to generate each of the sixteen Kharitonov plants with an associated Routh table for the resulting closed loop polynomial. Note that we must carry out computations parametrically in the controller gains K_1 and K_2. To illustrate, using the Kharitonov polynomials $N_2(s)$ and $D_1(s)$, we obtain the associated plant

$$P_{2,1}(s) = \frac{74s + 166}{s^4 + 4.6s^3 + 80.8s^2 + 30.1s - 0.1}$$

and, with PI compensator $C(s)$, the associated closed loop polynomial is found to be

$$p_{2,1}(s) = s^5 + 4.6s^4 + 80.8s^3 + (30.1 + 74K_1)s^2$$
$$+ (-0.1 + 166K_1 + 74K_2)s + 166K_2.$$

Now, using $p_{2,1}(s)$, we generate the Routh table

s^5	1	80.8	$-\gamma(K_1) + 74K_2$
s^4	4.6	$30.1 + 74K_1$	$166K_2$
s^3	$74.3 - 16.1K_1$	$-\gamma(K_1) + 37.9K_2$	0
s^2	$\alpha_1(K_1, K_2)/\alpha_2(K_1)$	$166K_2$	0
s^1	$\beta(K_1, K_2)/\alpha_1(K_1, K_2)$	0	0
s^0	$166K_2$	0	0

where

$$\alpha_1(K_1, K_2) = 2236.89 + 4249.99K_1 - 174.34K_2 - 1191.4K_1^2;$$
$$\alpha_2(K_1) \approx 74.3 - 16.1K_1;$$
$$\beta(K_1, K_2) \approx -223.689 + 370,899K_1 + 705,617K_1^2$$
$$-197,772K_1^3 - 831,606K_2 + 529,283K_1K_2$$
$$-88,182.9K_1^2K_2 - 6,607.49K_2^2;$$
$$\gamma(K_1) = 0.1 - 166K_1.$$

In a similar manner, one can generate Routh tables for the remaining fifteen Kharitonov plants. Using the sixteen Routh tables, the next step in the synthesis procedure is to enforce positivity for each of the first columns. This leads to inequalities involving K_1 and K_2; e.g., for the Routh table for the Kharitonov plant $P_{2,1}(s)$ above, positivity of the first column leads to the stability conditions

$$K_1 < 4.6; \quad K_2 > 0; \quad \frac{\alpha_1(K_1, K_2)}{\alpha_2(K_1)} > 0; \quad \frac{\beta(K_1, K_2)}{\alpha_1(K_1, K_2)} > 0.$$

We can now easily display the set of gains $\mathcal{K}_{2,1}$ satisfying the inequalities above; see Figure 11.5.1. In a similar manner, we can generate the set of stabilizing gains $\mathcal{K}_{i_1,i_2}$ for all remaining pairs $(i_1, i_2) \in \{1, 2, 3, 4\}$ and display the result graphically. To obtain the final result, we must enforce the requirement that (K_1, K_2) stabilizes all sixteen Kharitonov plants simultaneously. Hence, the desired set of stabilizing gains is given by

$$\mathcal{K} = \bigcap_{i_1, i_2} \mathcal{K}_{i_1, i_2}.$$

Any one of a wide variety of two-variable graphics routines can be used to display the set $\mathcal{K}$. To illustrate, for the range of gains

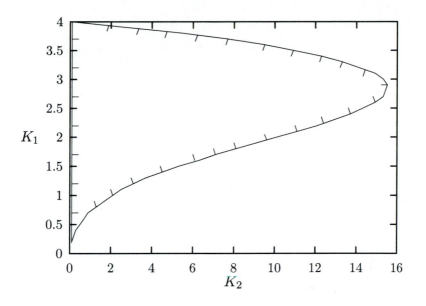

FIGURE 11.5.1   The Region $\mathcal{K}_{2,1}$ for Example 11.5.1

$0 \leq K_1 \leq 2$ and $0 \leq K_2 \leq 1.5$, the set of robust PI stabilizers is shown in Figure 11.5.2. Since this set is nonempty, $\mathcal{P}$ is robustly stabilizable. When stabilizing the interval plant $\mathcal{P}$, we can select any $(K_1, K_2) \in \mathcal{K}$. For example, a robust stabilizer is given by

$$C(s) = 0.9 + \frac{0.2}{s}.$$

Although we have already established that $C(s)$ is a robust stabilizer, it is of interest to provide an independent validation of this result via the Zero Exclusion Condition (Theorem 7.4.2). This task is simplified by using the octagonal value set characterization which is given in Section 10.6. Indeed, we first calculate the uncertain closed loop polynomial

$$p(s, q, r) = s^5 + r_3 s^4 + r_2 s^3 + (0.9q_1 + r_1)s^2 \\ + (0.9q_0 + 0.2q_1 + r_0)s + 0.2q_0.$$

Next, we generate the value set $p(j\omega, Q, R)$. An initial frequency sweep for 100 evenly spaced points in the range $0 \leq \omega \leq 7.5$ indicates that a low-frequency "zoom" is required to determine if the Zero Exclusion Condition is satisfied; see Figure 11.5.3. By concentrating

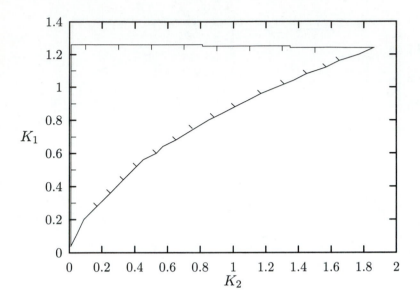

FIGURE 11.5.2   The Set of Robust Stabilizers $\mathcal{K}$ for Example 11.5.1

our attention on the range $0 \le \omega \le 1.2$, we use the second value set plot in Figure 11.5.4 to conclude that $0 \notin p(j\omega, Q, R)$ for all $\omega \ge 0$. Hence, we have verified that the Zero Exclusion Condition is satisfied and $C(s)$ is a robust stabilizer.

11.6   Machinery for Proof of Sixteen Plant Theorem

In this section, we develop some machinery to facilitate the proof of the Sixteen Plant Theorem. The reader interested solely in application of the theorem can proceed directly to Section 11.8.

11.6.1   Nonincreasing Phase Property

In this subsection, we concentrate on the one-parameter family of polynomials described by

$$p(s, \lambda) = f(s) + \lambda g(s),$$

where $f(s)$ and $g(s)$ are fixed polynomials and $\lambda \in \Lambda = [0, 1]$. Recalling Exercise 4.15.1, we cannot guarantee robust stability by simply checking the extremes $p(s, 0) = f(s)$ and $p(s, 1) = f(s) + g(s)$. However, under the strengthened hypothesis that $g(j\omega)$ has nonincreasing phase, we see below that such an extreme point result is ob-

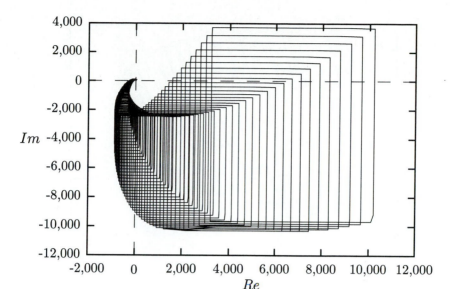

FIGURE 11.5.3   Crude Frequency Sweep for Example 11.5.1

tained. To this end, we begin by recalling that the value set $p(j\omega, \Lambda)$ is a straight line segment with endpoints $p(j\omega, 0)$ and $p(j\omega, 1)$; see Lemma 7.2.3. Furthermore, if $g(j\omega) \neq 0$, notice that the slope of $p(j\omega, \Lambda)$ is given by

$$m(\omega) = \frac{Im\ g(j\omega)}{Re\ g(j\omega)} = \tan\ \measuredangle g(j\omega).$$

If $Re\ g(j\omega) = 0$, then we interpret $p(j\omega, \Lambda)$ as a vertical line segment and if $Im\ g(j\omega) = 0$, we interpret $p(j\omega, \Lambda)$ as a horizontal line segment. Finally, for the case when $g(j\omega) = 0$, the value set $p(j\omega, \Lambda)$ degenerates to a single point; i.e., $p(j\omega, \Lambda) = \{f(j\omega)\}$. These situations are shown in Figure 11.6.1. The basic facts above are used in the proof of the lemma below. This lemma is due to Rantzer (1990) and Fu (1991).

LEMMA 11.6.2 (Nonincreasing Phase Property):   *Let $\mathcal{P}$ be a family of polynomials having invariant degree and described by*

$$p(s, \lambda) = f(s) + \lambda g(s),$$

where $f(s)$ and $g(s)$ are fixed polynomials and $\lambda \in \Lambda = [0, 1]$. Assuming that $\measuredangle g(j\omega)$ is nonincreasing, it follows that $\mathcal{P}$ is robustly stable if and only if $p(s, 0)$ and $p(s, 1)$ are stable.

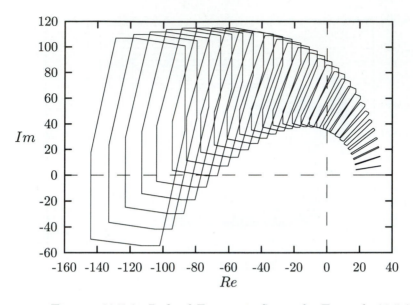

FIGURE 11.5.4   Refined Frequency Sweep for Example 11.5.1

PROOF: Since necessity is trivial, we concentrate on the proof of
sufficiency. We assume that $p(s, 0)$ and $p(s, 1)$ are stable and must
prove that the family $\mathcal{P}$ is robustly stable. Proceeding by contradiction, suppose $\mathcal{P}$ is not robustly stable. Then, by the Zero Exclusion
Condition (Theorem 7.4.2), there exists some frequency $\omega^* \geq 0$ such
that $0 \in p(j\omega^*, \Lambda)$. Since $p(s, 0)$ and $p(s, 1)$ are stable, the value
set $p(j\omega^*, \Lambda)$ cannot be a point. In accordance with the discussion
above, the value set $p(j\omega^*, \Lambda)$ is a line segment (perhaps vertical or
horizontal) which includes the origin $z = 0$ but not as an endpoint.

We consider the case when $p(j\omega^*, 0)$ and $p(j\omega^*, 1)$ lie in the
interior of Quadrants 1 and 3, respectively, and simply note that the
other possible geometries are handled in an identical manner. We
now define the two cones

$$\mathcal{C}_0 = \{z \in \mathbf{C} : \measuredangle p(j\omega^*, 0) < \measuredangle z < \frac{\pi}{2}\}$$

and

$$\mathcal{C}_1 = \{z \in \mathbf{C} : \measuredangle p(j\omega^*, 1) < \measuredangle z < \frac{3\pi}{2}\}$$

depicted in Figure 11.6.2. Using the stability of $p(s, 0)$ and $p(s, 1)$,
the Monotonic Angle Property (Lemma 5.7.6) and with continuity
of $\measuredangle p(j\omega, 0)$ and $\measuredangle p(j\omega, 1)$, we arrive at the following point: For

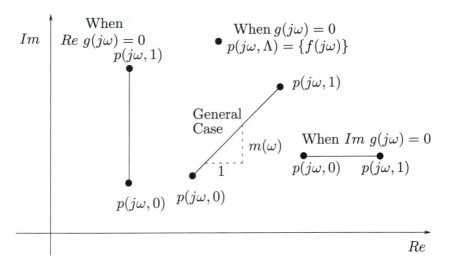

FIGURE 11.6.1   Value Set Possibilities for $p(j\omega, \Lambda)$

$\Delta\omega > 0$ sufficiently small, it must be true that $p(j(\omega^* + \Delta\omega), 0) \in C_0$ and $p(j(\omega^* + \Delta\omega), 1) \in C_1$. Notice that this condition implies that the slope of the value set satisfies $m(\omega^* + \Delta\omega) > m(\omega^*)$. On the other hand, since $\measuredangle g(j\omega)$ is nonincreasing, the slope $m(\omega) = \tan \measuredangle g(j\omega)$ must satisfy $m(\omega^* + \Delta\omega) \leq m(\omega^*)$. Hence, we have reached the contradiction which we seek. ∎

EXERCISE 11.6.3 (Results in the Literature):   To recover known results in the literature using Lemma 11.6.2, consider the two situations described below and argue that an extreme point result holds. (a) Suppose that $g(s)$ contains all even or all odd powers of s and specialize the lemma to obtain the result in Bialas and Garloff (1985). (b) Suppose that $g(s)$ is antistable (all roots in the strict right half plane) and specialize the lemma to obtain the extreme point result in Petersen (1990).

11.6.4   Transformations: Real Versus Complex

In this subsection, we state a basic lemma from the theory of polynomials. We see below that there is a fundamental relationship between stable real coefficient polynomials of order n and stable complex coefficient polynomials of order approximately equal to $n/2$. A nice proof of the lemma below is given in Jury (1974).

LEMMA 11.6.5 (Real Versus Complex Coefficients):   *Consider the*

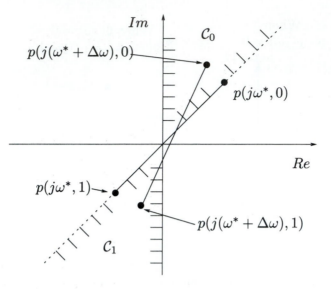

FIGURE 11.6.2   Illustration for the Proof of Lemma 11.6.2

real positive coefficient polynomial $p(s)$ decomposed into even and odd parts as

$$p(s) = p_{even}(s^2) + sp_{odd}(s^2).$$

Then the following three statements are equivalent:
(i) The real coefficient polynomial $p(s)$ is stable;
(ii) The complex coefficient polynomial

$$\tilde{p}_1(s) = p_{even}(js) + jp_{odd}(js)$$

is stable;
(iii) The complex coefficient polynomial

$$\tilde{p}_2(s) = p_{even}(-js) + sp_{odd}(-js)$$

is stable.

REMARKS 11.6.6 (Extreme Point Result):   In the lemma below, the usefulness of the transformation between real and complex coefficient polynomials is demonstrated. We obtain an extreme point result which facilitates the proof of the Sixteen Plant Theorem.

LEMMA 11.6.7 (A Stepping Stone):   *Let $f(s)$ be a polynomial of degree n and suppose that $g(s)$ is a polynomial of the form*

$$g(s) = (\alpha s + \beta)h(s)$$

with $h(s)$ being a polynomial of degree $k < n - 1$ having only even powers of s or only odd powers of s. Then, given any $\alpha, \beta \in \mathbf{R}$, the family of polynomials $\mathcal{P}$ described by

$$p(s, \lambda) = f(s) + \lambda g(s)$$

and $\lambda \in [0, 1]$ is robustly stable if and only if the two extremes $p(s, 0)$ and $p(s, 1)$ are stable.

PROOF: Since necessity is trivial, we proceed to establish sufficiency. Indeed, we assume that $p(s, 0)$ and $p(s, 1)$ are stable and must prove that $\mathcal{P}$ is robustly stable. In view of Part (b) of Exercise 11.6.3, we assume that α and β are both nonzero; otherwise λ enters into only all even order terms or all odd order terms and sufficiency follows immediately. In addition, we assume $h(s)$ has only even powers of s and work with $\tilde{p}_1(s)$ in Lemma 11.6.5; note that a nearly identical proof is used for the odd power case using $\tilde{p}_2(s)$ rather than $\tilde{p}_1(s)$ in the lemma. Finally, without loss of generality, we also assume $p(s, 0)$ and $p(s, 1)$ both have all positive coefficients. In this regard, recall that $p(s, 0)$ and $p(s, 1)$ are both assumed to be stable and have the same coefficient of s^n. Writing

$$f(s) = f_{even}(s^2) + s f_{odd}(s^2)$$

and

$$h(s) = h_{even}(s^2)$$

where $f(\cdot)$ and $h(\cdot)$ are polynomials, we obtain

$$
\begin{aligned}
p(s, \lambda) &= f_{even}(s^2) + s f_{odd}(s^2) + \lambda(\alpha s + \beta) h_{even}(s^2) \\
&= [f_{even}(s^2) + \lambda \beta h_{even}(s^2)] + s[f_{odd}(s^2) + \lambda \alpha h_{even}(s^2)].
\end{aligned}
$$

Observing that $p(s, \lambda)$ has positive coefficients for all $\lambda \in [0, 1]$, in accordance with Lemma 11.6.5, it suffices to prove that the complex coefficient polynomial

$$
\begin{aligned}
\tilde{p}_1(s, \lambda) &= [f_{even}(js) + \lambda \beta h_{even}(js)] + j[f_{odd}(js) + \lambda \alpha h_{even}(js)] \\
&= [f_{even}(js) + j f_{odd}(js)] + \lambda(\beta + \alpha j) h_{even}(js)
\end{aligned}
$$

is stable for all $\lambda \in [0, 1]$. To this end, we already know from the same lemma that $\tilde{p}_1(s, 0)$ and $\tilde{p}_1(s, 1)$ are stable.

The proof is now completed by contradiction: We assume that $\tilde{p}_1(s, \tilde{\lambda})$ is unstable for some $\tilde{\lambda} \in [0, 1]$ and note that the Zero Exclusion Condition (Theorem 7.4.2) also holds for complex coefficient polynomials; a nearly identical proof applies. Hence, there exists some $\omega^* \in \mathbf{R}$ and $\lambda^* \in (0,1)$ such that $\tilde{p}_1(j\omega^*, \lambda^*) = 0$. We first rule out the possibilities that $\tilde{p}_1(j\omega^*, 0) = 0$ or $\tilde{p}_1(j\omega^*, 1) = 0$ because this would contradict stability of $\tilde{p}_1(s, 0)$ and $\tilde{p}_1(s, 1)$.

It now follows that for each ω in some neighborhood Ω of ω^*, the value set $p(j\omega, \Lambda)$ is a line segment in the complex plane with endpoints $\tilde{p}_1(j\omega, 0)$ and $\tilde{p}_1(j\omega, 1)$ and constant slope $\tilde{m}(\omega) = \alpha/\beta$. To complete the proof, note that the origin $z = 0$ is not an endpoint of this line segment and moreover, since $\tilde{p}_1(s, 0)$ and $\tilde{p}_1(s, 1)$ are stable, the angles $\measuredangle \tilde{p}_1(j\omega, 0)$ and $\measuredangle \tilde{p}_1(j\omega, 1)$ are increasing functions of ω; note that we are using the fact that the Monotonic Angle Property (Lemma 5.7.6) remains valid for complex coefficient polynomials. The proof is now completed using an argument which is nearly identical to the one given in the proof of Lemma 11.6.7; i.e., using the Monotonic Angle Property of $\measuredangle \tilde{p}_1(j\omega, 0)$ and $\measuredangle \tilde{p}_1(j\omega, 1)$ and the fact that $0 \in p(j\omega^*, \Lambda)$, we can create two cones $\mathcal{C}_0$ and $\mathcal{C}_1$ as in Figure 11.6.2 and argue that for $|\omega - \omega^*|$ sufficiently small, the value set has slope $\tilde{m}(\omega) > \tilde{m}(\omega^*)$. This, however, contradicts the constancy of $\tilde{m}(\omega)$ for $\omega \in \Omega$. The proof is now complete. $\blacksquare$

11.7   Proof of the Sixteen Plant Theorem

Since the proof of necessity is trivial, we proceed directly to the proof of sufficiency. Indeed, we assume that $C(s)$ stabilizes the sixteen Kharitonov plants and must show that $C(s)$ robustly stabilizes the family $\mathcal{P}$. Since the closed loop polynomial

$$p(s, q, r) = K(s - z)N(s, q) + (s - p)D(s, r)$$

has degree $n + 1$ for all $q \in Q$ and $r \in R$, the Zero Exclusion Condition (Theorem 7.4.2) applies; i.e., to establish robust stability, it suffices to show that $0 \notin p(j\omega, Q, R)$ for all $\omega \geq 0$. Furthermore, in accordance with Theorem 10.5.1, it is sufficient to establish stability of at most thirty-two edge polynomials with two possible forms. The first form is

$$e(s, \lambda) = K(s - z)N_{i_1}(s) + (s - p)D_{i_2, i_3}(s, \lambda)$$

with $i_1 \in \{1, 2, 3, 4\}$, $(i_2, i_3) \in \{(1, 3), (1, 4), (2, 3), (2, 4)\}$ and

$$D_{i_2, i_3}(s, \lambda) = \lambda D_{i_2}(s) + (1 - \lambda)D_{i_3}(s).$$

The second form is

$$e(s, \lambda) = K(s - z)N_{i_1, i_2}(s) + (s - p)D_{i_3}(s, \lambda)$$

with $(i_1, i_2) \in \{(1,3), (1,4), (2,3), (2,4)\}$, $i_3 \in \{1, 2, 3, 4\}$ and

$$N_{i_1, i_2}(s, \lambda) = \lambda N_{i_1}(s) + (1 - \lambda) N_{i_2}(s).$$

To complete the proof of the theorem, we pick a typical edge polynomial $e(s, \lambda)$ above and argue that with $\Lambda = [0, 1]$, the standing assumption that the sixteen Kharitonov plants are stabilized guarantees stability for the entire edge family

$$\mathcal{E} = \{e(\cdot, \lambda) : \lambda \in \Lambda\}.$$

For illustrative purposes, we take $i_1 = 2$, $i_2 = 4$ and $i_3 = 3$ and note that the proof for all other edge combinations is carried out in an identical manner. Now, with

$$
\begin{aligned}
e(s, \lambda) &= K(s - z)N_{2,4}(s, \lambda) + (s - p)D_3(s) \\
&= K(s - z)[\lambda N_2(s) + (1 - \lambda)N_4(s)] + (s - p)D_3(s) \\
&= p_{4,3}(s) + \lambda K(s - z)[N_2(s) - N_4(s)],
\end{aligned}
$$

we make the following identifications with Lemma 11.6.5:

$$
\begin{aligned}
f(s) &\sim p_{4,3}(s); \\
h(s) &\sim N_2(s) - N_4(s); \\
\lambda(\alpha s + \beta) &\sim \lambda K(s - z).
\end{aligned}
$$

Since $C(s)$ stabilizes the sixteen Kharitonov plants, it is easy to verify that all preconditions of the lemma are satisfied. In particular, notice that $N_2(s) - N_4(s)$ has only even powers of s and $p(s, 0) = p_{4,3}(s)$ and $p(s, 1) = p_{2,3}(s)$ are stable. Therefore, the lemma guarantees that $e(s, \lambda)$ is stable for all $\lambda \in \Lambda = [0, 1]$. The proof of the theorem is now complete. ∎

11.8   Conclusion

Extreme point results presented in this chapter raise an interesting issue. Suppose that we remain within the realm of interval plants but we allow the compensator to be more general; i.e., we no longer restrict $C(s)$ to be first order. Then it is of interest to give conditions under which stabilization of some distinguished subset of the extreme plants implies stabilization of the entire interval family. In this regard, note that some sort of assumption must be imposed because of counterexamples of the sort given in Exercise 6.4.5. The question is: For the case of higher order compensators, how restrictive are the conditions under which an extreme point result holds? The results in the next chapter shed some light on this question.

Note, however, that for cases when higher order compensators lend themselves to extreme point results, we do not obtain a "nice" synthesis theory as in the case of first order compensators; such results are mainly useful in an analysis context. To elaborate on this point, note that when the number of parameters entering $C(s)$ is greater than two or three, the graphics approach described in this chapter is no longer valid. Although one can still use a finite number of Routh tables to generate inequality constraints on the compensator parameters, the finding of a feasible point (when one exists) amounts to solving a potentially difficult nonlinear program.

Notes and Related Literature

NRL 11.1 Historically, the result of Hollot and Yang (1990) paved the way for the Sixteen Plant Theorem. With the same setup as in the theorem, the following weaker result is established: To robustly stabilize the interval plant family, it is necessary and sufficient to stabilize the entire set of extreme plants. Recall that the number of extreme plants can be as high as $N_{ext} = 2^{m+n+1}$.

NRL 11.2 Lemma 11.6.7 is a minor extension of a result given in the paper by Hollot and Yang (1990).

NRL 11.3 As mentioned in Section 11.8, for the case of higher order compensators, it is unclear whether an extreme point result for interval plants is useful in a synthesis context. Except for some rather special cases (such as those involving minimum phase and one-sign high frequency gain assumptions), the robust stabilization problem for a finite plant collection is unsolved; for example, see Youla, Bongiorno and Lu (1974), Saeks and Murray (1982) and Vidyasagar and Viswanadham (1982).

NRL 11.4 An interesting embellishment of the Sixteen Plant Theorem involves further reduction in the number of plants required in the robust stability test. For example, under the strengthened hypothesis that the compensator $C(s)$ is either lead or lag with gain K of prescribed sign, only eight distinguished plants need to be tested; see the paper by Barmish, Hollot, Kraus and Tempo (1992) for additional details.

NRL 11.5 Although the Sixteen Plant Theorem was given in a robust stability context, an interesting robust performance result is obtainable as a byproduct: Consider a strictly proper interval plant family $\mathcal{P}$ as in Section 11.2 with robustly stable denominator family $\mathcal{D}$. Then, using the Sixteen Plant Theorem, it is easily

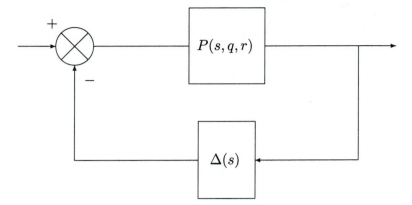

FIGURE 11.8.1   Configuration Associated with H^∞ Problem

shown that the *worst case* H^∞ norm is attained on one of the Sixteen Kharitonov plants; that is,

$$\max_{(q,r)\in Q\times R} \|P(s,q,r)\|_\infty = \max_{(i_1,i_2)\in\{1,2,3,4\}} \|P_{i_1,i_2}(s)\|_\infty.$$

This is the result given in Mori and Barnett (1988) and Chapellat, Dahleh and Bhattacharyya (1990). To prove this result using the Sixteen Plant Theorem, the key idea is to consider a feedback loop with unmodeled dynamics $\Delta(s)$ in the feedback path; see Figure 11.8.1. With $\Delta(s)$ proper, stable and rational, one can relate the H^∞ norm of $P(s,q,r)$ to a destabilizing perturbation $\Delta(s)$ which can be interpolated at the "critical frequency" via a first order compensator.

It is interesting to note that technical arguments used to establish this result do not easily generalize to *frequency weighted norms*; i.e., if $W(s)$ is proper, stable and rational, it is of interest to develop conditions under which the maximum

$$\mu_0 = \max_{(q,r)\in Q\times R} \|W(s)P(s,q,r)\|_\infty$$

is attained on one of the sixteen Kharitonov plants. The lack of a general extreme point result in this weighted case is related to the fact that a generalization of the Sixteen Plant Theorem is not immediate for higher order compensators. For example, for the interval plant

$$P(s,q) = \frac{[1,5000]}{s^2+25s}$$

of Hollot and Yang (1990) with compensator

$$C(s) = \frac{(s+3)(s+4)}{(s+0.1)(s+0.2)(s+75)}$$

connected as in Figure 11.2.1, it is easy to prove that $C(s)$ stabilizes the two extreme plants $P(s,1)$ and $P(s,5000)$, but the closed loop system obtained using $P(s,2)$ is unstable. It is also worth noting that even for simple controllers in the PID class, counterexamples of the sort above can be given.

Chapter 12

Rantzer's Growth Condition

Synopsis

In the preceding chapters, we encountered some special classes of polytopes of polynomials for which robust stability can be ascertained from stability of the extremes. This theme is more fully developed in this chapter. When we work with a typical edge polynomial described by $p(s, \lambda) = f(s) + \lambda g(s)$ and $\lambda \in [0, 1]$, the satisfaction of Rantzer's Growth Condition on the rate of change of the angle $\measuredangle g(j\omega)$ enables us to test for robust stability using the extreme polynomials $p(s, 0) = f(s)$ and $p(s, 1) = f(s) + g(s)$.

12.1 Introduction

When working with an affine linear uncertainty structure, we saw in Chapter 9 that the Edge Theorem holds under rather weak hypotheses. As we strengthened the hypotheses in Chapters 10 and 11, the results became progressively stronger. For example, given a simple feedback system involving an interval plant, we saw that the test for robust stability involves only thirty-two edges; see Theorem 10.5.1. By further restricting the compensator to be first order, we saw in Theorem 11.4.1 that the robust stability test turned out to involve only sixteen extreme plants.

In making the jump from an edge result to an extreme point result, we saw repeatedly that the following fundamental problem

196

arises: Let $f(s)$ and $g(s)$ be fixed polynomials and consider the one-parameter family $\mathcal{P}$ described by

$$p(s, \lambda) = f(s) + \lambda g(s)$$

and $\lambda \in [0, 1]$. Give conditions under which stability of the extremes $p(s, 0) = f(s)$ and $p(s, 1) = f(s) + g(s)$ implies robust stability of the family $\mathcal{P}$. In this chapter, we generalize the extreme point results developed thus far. Under mild regularity conditions, we see that if the angle $\angle g(j\omega)$ satisfies a certain growth condition, then robust stability of $\mathcal{P}$ is equivalent to stability of the extremes.

12.2   Convex Directions

The notion of a convex direction is instrumental to this chapter. In order to motivate this concept, we consider $p(s, \lambda)$ above and make one fundamental observation about the extreme point results attained in all previous chapters: In every case, if we recast the extreme point problem in the $f(s) + \lambda g(s)$ setting, the critical assumptions involve $g(s)$ but not $f(s)$. To make this point clear, we consider some examples and an exercise.

EXAMPLE 12.2.1 (Kharitonov's Problem):   By reducing Kharitonov's problem in Chapter 5 to an edge problem associated with uncertainty in the coefficient of s^k, it is straightforward to verify that we obtain an edge polynomial of the form

$$p(s, \lambda) = f(s) + \lambda s^k.$$

Hence, in terms of Kharitonov's framework, $f(s)$ can be rather general but the polynomial

$$g(s) = s^k$$

is restricted to having a special form.

EXAMPLE 12.2.2 (Nonincreasing Angle Property):   When we exposed the nonincreasing phase property in Section 11.6.1, $f(s)$ was rather general but a nonincreasing angle of $g(j\omega)$ was assumed.

EXAMPLE 12.2.3 (Sixteen Plant Theorem):   Lemma 11.6.2 was fundamental to the proof of the Sixteen Plant Theorem in Section 11.4. In the lemma, $f(s)$ was permitted to be any stable polynomial but $g(s)$ was constrained to be of the form

$$g(s) = (\alpha s + \beta)h(s)$$

with $\alpha, \beta \in \mathbf{R}$ and $h(s)$ being a polynomial satisfying

$$\deg h(s) < \deg f(s) - 1$$

and having only even powers of s or only odd powers of s.

EXERCISE 12.2.4 (Interval Plant and Compensator):   Consider an interval plant family $\mathcal{P}$ with compensator $C(s) = N_C(s)/D_C(s)$ connected in feedback configuration as in Figure 10.4.1.
(a) For the closed loop system, reduce the robust stability problem to an $f(s) + \lambda g(s)$ problem and show that $g(s)$ is of the form

$$g(s) = K s^k N_C(s)$$

or

$$g(s) = K s^k D_C(s).$$

(b) Using Theorem 10.5.1, argue that it suffices to study no more than thirty-two $f(s) + \lambda g(s)$ problems with $g(s)$ of the form

$$g(s) = h(s) N_C(s)$$

or

$$g(s) = h(s) D_C(s)$$

with $h(s)$ having only even powers of s or only odd powers of s.
(c) For the more general case when the plant numerator and denominator have affine linear uncertainty structures, describe the appropriate polynomial $g(s)$ to be studied.

REMARKS 12.2.5 (Preparing for a Definition):   In view of the exercise above, extreme point results involving a general $f(s)$ but a specific class of $g(s)$ have interpretations in a feedback context. That is, a condition on $g(s)$ implicitly describes a class of compensators for which an extreme point result holds. In the theory to follow, we view $g(s)$ as a direction in a space of stable polynomials and seek directions which are "nice" in the sense that an extreme point result holds. These nice directions are now defined more formally.

DEFINITION 12.2.6 (Convex Direction):   A monic polynomial $g(s)$ is said to be a *convex direction* (for the space of stable n-th order polynomials) if the following condition is satisfied: Given any stable n-th order polynomial $f(s)$ such that $f(s) + g(s)$ is also stable and $\deg(f(s) + \lambda g(s)) = n$ for all $\lambda \in [0, 1]$, it follows that the polynomial $f(s) + \lambda g(s)$ is stable for all $\lambda \in [0, 1]$.

REMARKS 12.2.7 (Interpretation): Note that the monicity requirement in the definition above is introduced solely for convenience; it
facilitates notation in the sequel. When dealing with an edge polyno-
mial $p(s, \lambda) = f(s) + \lambda g(s)$ with nonmonic $g(s) = \sum_{i=0}^{m} a_i s^i$, robust
stability can be ascertained by scaling $f(s)$ and $g(s)$ by the factor
a_m to induce monicity.

The convex direction concept is depicted graphically in Figure 12.2.1. From the figure, it is apparent that $g_1(s)$ is a convex

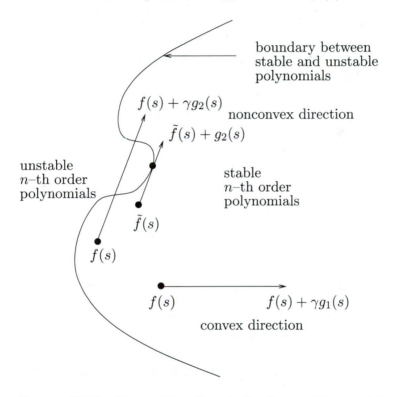

FIGURE 12.2.1   Convex Directions in the Space of Polynomials

direction because $f(s) + \gamma g_1(s)$ remains within the stable set for all
$\gamma \geq 0$ and all starting points $f(s)$. On the other hand, for the case
of $g_2(s)$, $f(s) + \gamma g_2(s)$ can become unstable for a range of $\gamma > 0$.
Hence, $g_2(s)$ cannot be a convex direction because we can find a new
starting point $\tilde{f}(s)$ having the property that $\tilde{f}(s)$ and $\tilde{f}(s) + g_2(s)$
are stable but $\tilde{f}(s) + \lambda g_2(s)$ is unstable for some $\lambda \in (0, 1)$. To
see intuitively how $\tilde{f}(s)$ is obtained, let $[\gamma_1, \gamma_2]$ denote the range of
instability for γ. Notice that if $\gamma_2 < 1$, we can take $\tilde{f}(s) = f(s)$
and induce instability for $\lambda = \gamma_2$. On the other hand, if $\gamma_2 > 1$,

by parallel translation of the original $f(s) + \lambda g(s)$ line, we obtain a new stable starting point $\tilde{f}(s)$ for which $\tilde{f}(s) + g_2(s)$ is stable but $\tilde{f}(s) + \lambda g(s)$ is unstable for some $\lambda \in (0,1)$.

EXERCISE 12.2.8 (First Order Case):   Argue that a monic first order polynomial $g(s)$ defines a convex direction in the space of polynomials of order $n \geq 1$. *Hint*: See the machinery associated with the proof of the Sixteen Plant Theorem in Section 11.7.

EXERCISE 12.2.9 (Second Order Case):   Using the fact that a monic second order polynomial is stable if and only if all coefficients have the same sign, argue that every monic polynomial $g(s) = s^2 + a_1 s + a_0$ defines a convex direction.

12.3   Rantzer's Growth Condition

In the theorem below, a complete characterization of convex directions is provided. We relegate the proof to the next two sections.

THEOREM 12.3.1 (Rantzer (1992a)):   *A polynomial $g(s)$ is a convex direction if and only if the growth condition*

$$\frac{d}{d\omega} \angle g(j\omega) \leq \left| \frac{\sin 2\angle g(j\omega)}{2\omega} \right|$$

is satisfied for all frequencies $\omega > 0$ such that $g(j\omega) \neq 0$.

REMARKS 12.3.2 (Interpretation):   Notice that the theorem excludes consideration of frequencies $\omega \in \mathbf{R}$ for which $g(j\omega) = 0$. The condition $g(j\omega) \neq 0$ also guarantees that the expression for $\angle g(j\omega)$ makes sense; i.e., if $g(j\omega) = 0$, the angle of $g(j\omega)$ is ambiguous.

EXAMPLE 12.3.3 (Checking the Growth Condition):   For the polynomial

$$g(s) = s^4 - 2s^3 - 13s^2 + 14s + 24,$$

we first generate

$$\frac{d \angle g(j\omega)}{d\omega} = \frac{-2\omega^6 - 16\omega^4 - 38\omega^2 + 336}{(\omega^4 + 13\omega^2 + 24)^2 + (2\omega^3 + 14\omega)^2}$$

and

$$\frac{\sin 2\angle g(j\omega)}{2\omega} = \frac{2\omega^6 + 40\omega^4 + 230\omega^2 + 336}{(\omega^4 + 13\omega^2 + 24)^2 + (2\omega^3 + 14\omega)^2}.$$

By plotting the two quantities above for $\omega \geq 0$, it is straightforward to verify that the growth condition is satisfied. By carrying out the plot, it is also easy to see that we eventually encounter a cutoff frequency $\omega_c > 0$ above which

$$\frac{d \; \sphericalangle g(j\omega)}{d\omega} < 0.$$

Hence, the growth condition is automatically satisfied for $\omega > \omega_c$. We can obtain an "estimate" for ω_c by computing the maximum positive root of the numerator $2\omega^6 + 16\omega^4 + 38\omega^2 - 336$ associated with the rate of change of $\sphericalangle g(j\omega)$. We obtain $\omega_c \approx 1.7677$ and can simplify the computation by studying the growth condition for frequencies $\omega \in [0, \omega_c)$.

EXERCISE 12.3.4 (Checking the Growth Condition): Determine if $g(s) = s^5 - s^4 + s^3 + s^2 + s - 1$ is a convex direction.

EXERCISE 12.3.5 (Consistency with Previous Results): The objective of this exercise is to demonstrate that the growth condition is consistent with other results already developed in this text. In each case below, show that $g(s)$ is a convex direction.
(a) Reconciliation with Kharitonov's Theorem: Suppose that $g(s)$ contains either all even powers of s or all odd powers of s.
(b) Reconciliation with Nonincreasing Angle Property: Suppose that $\sphericalangle g(j\omega)$ is nonincreasing.
(c) Reconciliation with Sixteen Plant Theorem: Suppose that $g(s)$ is of the form $g(s) = (s + \alpha)h(s)$ with $h(s)$ containing either all even powers of s or all odd powers of s.

EXERCISE 12.3.6 (Extension of Sixteen Plant Theorem): The objective of this exercise is to extend the domain of applicability of the Sixteen Plant Theorem given in Section 11.4. Indeed, consider an interval plant family $\mathcal{P}$ satisfying the conditions of the theorem with proper compensator

$$C(s) = \frac{N_C(s)}{D_C(s)}$$

which is no longer required to be first order. Instead, assume that $N_C(s)$ and $D_C(s)$ satisfy Rantzer's Growth Condition. Now, prove that $C(s)$ robustly stabilizes $\mathcal{P}$ if and only if $C(s)$ stabilizes each of the sixteen Kharitonov plants. *Hint*: First prove that if $g(s)$ is a convex direction and $h(s)$ is a monic polynomial containing either all even powers of s or all odd powers of s, then

$$\tilde{g}(s) = g(s)h(s)$$

is also a convex direction. Subsequently, in mimicking the proof of the Sixteen Plant Theorem, observe that each of the thirty-two distinguished edges (see Theorem 10.5.1) are expressible in the form

$$p(s, \lambda) = f(s) + \lambda N_C(s)h(s)$$

or

$$p(s, \lambda) = f(s) + \lambda D_C(s)h(s)$$

with $f(s)$ being stable and $h(s)$ as above.

12.4   Machinery for Rantzer's Growth Condition

In this section, we develop some rather technical machinery which facilitates the proof of Theorem 12.3.1. This section and the next can be skipped by the reader interested primarily in application of the result. We first introduce some notation and then provide two lemmas which conveniently describe the rate of change of the angle of a polynomial.

NOTATION 12.4.1 (Angles and Their Derivatives):   To avoid cumbersome notation associated with the calculus of phase derivatives, we adopt the following notational convention: Given a fixed polynomial $p(s)$ and a frequency $\omega \geq 0$ such that $p(j\omega) \neq 0$, we take

$$\theta_p(\omega) = \sphericalangle p(j\omega)$$

and

$$\theta_p'(\omega) = \frac{d}{d\omega} \sphericalangle p(j\omega).$$

Having defined these quantities, we use a compact notion for description of phase derivatives evaluated at a given frequency $\omega_0 > 0$; that is, we take

$$\theta_p'(\omega_0) = \frac{d}{d\omega} \sphericalangle p(j\omega)\Big|_{\omega=\omega_0}$$

to be the evaluation of $\theta_p'(\omega)$ at $\omega = \omega_0$.

LEMMA 12.4.2 (Rate of Change of Angle):   *Given a polynomial $p(s)$ and a frequency $\omega \geq 0$ such that $p(j\omega) \neq 0$, it follows that*

$$\theta_p'(\omega) = Im \left[\frac{p'(j\omega)}{p(j\omega)}\right],$$

where $p'(j\omega)$ denotes the derivative of $p(j\omega)$; i.e.,

$$p'(j\omega) = \frac{d}{d\omega}Re\ p(j\omega) + j\frac{d}{d\omega}Im\ p(j\omega).$$

PROOF: Using the shorthand notation

$$R(\omega) = Re\ p(j\omega)$$

and

$$I(\omega) = Im\ p(j\omega)$$

with associated derivatives $R'(\omega)$ and $I'(\omega)$, respectively, we write

$$p'(j\omega) = R'(\omega) + jI'(\omega).$$

Next, we carry out a straightforward calculation to obtain

$$\theta_p'(\omega) = \frac{R(\omega)I'(\omega) - I(\omega)R'(\omega)}{R^2(\omega) + I^2(\omega)}.$$

Comparing the expressions for $p'(j\omega)$ and $\theta_p'(j\omega)$ above with

$$\frac{p'(j\omega)}{p(j\omega)} = \frac{R'(\omega) + jI'(\omega)}{R(\omega) + jI(\omega)}$$

$$= \frac{R(\omega)R'(\omega) + I(\omega)I'(\omega) + j[R(\omega)I'(\omega) - I(\omega)R'(\omega)]}{R^2(\omega) + I^2(\omega)},$$

it is easy to verify that

$$\theta_p'(j\omega) = Im\left[\frac{p'(j\omega)}{p(j\omega)}\right]. \blacksquare$$

LEMMA 12.4.3 (Angle Formula for Convex Combinations):   *Given two polynomials $f(s)$ and $g(s)$, a scalar $\lambda^* \in (0,1)$ and a frequency $\omega^* > 0$ such that $f(j\omega^*) \neq 0$, $g(j\omega^*) \neq 0$, $f(j\omega^*) + g(j\omega^*) \neq 0$ and $f(j\omega^*) + \lambda^* g(j\omega^*) = 0$, it follows that*

$$\theta_g'(\omega^*) = \lambda^*\theta_f'(\omega^*) + (1 - \lambda^*)\theta_{f+g}'(\omega^*).$$

PROOF: Let $\omega > 0$ be given such that $f(j\omega) \neq 0$, $g(j\omega) \neq 0$ and $f(j\omega) + g(j\omega) \neq 0$. For $\lambda \in (0,1)$, we can express $g'(j\omega)/g(j\omega)$ in the convenient form

$$\frac{g'(j\omega)}{g(j\omega)} = \lambda \frac{f'(j\omega)}{\lambda f(j\omega) - \lambda(f(j\omega) + g(j\omega))}$$

$$+ (1 - \lambda)\frac{f'(j\omega) + g'(j\omega)}{(1 - \lambda)(f(j\omega) + g(j\omega)) - (1 - \lambda)f(j\omega)}.$$

Now, specializing to $\omega = \omega^*$ and $\lambda = \lambda^*$ and invoking Lemma 12.4.2, we obtain

$$\theta'_g(\omega^*) = Im\left[\frac{g'(j\omega)}{g(j\omega^*)}\right]$$

$$= \lambda\left[\frac{f'(j\omega^*)}{f(j\omega^*)}\right] + (1 - \lambda)Im\left[\frac{f'(j\omega^*) + g'(j\omega^*)}{f(j\omega^*) + g(j\omega^*)}\right]$$

$$= \lambda^*\theta'_f(\omega^*) + (1 - \lambda^*)\theta'_{f+g}(\omega^*),$$

which is the stated result of the lemma. ∎

12.4.4   Two Phase Derivative Minimization Problems

We now formulate and solve two minimization problems whose solutions are instrumental to the proof of Theorem 12.3.1. The data describing the first problem consists of an integer $d \geq 1$, an angle $\theta \in [0, 2\pi]$ and a fixed frequency $\omega = \omega_0 > 0$. We seek

$$\mu^* = \inf_{p \in \mathcal{P}} \theta'_p(\omega_0),$$

where the infimum above is taken over the family $\mathcal{P}$ consisting of all stable polynomials $p(s)$ of degree d such that $\sphericalangle p(j\omega_0) = \theta$. For simplicity, we denote this requirement in the sequel by writing

$$\sphericalangle p(j\omega_0) = \theta.$$

We refer to the problem of finding μ^* as the *Phase Derivative Minimization Problem.*

The second problem is the same as the first except for the fact that $\mathcal{P}_R$ consists of all polynomials of the form

$$p(s) = s^k r(s)$$

with $k > 0$ being any positive integer and $r(s)$ being a stable polynomial of degree d. Since the constraint set $\mathcal{P}_R$ is a superset of the constraint set $\mathcal{P}$ used for the first problem, we refer to this second problem as the *Relaxed Phase Derivative Minimization Problem.* We let μ_R^* denote the infimal value for this problem and note that

$$\mu_R^* \leq \mu^*.$$

LEMMA 12.4.5 (Solution for First Order Case):   *For $d = 1$ and $\pi/2 \leq \theta \leq 2\pi$, the Phase Derivative Minimization Problem has no feasible points. For $0 < \theta < \pi/2$, an optimal solution is given by*

$$p^*(s) = K(s + \omega_0 \cot \theta)$$

with $K > 0$ being an arbitrary constant. Furthermore, the minimum value is

$$\mu^* = \frac{\sin 2\theta}{2\omega_0}.$$

PROOF: Since a stable polynomial $p(s)$ has an angle which satisfies $0 < \theta_p(\omega) < \pi/2$ for all $\omega > 0$, the Phase Derivative Minimization Problem has no feasible points for $\pi/2 \leq \theta \leq 2\pi$. We now take $0 < \theta < \pi/2$ and consider candidate polynomials of the form

$$p(s) = K(s + \alpha)$$

with $\alpha > 0$ for stability and $K \neq 0$ to guarantee $d = 1$. In this case, the constraint $\theta_p(\omega_0) = \theta$ forces

$$\alpha = \omega_0 \cot \theta.$$

Now, differentiating the angle of $p^*(j\omega)$, we obtain

$$\mu^* = \theta'_{p*}(\omega_0)$$

$$= \frac{d}{d\omega} \tan^{-1} \left[\frac{\omega}{\omega_0 \cot \theta} \right] \Bigg|_{\omega = \omega_0}$$

Using the differentiation formula

$$\frac{d}{dx} \tan^{-1} u = \frac{1}{1 + u^2} \frac{du}{dx},$$

a straightforward calculation yields

$$\mu^* = \frac{\sin 2\theta}{2\omega_0}. \quad \blacksquare$$

EXERCISE 12.4.6 (Minor Extension):   For $d = 1$, prove that the Relaxed Phase Derivative Minimization Problem has minimum value given by

$$\mu^* = \left| \frac{\sin 2\theta}{2\omega_0} \right|.$$

Describe elements $p^*(s)$ in $\mathcal{P}_R$ which achieve the minimum.

EXERCISE 12.4.7 (Solution for Special Case):   For the special case when θ is an integer multiple of $\pi/2$, show that the infimal value for the Phase Derivative Minimization Problem and its relaxed version is given by

$$\mu^* = \mu_R^* = 0.$$

Hint: Construct a sequence of stable polynomials $\{p_k(s)\}_{k=1}^\infty$ such that $\measuredangle p_k(j\omega_0) = m\pi/2$ and $\theta'_{p_k}(\omega_0) \to 0$.

LEMMA 12.4.8 (Lower Bound for Higher Order Case):   *For arbitrary degree $d > 1$, any admissible polynomial $p \in \mathcal{P}_R$ for the Relaxed Phase Derivative Minimization Problem satisfies*

$$\theta'_p(\omega_0) > \left| \frac{\sin 2\theta}{2\omega_0} \right|.$$

Hence, a lower bound for the infimal values of the Phase Derivative Minimization Problem and its relaxed version is given by

$$\mu^* \geq \mu_R^* \geq \left| \frac{\sin 2\theta}{2\omega_0} \right|,$$

which is never attained.

PROOF: In accordance with the problem formulation, we consider candidate polynomials of the form

$$p(s) = s^k r(s)$$

with $r(s)$ being stable with degree $d \geq 1$. We proceed by induction on the degree of $r(s)$ and use the results in Exercise 12.4.6 for $d = 1$. Therefore, to prove the lemma, we assume that the desired inequality for μ^* holds, but not strictly, for degree $d = n$. We must prove that it holds strictly for degree $d = n + 1$. Indeed, given any admissible $p(s)$ in $\mathcal{P}_R$ associated with $r(s)$ having degree $d = n + 1$, we can write

$$r(s) = (s^2 + \alpha_1 s + \alpha_0)t(s)$$

with $\alpha_1 > 0$, $\alpha_0 > 0$ and $t(s)$ being a stable polynomial of degree $d_t = n - 1$. Now, with

$$p(s) = s^k(s^2 + \alpha_1 s + \alpha_0)t(s),$$

we consider three cases.

Case 1: If $\omega_0^2 < \alpha_0$, we define a polynomial of one lower degree by

$$p_1(s) = s^k(\alpha_1 s + \alpha_0 - \omega_0^2)t(s).$$

It is straightforward to verify that $\theta_{p_1}(\omega_0) = \theta_p(\omega_0)$ and by straightforward differentiation,

$$\theta'_{p_1} = \frac{\alpha_1(\alpha_0 - \omega_0^2)}{(\alpha_0 - \omega_0^2)^2 + \alpha_1^2 \omega_0^2} + \theta'_t(\omega_0).$$

Now, by the inductive hypothesis,

$$\theta'_{p_1}(\omega_0) \geq \left| \frac{\sin 2\theta}{2\omega_0} \right|.$$

Now, to complete the proof for this case, it suffices to show that $\theta'_p(\omega_0) > \theta'_{p_1}(\omega_0)$. To this end, we compare the left and right-hand sides above. Indeed, by straightforward calculation, we arrive at the desired conclusion from the chain of inequalities

$$\theta'_p(\omega_0) = \frac{\alpha_1(\alpha_0 + \omega_0^2)}{(\alpha_0 - \omega_0^2)^2 + \alpha_1^2 \omega_0^2} + \theta'_t(\omega_0)$$

$$> \frac{\alpha_1(\alpha_0 - \omega_0^2)}{(\alpha_0 - \omega_0^2)^2 + \alpha_1^2 \omega_0^2} + \theta'_t(\omega_0)$$

$$= \theta'_{p_1}(\omega_0).$$

Case 2: If $\omega_0^2 > \alpha_0$, we define the polynomial

$$p_2(s) = s^{k+1}\left[(1 - \frac{\alpha_0}{\omega_0^2})s + \alpha_1\right]t(s)$$

and it is straightforward to verify that $\theta_{p_2}(\omega_0) = \theta_p(\omega_0)$. Furthermore, by the inductive hypothesis,

$$\theta'_{p_2}(\omega_0) \geq \left| \frac{\sin 2\theta}{2\omega_0} \right|.$$

Hence, to complete the proof for this case, it suffices to show that $\theta'_p(\omega_0) > \theta'_{p_2}(\omega_0)$. Now, using a chain of inequalities which is nearly identical to the one used in Case 1, we obtain

$$\theta'_p(\omega_0) = \frac{\alpha_1(\alpha_0 + \omega_0^2)}{(\alpha_0 - \omega_0^2)^2 + \alpha_1^2\omega_0^2} + \theta'_t(\omega_0)$$

$$> \frac{\alpha_1(\omega_0^2 - \alpha_0)}{(\alpha_0 - \omega_0^2)^2 + \alpha_1^2\omega_0^2} + \theta'_t(\omega_0)$$

$$= \theta'_{p_2}(\omega_0).$$

Case 3: If $\omega_0^2 = \alpha_0$, we work with the polynomial

$$p_3(s) = \alpha_1 s^{k+1} t(s),$$

and, analogous to Case 2, we have $\theta_{p_3}(\omega_0) = \theta_p(\omega_0)$. Since $p_3(s)$ satisfies the inductive hypothesis, it remains to prove that the inequality $\theta'_p(\omega_0) > \theta'_{p_3}(\omega_0)$ holds. Indeed, using a straightforward calculation, we arrive at the desired conclusion by noting that

$$\theta'_p(\omega_0) = \frac{2}{\alpha_1} + \theta'_t(\omega_0) > \theta'_t(\omega_0) = \theta_{p_3}(\omega_0).$$

This completes the proof of the lemma. ∎

LEMMA 12.4.9 (The Infimal Value):   *The infimal value for the Phase Derivative Minimization Problem and its relaxed version is given by*

$$\mu^* = \mu_R^* = \left| \frac{\sin 2\theta}{2\omega_0} \right|.$$

PROOF: In view of Lemma 12.4.8, it suffices to establish the equality above for μ^*; i.e., the result for μ_R^* then follows automatically. Furthermore, we already know that

$$\mu^* \geq \left| \frac{\sin 2\theta}{2\omega_0} \right|$$

and for $d = 1$, we also know that the desired result holds with equality; see Lemma 12.4.5. To complete the proof, we consider $d \geq 2$ and our objective is to exhibit a sequence of stable polynomials $\{p_k(s)\}_{k=1}^{\infty}$ such that $\theta_{p_k}(\omega_0) = \theta$ for all k and

$$\lim_{k \to \infty} \theta'_{p_k}(\omega_0) = \left| \frac{\sin 2\theta}{2\omega_0} \right|.$$

For simplicity, we break the remainder of the proof into three cases and exclude consideration of $\theta = m\pi/2$ for $m = 0, 1, 2, 3$; see Exercise 12.4.7 for the analysis of this case.

Case 1: If $0 < \theta < \pi/2$, we construct the desired polynomial sequence by exploiting the solution for $d = 1$; that is, with scalar $\alpha = \omega_0 \cot \theta$, we know that $p(s) = s + \alpha$ is a stable polynomial such that $\theta_p(\omega_0) = \theta$ and $\theta'_p(\omega_0) = \mu^*$. Now, taking $\alpha_k = \alpha + 1/k$, we define the sequence members

$$p_k(s) = (s + \alpha_k)(\delta_k s + 1)^{d-1},$$

where $\delta_k > 0$ is selected (as a function of α_k) such that $\theta_{p_k}(\omega_0) = \theta$. Note that the existence of δ_k is guaranteed because the equation

$$\measuredangle(j\omega_0 + \alpha_k) + (d-1)\measuredangle(j\delta_k\omega_0 + 1) = \theta$$

has a solution for $\delta > 0$ suitably small. Now, since $\alpha_k \to \alpha$ as $k \to \infty$, it follows that $\delta_k \to 0$ as $k \to \infty$ to preserve satisfaction of the requirement that $\theta_{p_k}(\omega_0) = \theta$. The proof is now completed by exploiting continuity of the phase derivative; i.e.,

$$\lim_{k \to \infty} \theta'_{p_k}(\omega_0) = \left. \frac{d}{d\omega}(\measuredangle j\omega + \alpha)\right|_{\omega = \omega_0} = \left| \frac{\sin 2\theta}{2a} \right|.$$

Case 2: If $2 \le d \le 4$ and $\pi(d-1)/2 < \theta < \pi d/2$, then with $\alpha_k = \alpha + 1/k$, we define the sequence members

$$p_k(s) = (\alpha_k s + 1)(s + \delta_k)^{d-1},$$

where δ_k is selected (as a function of α_k) such that $\theta_{p_k}(\omega_0) = \theta$. The remainder of the proof now runs along the same lines as in Case 1.

Case 3: If $d \ge 3$, $\pi/2 < \theta < \min\{2\pi, \pi(d-1)/2\}$ and $\theta \ne m\pi/2$ for $m = 2, 3, 4$, we define the sequence members

$$p_k(s) = (s + \alpha)(s + \gamma_k)^\nu(\delta_k s + 1)^{d-\nu-1},$$

where $\alpha = \omega_0|\tan\theta|$, $1 \le \nu \le d - 2$ and $\gamma_k \to 0$ and $\delta_k \to 0$ are selected so that $\theta_{p_k}(\omega_0) = \theta$ for all k. Once again, the proof is completed in the same manner as in Case 1. ∎

12.4.10   Construction of Nonconvex Directions

In this subsection, we provide a result which is used to prove that a convex direction $g(s)$ necessarily satisfies the growth condition.

LEMMA 12.4.11 (Perturbed Polynomial):   *Suppose $g(s)$ is a polynomial of degree $d \geq 2$ such that at frequency $\omega_0 > 0$, $g(j\omega_0) \neq 0$ and*

$$\theta_g'(\omega_0) > \left| \frac{\sin 2\theta_g(\omega_0)}{2\omega_0} \right|.$$

Then, given any $n \geq \max\{4, d\}$, there exists a polynomial $f(s)$ of degree $n - 2$ such that

$$p_\epsilon(s) = (s^2 + \omega_0^2)f(s) + \epsilon g(s)$$

is stable for ϵ suitably small.

PROOF: Considering the Phase Derivative Minimization Problem with $\theta = \theta_g(\omega_0)$, we note that Lemma 12.4.9 indicates that there exists a stable polynomial $f(s)$ of degree $n-2$ such that $\theta_f(\omega_0) = \theta_g(\omega_0)$ or $\theta_f(\omega_0) = -\theta_g(\omega_0)$ and

$$\theta_g'(\omega_0) > \theta_f'(\omega_0) > \left| \frac{\sin 2\theta_g(\omega_0)}{2\omega_0} \right|.$$

Note that if the degree of $g(s)$ is greater than that of $f(s)$, we have $\theta_f(\omega_0) = \theta_g(\omega_0) - \pi$, and if the degree of $f(s)$ is greater than or equal to that of $g(s)$, we have $\theta_f(\omega_0) = \theta_g(\omega_0)$.

   To prove that $f(s)$ satisfies the requirements of the lemma, we use a root locus argument; i.e., we consider a unity feedback system with open loop transfer function

$$P_\epsilon(s) = \frac{\epsilon g(s)}{(s^2 + \omega_0^2)f(s)}.$$

Notice that the root locus with respect to ϵ tells us about the desired behavior of $p_\epsilon(s)$. We first observe that as $\epsilon \to 0$, the root locus branches are close to the open loop poles. Since $f(s)$ is stable, there are two branches near the points $s = \pm j\omega_0$ and the remaining branches are in the strict left half plane. Therefore, to complete the proof, we develop an ϵ-parameterization of the potentially bad pair of conjugate roots emanating from $s = \pm j\omega_0$; by continuous dependence of the roots on ϵ (see Lemma 4.8.2), we are assured that the remaining roots are in the strict left half plane for ϵ suitably small.

   We now consider the mapping $\epsilon \mapsto z(\epsilon)$ defined implicitly by the conditions $z(0) = j\omega_0$ and $p_\epsilon(z(\epsilon)) = 0$. Equivalently, letting

$$h(s) = (s + j\omega_0)f(s),$$

the equation $p_\epsilon(z(\epsilon)) = 0$ is changed to

$$p_\epsilon(z(\epsilon)) = (z(\epsilon) - j\omega_0)h(z(\epsilon)) + \epsilon g(z(\epsilon)) = 0.$$

To complete the proof, it suffices to show that

$$Re \ z'(0) = \frac{d}{d\epsilon} Re \ z(\epsilon)\Big|_{\epsilon=0} = 0$$

and

$$Re \ z''(0) = \frac{d^2}{d\epsilon^2} Re \ z(\epsilon)\Big|_{\epsilon=0} < 0.$$

In other words, the satisfaction of these two conditions guarantees that $z(\epsilon)$ lies in the strict left half plane for ϵ sufficiently small.

To study the behavior of $z(\epsilon)$, we differentiate both sides of the equation $p(z(\epsilon)) = 0$ and obtain

$$z'(\epsilon) = -\frac{g(z(\epsilon))}{h(z(\epsilon)) + (z(\epsilon) - j\omega_0)h'(z(\epsilon)) + \epsilon g'(z(\epsilon))}.$$

Setting $\epsilon = 0$, it follows that

$$z'(0) = -\frac{g(j\omega_0)}{h(j\omega_0)}.$$

Now, to prove that $Re \ z'(0) = 0$, we claim that $z'(0)$ is purely imaginary. To prove this claim, we recall that $\theta_f(\omega_0) = \theta_g(\omega_0)$ or $\theta_f(\omega_0) = \theta_g(\omega_0) - \pi$. Using the formula for $z'(0)$ above, it follows that

$$z'(0) = j\frac{|g(j\omega_0)|}{2\omega_0|f(j\omega_0)|}$$

or

$$z'(0) = -j\frac{|g(j\omega_0)|}{2\omega_0|f(j\omega_0)|}.$$

In either event, $z'(0)$ is purely imaginary.

The next objective is to prove that $Re \ z''(0) < 0$. To this end, a lengthy differentiation of $z'(\epsilon)$ yields the complicated formula

$$z''(\epsilon) = \frac{g(z(\epsilon))[h'(z(\epsilon))z'(\epsilon) + z'(\epsilon)h'(z(\epsilon)) + (z(\epsilon) - j\omega_0)h''(z(\epsilon))z'(\epsilon)]}{[h(z(\epsilon)) + (z(\epsilon) - j\omega_0)h'(z(\epsilon)) + \epsilon g'(z(\epsilon))]^2}$$

$$-\frac{g'(z(\epsilon))z'(\epsilon)[h(z(\epsilon)) + (z(\epsilon) - j\omega_0)h'(z(\epsilon)) + \epsilon g'(z(\epsilon))]}{[h(z(\epsilon)) + (z(\epsilon) - j\omega_0)h'(z(\epsilon)) + \epsilon g'(z(\epsilon))]^2}$$

$$+\frac{g(z(\epsilon))[g'(z(\epsilon)) + \epsilon g''(z(\epsilon))z'(\epsilon)]}{[h(z(\epsilon)) + (z(\epsilon) - j\omega_0)h'(z(\epsilon)) + \epsilon g'(z(\epsilon))]^2}.$$

Setting $\epsilon = 0$ and substituting the expression for $z'(0)$ found above, after some straightforward algebra, we obtain

$$z''(0) = 2j \frac{g^2(j\omega_0)}{h^2(j\omega_0)} \left(\frac{h'(j\omega_0)}{h(j\omega_0)} - \frac{g'(j\omega_0)}{g(j\omega_0)} \right).$$

Taking the real part of $z''(0)$ and invoking Lemma 12.4.2, it now follows that

$$Re \ z''(0) = \frac{2g^2(j\omega_0)}{h^2(j\omega_0)} \left(\theta_g'(\omega_0) - \theta_h'(\omega_0) \right).$$

To complete the proof of the lemma, recall that $g(j\omega_0)/h(j\omega_0)$ is purely imaginary and observe that $\theta_g'(\omega_0) > \theta_h'(\omega_0)$ follows from $\theta_g'(\omega) > \theta_f'(\omega)$ and $\theta_f'(\omega_0) = \theta_h'(\omega)$. Using these facts and the expression for $Re \ z''(0)$ above, it follows immediately that $Re \ z''(0) < 0$. ∎

12.5 Proof of the Theorem

We now proceed to prove Theorem 12.3.1. To establish sufficiency, we assume that the growth condition is satisfied and must prove that $g(s)$ is a convex direction. In view of the fact that all directions are trivially convex for $d = 1$ (Exercise 12.3.5), we assume that $d \geq 2$ and proceed by contradiction. Indeed, if $g(s)$ is not a convex direction, then there exists a stable polynomial $f(s)$ of order $n \geq d$ such that $f(s) + g(s)$ is stable, $p(s, \lambda) = f(s) + \lambda g(s)$ has order n for all $\lambda \in (0, 1)$ and $p(s, \lambda^*)$ is unstable for some $\lambda^* \in (0, 1)$. Applying the Zero Exclusion Condition (see Theorem 7.4.2), there exists some $\lambda_0 \in (0, 1)$ and $\omega_0 > 0$ such that the condition $p(j\omega_0, \lambda_0) = 0$ holds. Hence, with $\Lambda = [0, 1]$, the value set $p(j\omega_0, \Lambda)$ is a straight line segment which includes $z = 0$ but not as an endpoint; see Figure 12.5.1 for the case $0 \leq \theta_g(\omega_0) \leq \pi$. Without loss of generality, we complete the proof for the case $0 \leq \theta_g(\omega_0) \leq \pi$, noting that a nearly identical proof is used if $-\pi < \theta_g(\omega_0) < 0$. From the figure, we see that $\theta_{f+g}(\omega_0) = \theta_g(\omega_0)$ and $\theta_f(\omega_0) = \theta_g(\omega_0) - \pi$.

Applying Lemma 12.4.8 to both $f(s)$ and $f(s) + g(s)$, we obtain

$$\theta_f'(\omega_0) > \left| \frac{\sin 2(\pi - \theta_g(\omega_0))}{2\omega_0} \right| = \left| \frac{\sin 2\theta_g(\omega_0)}{2\omega_0} \right|$$

and

$$\theta_{f+g}'(\omega) > \left| \frac{\sin 2\theta_g(\omega_0)}{2\omega_0} \right|.$$

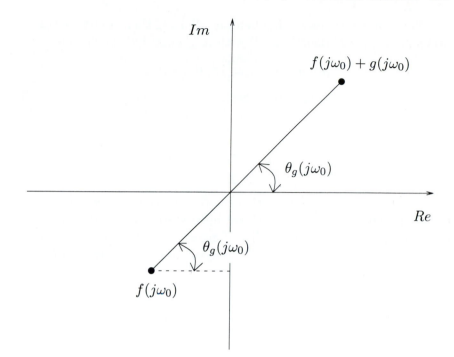

Im

$f(j\omega_0) + g(j\omega_0)$

$\theta_g(j\omega_0)$

Re

$\theta_g(j\omega_0)$

$f(j\omega_0)$

FIGURE 12.5.1   Value Set Geometry for Proof of Sufficiency

However, in view of Lemma 12.4.3, we also know that

$$\theta'_g(\omega_0) = \lambda^* \theta'_f(\omega_0) + (1 - \lambda^*) \theta'_{f+g}(\omega).$$

Combining this equality with the two inequalities above for $\theta'_f(\omega_0)$ and $\theta'_{f+g}(\omega_0)$, it follows that

$$\theta'_g(\omega_0) > \lambda^* \left| \frac{\sin\ 2\theta_g(\omega_0)}{2\omega_0} \right| + (1 - \lambda^*) \left| \frac{\sin\ 2\theta_g(\omega_0)}{\omega_0} \right| = \left| \frac{\sin\ 2\theta_g(\omega_0)}{2\omega_0} \right|.$$

This is the contradiction we seek.

   We now proceed to establish necessity. That is, we assume that $g(s)$ is a convex direction and must prove that the growth condition is satisfied. As in the proof of sufficiency, we take $d = \deg g(s) \geq 2$. Proceeding by contradiction, we assume that

$$\theta'_g(\omega_0) > \left| \frac{\sin\ 2\theta_g(\omega_0)}{2\omega_0} \right|$$

for some $\omega_0 > 0$ such that $\theta_g(\omega_0) \neq 0$.

Now, in accordance with Lemma 12.4.11, there exists a stable polynomial $f(s)$ of degree $n - 2$ with $n \geq \max\{4, d\}$ such that

$$p_\epsilon(s) = (s^2 + \omega_0^2)f(s) + \epsilon g(s)$$

is stable for ϵ sufficiently small. Selecting some fixed $\epsilon > 0$ for which the polynomials $p_\epsilon(s)$ and $p_{-\epsilon}(s)$ are both stable, we define a family of polynomials by

$$p_\epsilon(s, \lambda) = \frac{(s^2 + \omega_0^2)f(s) - \epsilon g(s)}{2\epsilon} + \lambda g(s)$$

and $\lambda \in [0, 1]$. Observe that this family has the property that the two extremes $p(s, 0)$ and $p(s, 1)$ are both stable. Furthermore, with the parameter ϵ reduced further if necessary, the construction of $f(s)$ guarantees that this family of polynomials has invariant degree. Noting, however, that

$$2\epsilon p_\epsilon(s, 0.5) = (s^2 + \omega_0^2)f(s),$$

it follows that $p_\epsilon(s, 0.5)$ is unstable. This contradicts the fact that $g(s)$ is a convex direction. ∎

12.6   Diamond Families: An Illustrative Application

In this section, we demonstrate the power of the growth condition by considering a problem which can be viewed as "dual" to Kharitonov's problem. Motivated by classical duality between ℓ^∞ and ℓ^1 for finite sequences, we endow the space of uncertain parameters with the ℓ^1 norm; i.e., if $q = (q_0, q_1, \ldots, q_n)$, we take

$$\|q\|_1 = \sum_{i=0}^{n} |q_i|.$$

This choice of norm motivates the definition below.

DEFINITION 12.6.1 (Diamond Polynomial Family):   A *diamond polynomial family* $\mathcal{P}$ is described by an uncertain polynomial of the form

$$p(s, q) = q_0 + q_1 s + q_2 s^2 + \cdots + q_{n-1}s^{n-1} + q_n s^n$$

with coefficients q_i known to lie in the $(n+1)$-dimensional diamond with center $q^* = (q_0^*, q_1^*, \ldots, q_n^*)$ and radius $r > 0$; i.e., admissible coefficients $q = (q_0, q_1, \ldots, q_n)$ satisfy

$$|q_0 - q_0^*| + |q_1 - q_1^*| + \cdots + |q_n - q_n^*| \leq r.$$

Letting Q denote this uncertainty bounding set, the resulting polynomial family is $\mathcal{P} = \{p(\cdot, q) : q \in Q\}$.

EXERCISE 12.6.2 (Value Set and Edges):   For the diamond polynomial family $\mathcal{P}$ above, assume that the order n is even and prove that for $\omega > 0$, the value set $p(j\omega, Q)$ is a diamond in the complex plane described by

$$p(j\omega, Q) = \operatorname{conv}\{v_1(\omega), v_2(\omega), v_3(\omega), v_4(\omega)\},$$

where

$$
\begin{aligned}
v_1(\omega) &= p(j\omega, q^*) + r \max\{1, \omega^n\}; \\
v_2(\omega) &= p(j\omega, q^*) - r \max\{1, \omega^n\}; \\
v_3(\omega) &= p(j\omega, q^*) + jr \max\{\omega, \omega^{n-1}\}; \\
v_4(\omega) &= p(j\omega, q^*) - jr \max\{\omega, \omega^{n-1}\}.
\end{aligned}
$$

Now, describe the value set for the case when the order n is odd.

EXERCISE 12.6.3 (Eight Distinguished Edges):   With setup as in Exercise 12.6.2, use the Edge Theorem (see Section 9.3) in conjunction with the value set description for $p(j\omega, Q)$ in the exercise above to prove the result in Tempo (1990); i.e., assume that $|q_n^*| > r$ and argue that $\mathcal{P}$ is robustly stable if and only if the eight edge polynomials

$$
\begin{aligned}
e_1(s, \lambda) &= p(s, q^*) - \lambda r - (1 - \lambda)rs; \\
e_2(s, \lambda) &= p(s, q^*) + \lambda r - (1 - \lambda)rs; \\
e_3(s, \lambda) &= p(s, q^*) + \lambda r + (1 - \lambda)rs; \\
e_4(s, \lambda) &= p(s, q^*) - \lambda r + (1 - \lambda)rs; \\
e_5(s, \lambda) &= p(s, q^*) - \lambda r s^{n-1} - (1 - \lambda)rs^n; \\
e_6(s, \lambda) &= p(s, q^*) + \lambda r s^{n-1} - (1 - \lambda)rs^n; \\
e_7(s, \lambda) &= p(s, q^*) + \lambda r s^{n-1} + (1 - \lambda)rs^n; \\
e_8(s, \lambda) &= p(s, q^*) - \lambda r s^{n-1} + (1 - \lambda)rs^n
\end{aligned}
$$

are stable for all $\lambda \in [0, 1]$.

EXERCISE 12.6.4 (Extreme Point Result):   Using the eight edge polynomials in the exercise above, apply Theorem 12.3.1 and arrive at the result of Barmish, Tempo, Hollot and Kang (1992); i.e., $\mathcal{P}$ is robustly stable if and only if the eight extreme polynomials

$$p_1(s) = p(s, q^*) + r;$$
$$p_2(s) = p(s, q^*) - r;$$
$$p_3(s) = p(s, q^*) + rs;$$
$$p_4(s) = p(s, q^*) - rs;$$
$$p_5(s) = p(s, q^*) + rs^{n-1};$$
$$p_6(s) = p(s, q^*) - rs^{n-1};$$
$$p_7(s) = p(s, q^*) + rs^n;$$
$$p_8(s) = p(s, q^*) - rs^n$$

are stable.

EXERCISE 12.6.5 (Extreme Points for a Diamond Family):   Consider the diamond family $\mathcal{P}$ of the fourth order polynomials described by center $q^* = (q_0^*, q_1^*, q_2^*, q_3^*, q_4^*) = (3.49, 7.98, 6.49, 3.00, 1.00)$ and radius $r = 0.5$. Determine if $\mathcal{P}$ is robustly stable.

EXERCISE 12.6.6 (Robustness Margin):   With the setup as in Exercise 12.6.5, replace the radius $r = 0.5$ by a variable parameter $r \geq 0$. Denoting the resulting family of polynomials by $\mathcal{P}_r$, take r_{max} to be the supremal value of the radius r for which robust stability of $\mathcal{P}_r$ is preserved. Verify that $r_{max} \approx 0.9467$.

12.7   Conclusion

The main objective of this chapter was to enrich the class of polytopes lending themselves to extreme point results for robust stability. This was accomplished via Rantzer's Growth Condition. In the next chapter, we continue to emphasize extreme points. However, instead of concentrating on uncertainty structure, we concentrate on the desired root location region $\mathcal{D}$. The fundamental question is: Under what conditions on $\mathcal{D}$ do we obtain extreme point results for interval polynomial families? We see that a convexity condition involving the reciprocal set $1/\mathcal{D}$ plays an important role.

Notes and Related Literature

NRL 12.1 Most of the technical concepts in this chapter associated with the growth condition in Theorem 12.3.1 are due to Rantzer (1992a).

NRL 12.2 The study of robust stability in the diamond framework was first suggested by Tempo (1990).

NRL 12.3 It is interesting to note that the extreme point result of Exercise 12.6.4 does not hold for *weighted diamonds*. To demonstrate this point, we consider the weighted diamond family of polynomials described as follows: The *center polynomial* is given by $p(s, q^*) = s^3 + 2s^2 + 2.201s + 4$ and the uncertainty bounding set is described by $10|q_3 - q_3^*| + 100|q_2 - q_2^*| + 100|q_1 - q_1^*| + 2.5|q_0 - q_0^*| \leq 1$. It is easily verified that all extreme polynomials $p(s, q^i)$ are stable but the polynomial $p^*(s) = 1.05s^3 + 2s^2 + 2.201s + 4.2$ is an unstable member of this diamond family; see the dissertation by Kang (1992) for further details.

NRL 12.4 Contrary to most results on robust stability, an extreme point result does *not* hold for certain types of diamond families with complex polynomials; i.e., real and complex coefficient families are fundamentally different in this regard. Indeed, we consider $p(s, u, v) = \sum_{i=0}^{n}(u_i + jv_i)s^i$, where $u = (u_0, u_1, \ldots, u_n)$ and $v = (v_0, v_1, \ldots, v_n)$ lie in unit diamonds U and V, respectively. Now, for the second order case with diamond centers given by $u_0^* = -4.3176$, $u_1^* = 0.0111$, $u_2^* = 1.2272$, $v_0^* = 1.8398$, $v_1^* = 15.1285$ and $v_2^* = 6.3118$, it is straightforward to verify that there are at most 36 extreme polynomials which are all stable. However, the polynomial $p^*(s) = p(s, u^*, v^*) + 0.5s - (0.5 + j)$ is a member of the family and has two roots given by $s_1 \approx -2.3249 - j0.0440$ and $s_2 \approx 0.0002 - j0.3271$. Hence, the family is not robustly stable even though all the extremes are stable; see the paper by Barmish, Tempo, Hollot and Kang (1992) for further details.

NRL 12.5 There is also an extension of Rantzer's Growth Condition which is applicable to delay systems. In the paper by Kharitonov and Zhabko (1992), convex combinations of quasipolynomials are considered and a *generalized growth condition* is used to characterize convex directions.

Chapter 13

Schur Stability and Kharitonov Regions

Synopsis

For the problem of robust Schur stability of interval polynomials, there has been considerable attention devoted to the attainment of Kharitonov-like results. Although it can be argued that no result to date compares with Kharitonov's Theorem in its elegance and simplicity, a number of useful robustness criteria have nevertheless emerged. The first part of this chapter covers some developments along these lines and motivates the study of weak Kharitonov regions. These are regions $\mathcal{D}$ in the complex plane for which $\mathcal{D}$-stability of all the extreme polynomials implies robust $\mathcal{D}$-stability.

13.1 Introduction

When studying robust stability of discrete-time systems, the following question immediately comes to mind: Is there a simple and elegant robust Schur stability criterion which might be appropriately called the *discrete-time analogue* of Kharitonov's Theorem? In this regard, we consider the interval polynomial

$$p(z, q) = \sum_{i=0}^{n} [q_i^-, q_i^+] z^i$$

with argument z instead of s to emphasize that the open unit disc is the desired root location $\mathcal{D}$ of concern. As usual, we let $\{q^i\}$ denote

218

the finite set of extreme points (2^{n+1} at most) of the uncertainty bounding set Q and pose a basic question: Can we identify a distinguished subset $\{\bar{q}^i\}$ of $\{q^i\}$ which can be used to guarantee robust Schur stability? That is, is there a subset $\{\bar{q}^i\}$ of $\{q^i\}$ having the property that Schur stability of each $p(z, \bar{q}^i)$ implies that the interval polynomial family $\mathcal{P} = \{p(\cdot, q) : q \in Q\}$ is robustly Schur stable? This chapter begins by answering the most general version of this question in the negative. Subsequently, we proceed to deal with a number of special cases for which a positive answer can be given.

The question about an extreme point solution for the robust Schur stability problem is a special case of a more general question: Given an interval polynomial family $\mathcal{P}$, for what class of $\mathcal{D}$ regions can we establish that $\mathcal{D}$-stability of a subset of the extremes implies robust $\mathcal{D}$-stability? After dealing with the case when $\mathcal{D}$ is the open unit disc, we consider the generalization to large classes of $\mathcal{D}$ regions.

EXAMPLE 13.1.1 (No General Extreme Point Result):   To demonstrate that Kharitonov's Theorem does not generalize to the Schur stability case in the obvious way, we consider the interval polynomial

$$p(z, q) = z^4 + \left[-\frac{17}{8}, \frac{17}{8} \right] z^3 + \frac{3}{2} z^2 - \frac{1}{3}$$

of Bose and Zeheb (1986). For $q = -17/8$, it is easy to verify that the extreme polynomial $p(z, -17/8)$ has four roots $z_{1,2} \approx 0.786 \pm j0.596$, $z_3 \approx 0.924$ and $z_4 \approx -0.371$ which are all interior to the unit disc; a similar conclusion is reached for $p(z, 17/8)$. That is, its four roots are given by $z_{1,2} \approx 0.786 \pm j0.596$, $z_3 \approx -0.924$ and $z_4 \approx 0.371$. However, a final computation reveals that the intermediate polynomial $p(z, 0)$ has roots $z_{1,2} \approx \pm j1.303$, $z_{3,4} \approx \pm 0.4433$ which are not all inside the unit disc.

REMARKS 13.1.2 (Low Order Polynomials):   Later in the chapter, we establish that examples of the sort given above cannot be given for polynomials having order $n \leq 3$; i.e., a discrete-time interval polynomial of order $n \leq 3$ is robustly Schur stable if and only if the extreme polynomials are Schur stable.

13.2   Low Order Coefficient Uncertainty

Although Example 13.1.1 rules out a general Kharitonov-like result for the robust Schur stability problem, there is one special case for which strong results can be given. Namely, when the coefficient of z^i

is fixed for $i > \lfloor \frac{n}{2} \rfloor$, an extreme point result emerges; by $\lfloor \frac{n}{2} \rfloor$ above, we mean the largest integer less than or equal to $n/2$. To simplify the proof of the technical results to follow, without loss of generality, we make a simplifying assumption.

DEFINITION 13.2.1 (Nontriviality Condition): The discrete-time interval polynomial

$$p(z, q) = \sum_{i=0}^{n} [q_i^-, q_i^+] z^i$$

is said to satisfy a *Nontriviality Condition* if the two quantities

$$\mu_0^- = \sum_{i=0}^{n} q_i^-$$

and

$$\mu_0^+ = \sum_{i=0}^{n} q_i^+$$

are either both positive or both negative.

REMARKS 13.2.2 (No Loss of Generality): To see that there is no loss of generality associated with the imposition of the Nontriviality Condition, notice that if μ_0^- and μ_0^+ have opposite signs, it follows that

$$\sum_{i=0}^{n} q_i^* = 0$$

for some q^*. Hence, $p(1, q^*) = 0$, which rules out robust Schur stability; i.e., $z = 1$ is a root of $p(s, q^*)$. Although not needed in the sequel, a similar nontriviality condition can be enforced with respect to $z = -1$. In other words, there would be no loss of generality in assuming that the two quantities

$$\nu_0^- = \sum_{\substack{i=0 \\ i \ even}}^{n} q_i^- + \sum_{i \ odd}^{n} q_i^+$$

and

$$\nu_0^+ = \sum_{\substack{i=0 \\ i \ even}}^{n} q_i^+ + \sum_{i \ odd}^{n} q_i^-$$

are either both positive or both negative.

THEOREM 13.2.3 (Hollot and Bartlett, (1986)):   *Consider the interval polynomial family $\mathcal{P}$ described by $p(z,q) = \sum_{i=0}^{n}[q_i^-, q_i^+]z^i$ with uncertainty bounding set Q having extreme point set $\{q^i\}$. In addition, assume that the Nontriviality Condition is satisfied and $q_i^- = q_i^+$ for $i = \lfloor \frac{n}{2} \rfloor + 1, \lfloor \frac{n}{2} \rfloor + 2, \dots, n$ where $\lfloor \frac{n}{2} \rfloor$ denotes the largest integer less than or equal to $\frac{n}{2}$. Then $\mathcal{P}$ is robustly Schur stable if and only if each of the extreme polynomials $p(z, q^i)$ is Schur stable.*

PROOF: The proof of necessity is trivial because robust Schur stability of $\mathcal{P}$ implies Schur stability of each $p(z, q^i)$; i.e., $p(z, q^i) \in \mathcal{P}$. To establish sufficiency, we use the well-known bilinear transformation to convert a unit disc problem into a left half plane problem. Namely, by defining the uncertain polynomial

$$\tilde{p}(s,q) = (s-1)^n p\left(\frac{s+1}{s-1}, q\right),$$

it follows that the original interval polynomial family $\mathcal{P}$ is robustly Schur stable if and only if the transformed family of polynomials $\tilde{\mathcal{P}} = \{\tilde{p}(\cdot, q) : q \in Q\}$ is robustly stable (strict left half plane). Furthermore, we observe that the transformed family of polynomials $\tilde{\mathcal{P}}$ has invariant degree because the coefficient of s^n is $\sum_{i=0}^{n} q_i$; i.e., the Nontriviality Condition guarantees that this sum cannot vanish.

We now assume that each of the extreme polynomials $\tilde{p}(s, q^i)$ is stable and it must be shown that $\tilde{\mathcal{P}}$ is robustly stable. Proceeding by contradiction, suppose $\tilde{\mathcal{P}}$ is not robustly stable. Since $\tilde{p}(s, q)$ has an affine linear uncertainty structure, the Edge Theorem (see Section 9.3) implies that there exists some $k \leq \lfloor \frac{n}{2} \rfloor$ and extremal settings $q_i = q_i^{\pm} = q_i^-$ or q_i^+ for $i \neq k$ such that the edge polynomial

$$p_k(s, q_k) = q_k(s+1)^k(s-1)^{n-k} + \sum_{i \neq k} q_i^{\pm}(s+1)^i(s-1)^{n-i}$$

is stable for $q_k = q_k^-$ and $q_k = q_k^+$ but unstable for some $q_k \in (q_k^-, q_k^+)$. Now, by the Zero Exclusion Condition (see Theorem 7.4.2), there exists some $q_k^* \in (q_k^-, q_k^+)$ and some $\omega^* \geq 0$ such that $p_k(j\omega^*, q_k^*) = 0$. Equivalently, the origin, $z = 0$, lies in the interior of the straight line segment joining $p_k(j\omega^*, q_k^-)$ and $p_k(j\omega^*, q_k^+)$.

Without loss of generality, we take $0 \leq \measuredangle p(j\omega^*, q_k^+) < \pi/2$ and $\pi < \measuredangle p(j\omega^*, q_k^-) < 3\pi/2$ as indicated in Figure 13.2.1; the argument to follow is not specific to these quadrants. Now, recalling Example 7.2.2, the slope of this value set line segment is

$$m_k(\omega) = \frac{Im\ (j\omega + 1)^k(j\omega - 1)^{n-k}}{Re\ (j\omega + 1)^k(j\omega - 1)^{n-k}}.$$

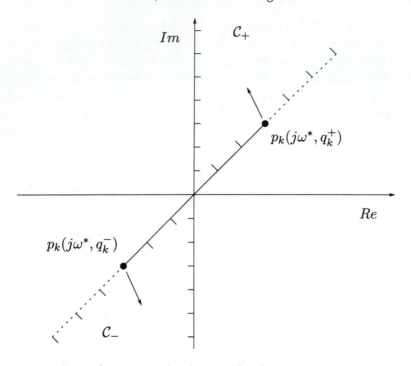

FIGURE 13.2.1   Value Set for the Proof of Theorem 13.2.3

Claim: The slope $m_k(\omega)$ is nonincreasing with respect to $\omega \geq 0$. To prove this claim, we define the complex frequency function

$$z_k(\omega) = (j\omega + 1)^k (j\omega - 1)^{n-k}$$

whose angle is

$$\phi_k(\omega) = k \tan^{-1} \omega + (n - k)(\pi - \tan^{-1} \omega)$$
$$= (2k - n) \tan^{-1} \omega + n\pi.$$

Since

$$m_k(\omega) = \tan \phi_k(\omega)$$

and the tangent function is monotonic, we need only show that

$$\frac{d\phi_k(\omega)}{d\omega} \leq 0.$$

Indeed, differentiating above, we obtain

$$\frac{d\phi_k(\omega)}{d\omega} = \frac{2k - n}{1 + \omega^2}.$$

Now, since $k \leq \lfloor \frac{n}{2} \rfloor$, we conclude that

$$\frac{d\phi_k(\omega)}{d\omega} \leq 0.$$

Hence, the claim is established.

To complete the proof of the theorem, we apply the result in the claim above to the critical frequency $\omega = \omega^*$. To arrive at the desired contradiction, we first exploit stability of $p_k(s, q_k^-)$ and $p_k(s, q_k^+)$. That is, for $\Delta\omega > 0$ sufficiently small, the Monotonic Angle Property (see Lemma 5.7.6) forces $p_k(j(\omega^* + \Delta\omega), q_k^+)$ into the interior of the cone

$$C_+ = \{z : \angle p_k(j\omega^*, q_k^+) < \angle z < \frac{\pi}{2}\}$$

and $p_k(j(\omega^* + \Delta\omega), q_k^-)$ into the interior of the cone

$$C_- = \{z : \angle p_k(j\omega^*, q_k^-) < \angle z < \frac{3\pi}{2}\}.$$

However, these new locations for $p_k(j\omega, q_k^-)$ and $p_k(j\omega, q_k^+)$ are inconsistent with the fact that $m_k(\omega)$ is nonincreasing; i.e., the straight line joining any two points in the interiors of C_- and C_+ has slope greater than $m_k(\omega^*)$. Having arrived at this contradiction, the proof of the theorem is now complete. ∎

13.3   Low Order Polynomials

With Theorem 13.2.3 in hand, we now consider the special case of interval polynomials having order $n \leq 3$. First, we address the simplest cases, $n = 1$ and $n = 2$, in the exercise below.

EXERCISE 13.3.1 (First and Second Order Cases):   For monic interval polynomials of order $n = 1$ and $n = 2$, argue that robust Schur stability is equivalent to Schur stability of the extremes by explicitly displaying the roots.

REMARKS 13.3.2 (Third Order Case):   We now argue that the third order case can be reduced to a case which can be handled by Theorem 13.2.3. Indeed, for the third order interval polynomial

$$p(z, q) = z^3 + [q_2^-, q_2^+]z^2 + [q_1^-, q_1^+]z + [q_0^-, q_0^+]$$

with uncertainty bounding set Q, we can write down conditions for Schur stability expressed parametrically in q; e.g., by applying the

positive innerwise criterion of Jury (1974), robust Schur stability is
guaranteed if and only if the inequalities

$$1 + q_0 + q_1 + q_2 > 0;$$
$$1 - q_2 + q_1 - q_0 > 0;$$
$$1 - q_0^2 - q_1 + q_0 q_2 > 0;$$
$$1 + q_3 > 0;$$
$$1 - q_3 > 0$$

are satisfied.

Now, observe that if q_0 and q_1 are fixed and q_2 is allowed to vary,
the inequalities above are affine linear with respect to $q_2^- \leq q_2 \leq q_2^+$.
Using the basic fact that a linear function $f(q_2)$ over an interval
$[q_2^-, q_2^+]$ remains positive if and only if $f(q_2^-) > 0$ and $f(q_2^+) > 0$, we
arrive at the following point: Robust Schur stability is guaranteed if
and only if $p(z, q)$ is Schur stable for all q of the form $q = (q_0, q_1, q_2^-)$
or $q = (q_0, q_1, q_2^+)$ with $q_0^- \leq q_0 \leq q_0^+$ and $q_1^- \leq q_1 \leq q_1^+$. We
have now reduced the problem to a point which permits application
of Theorem 13.2.3; i.e., since only coefficients of order one and two
are uncertain, we conclude that Schur stability of the extremes is
equivalent to robust Schur stability. There are at most eight such
extremes.

13.4   Weak and Strong Kharitonov Regions

Questions involving the attainment of extreme point results for robust Schur stability of interval polynomials can be phrased in a much
more general context: Given an interval polynomial family $\mathcal{P}$ with
set of extremes $\{p(s, q^i)\}$ and a desired root location region $\mathcal{D}$, under what conditions does $\mathcal{D}$-stability for each of the extremes $p(s, q^i)$
imply robust $\mathcal{D}$-stability of $\mathcal{P}$? To address this question in a precise
manner, we require some preliminaries.

DEFINITION 13.4.1 (Weak and Strong Kharitonov Regions):   An
open set $\mathcal{D} \subseteq \mathbf{C}$ in the complex plane is said to be a *weak Kharitonov
region* if the following condition is satisfied: For any given interval
polynomial family $\mathcal{P} = \{p(\cdot, q) : q \in Q\}$ having invariant degree,
stability of $p(s, q^i)$ for each extreme point q^i of Q implies robust
$\mathcal{D}$-stability of $\mathcal{P}$. We say that $\mathcal{D}$ is a *strong Kharitonov region* if
there exists a finite index set $I(\mathcal{D}) \subseteq \{1, 2, 3, \ldots, N\}$ having the
following property: Given any interval polynomial family $\mathcal{P}$ with
invariant degree, there exists a labeling of the extreme polynomi-

als $\{p(s, q^i)\}$ such that $\mathcal{D}$-stability of $p(s, q^i)$ for $i \in I(\mathcal{D})$ implies robust $\mathcal{D}$-stability of $\mathcal{P}$. The understanding above is that the labeling scheme for the extreme polynomials may depend on the order n of the interval polynomial family, but the cardinality N of $I(\mathcal{D})$ is independent of n, $p(\cdot, q)$ and Q; N only depends on the region $\mathcal{D}$.

REMARKS 13.4.2 (Elaboration): To elaborate on the definition above, observe that the strict left half plane is a strong Kharitonov region because the requirements of Definition 13.4.1 are satisfied by taking $I(\mathcal{D}) = \{1, 2, 3, 4\}$. From the point of view of computational complexity, it is obvious that results involving strong Kharitonov regions are quite powerful when the cardinality of $I(\mathcal{D})$ is small. In contrast, weak Kharitonov regions have the undesirable property that the number of extremes to be tested increases exponentially with respect to the degree n.

13.5   Characterization of Weak Kharitonov Regions

In this section, the main objective is to provide the rather general characterization of weak Kharitonov regions due to Rantzer (1992b).

DEFINITION 13.5.1 (Regularity): A region $\mathcal{D} \subseteq \mathbf{C}$ is said to be *regular* if it is open and simply connected with boundary $\partial \mathcal{D}$, which can be directed in a positively oriented manner with an associated piecewise $\mathbf{C}^2$ boundary sweeping $\Phi_{\mathcal{D}} : I \to \partial \mathcal{D}$; see Section 7.5 for an introduction to boundary sweeping functions. The proof of the theorem below is relegated to the next two sections.

THEOREM 13.5.2 (Rantzer (1992b)): *Suppose $\mathcal{D} \subseteq \mathbf{C}$ is regular. Then $\mathcal{D}$ is a weak Kharitonov region if both $\mathcal{D}$ and its reciprocal,*

$$\frac{1}{\mathcal{D}} = \{z \in \mathbf{C} : zd = 1 \text{ for some } d \in \mathcal{D}\},$$

are convex.

REMARKS 13.5.3 (Real Versus Complex Coefficients): The reader is reminded that the standing assumption in this text is that unless otherwise stated, all polynomials have real coefficients. We emphasize this point because the theorem above provides an example of a result for which the real and complex versions are different. In the real coefficient case, convexity of $\mathcal{D}$ and $1/\mathcal{D}$ is only sufficient whereas in the complex coefficient case, this condition is both necessary and sufficient. Before proceeding toward the proof of the theorem, we demonstrate the power of the convexity condition in two ways. First, we specialize the theorem to recover known results. Second, we demonstrate how the theorem is used to determine interesting weak Kharitonov regions. This is accomplished via a sequence of examples and exercises.

EXAMPLE 13.5.4 (Strict Left Half Plane): If $\mathcal{D}$ is the strict left half plane, then $1/\mathcal{D} = \mathcal{D}$. Hence, both $\mathcal{D}$ and $1/\mathcal{D}$ are convex and we arrive at a conclusion which is consistent (also weaker) than Kharitonov's Theorem. That is, the strict left half plane is a weak Kharitonov region.

EXAMPLE 13.5.5 (Shifted Half Plane): For degree of stability problems, a scalar $\sigma > 0$ is specified and the shifted half plane

$$\mathcal{D} = \{z \in \mathbf{C} : Re\, z < -\sigma\}$$

is the root location region of interest. Now, straightforward calculation indicates that $z \in 1/\mathcal{D}$ if and only if

$$\left| z + \frac{1}{2\sigma} \right| < \frac{1}{2\sigma}.$$

From the description of $1/\mathcal{D}$ above and the sketch in Figure 13.5.1, it is again obvious that both $\mathcal{D}$ and $1/\mathcal{D}$ are convex. We conclude that $\mathcal{D}$ is a weak Kharitonov region.

EXAMPLE 13.5.6 (Unit Disc): For robust Schur stability problems, we take $\mathcal{D}$ to be the interior of the unit disc and obtain

$$\frac{1}{\mathcal{D}} = \{z \in \mathbf{C} : |z| \geq 1\}.$$

The nonconvexity of this region is consistent with the absence of extreme point results in the Schur case; see Section 13.1.

EXAMPLE 13.5.7 (Damping Cone): When both stability and damping is of concern, we take

$$\mathcal{D} = \{z \in \mathbf{C} : \pi - \phi < \measuredangle z < \pi + \phi\},$$

where $\phi \in (0, \pi/2)$ is the so-called damping angle. Since $1/\mathcal{D} = \mathcal{D}$, it follows that both $\mathcal{D}$ and $1/\mathcal{D}$ are convex. Hence, we conclude that $\mathcal{D}$ is a weak Kharitonov region.

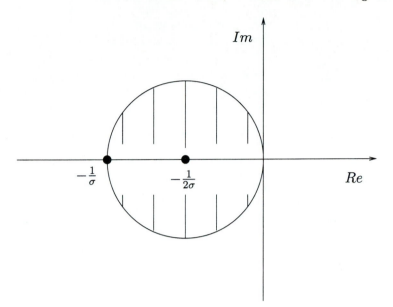

FIGURE 13.5.1   The Region $\mathcal{D}^{-1}$ for Example 13.5.5

EXERCISE 13.5.8 (Shifted Disc):   In the theory of delta transformation, it is important to know whether or not the roots of a given polynomial lie in the interior of a disc $\mathcal{D}$ which is wholly contained in the strict left half plane.
(a) Show that such a set $\mathcal{D}$ is a weak Kharitonov region.
(b) For the more general case when $\mathcal{D}$ is the interior of a disc with arbitrary center and radius, argue that $1/\mathcal{D}$ is convex if and only if the condition

$$0 \notin \text{int } \mathcal{D}$$

is satisfied.

EXERCISE 13.5.9 (Intersection of Weak Kharitonov Regions):   Let $\mathcal{D}_1$ and $\mathcal{D}_2$ be two open convex regions in $\mathbf{C}$ such that $1/\mathcal{D}_i$ is convex for $i = 1, 2$. Letting

$$\mathcal{D} = \mathcal{D}_1 \cap \mathcal{D}_2,$$

show that $1/\mathcal{D}$ must be convex. Use this result to prove that the region $\mathcal{D}$ in Figure 13.5.2 is a weak Kharitonov region. This region reflects concerns for both damping and degree of stability.

EXERCISE 13.5.10 (General Characterization):   If $\mathcal{D} \subseteq \mathbf{C}$ is regular, argue that $\mathcal{D}$ and $1/\mathcal{D}$ are both convex if and only if $\mathcal{D}$ is an

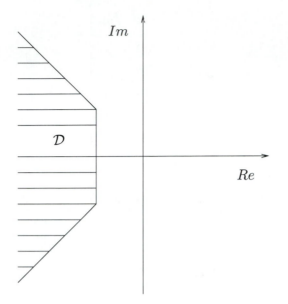

FIGURE 13.5.2 $\mathcal{D}$ Region for Exercise 13.5.9

intersection of open discs and halfplanes which exclude zero; this intersection is not necessarily finite.

13.6   Machinery for Proof of the Theorem

The objective in this section is to develop a number of technical results which are instrumental to the proof of Theorem 13.5.2. Both this section and the next can be skipped by the reader interested primarily in application of the results.

13.6.1   Convention for Functions and Their Derivatives

Suppose that $\mathcal{D} \subseteq \mathbf{C}$ is regular with piecewise C^2 boundary sweeping function $\Phi_{\mathcal{D}} : I \to \partial\mathcal{D}$ and $p(s)$ is a polynomial. Then throughout this section we encounter many complex functions of the generalized frequency $\delta \in I$. If $F : I \to \mathbf{C}$ is a piecewise C^2 complex function of δ, whenever notationally convenient we use $F'(\delta)$ to denote the derivative of $F(\delta)$. In the sequel, $F'(\delta)$ can be taken as either the left or right derivative whenever ambiguity arises. For example, if $\delta = \delta^*$ is a generalized frequency where $F'(\delta)$ does not exist, we can take $F'(\delta^*)$ to be the left derivative

$$F'(\delta^*) = F'_-(\delta^*) = \lim_{h\downarrow 0} \frac{F(\delta^*) - F(\delta^* - h)}{h}$$

or the right derivative

$$F'(\delta^*) = F'_+(\delta^*) = \lim_{h\downarrow 0} \frac{F(\delta^* + h) - F(\delta^*)}{h}.$$

Finally, we draw attention to a shorthand notation for evaluation of derivatives; i.e.,

$$F'(\delta^*) = F'(\delta)|_{\delta=\delta^*}.$$

In some cases, $F(\delta)$ corresponds to the evaluation of a polynomial $p(s)$ for $s = \Phi_{\mathcal{D}}(\delta)$. With $F(\delta) = p(\Phi_{\mathcal{D}}(\delta))$, we use the shorthand notation

$$\theta_p(\delta) = 4p(\Phi_{\mathcal{D}}(\delta)).$$

EXAMPLE 13.6.2 (The Generalized Frequency δ Is an Endpoint): Suppose that $\mathcal{D}$ is the interior of the unit disc and we use the boundary sweeping function $\Phi_{\mathcal{D}}(\delta) = \cos 2\pi\delta + j \sin 2\pi\delta$ with $\delta \in [0, 1]$. Then, for $\delta = 0$, if we use the notation $\Phi'_{\mathcal{D}}(0)$, the understanding is that the right derivative, $\Phi'_{\mathcal{D}}(0) = \Phi'_{\mathcal{D},+}(0) = 2\pi j$ is intended. For $\delta = 1$, if we use the notation $\Phi'_{\mathcal{D}}(1)$, the understanding is that the left derivative $\Phi'_{\mathcal{D}}(1) = \Phi'_{\mathcal{D},-}(1) = 2\pi j$ is intended.

EXAMPLE 13.6.3 (Nondifferentiable Point):   Consider the damping cone

$$\mathcal{D} = \{z \in \mathbf{C} : \frac{3\pi}{4} < 4z < \frac{5\pi}{4}\}$$

with boundary sweeping function

$$\Phi_{\mathcal{D}}(\delta) = \begin{cases} -\delta e^{j\frac{5\pi}{4}} & \text{if } \delta < 0; \\ \delta e^{j\frac{\pi}{4}} & \text{if } \delta \geq 0. \end{cases}$$

Then for $\delta = 0$, we can take $\Phi'_{\mathcal{D}}(0)$ to be either the left derivative,

$$\Phi'_{\mathcal{D}}(0) = \Phi_{\mathcal{D},-}(0) = -e^{j\frac{\pi}{4}},$$

or instead we can use the right derivative,

$$\Phi'_{\mathcal{D}}(0) = \Phi_{\mathcal{D},+}(0) = e^{j\frac{\pi}{4}}.$$

EXERCISE 13.6.4 (Shifted Disc):   Consider the shifted disc

$$\mathcal{D} = \{z \in \mathbf{C} : |z + 1| < 1\}$$

with boundary sweeping function $\Phi_{\mathcal{D}}(\delta) = -1 - \cos\ 2\pi\delta - j\sin\ 2\pi\delta$. With $\theta(\delta) = \angle\Phi_{\mathcal{D}}(\delta)$, describe the one-sided derivatives $\theta'_+(0)$, $\theta'_-(1)$ and the total derivative $\theta'(\delta)$ for $\delta \in (0, 1)$.

REMARKS 13.6.5 (Angle Considerations):   An important concept entering into the proof of Theorem 13.5.2 is a generalization of the Monotonic Angle Property; see Lemma 5.7.6. Indeed, if $\mathcal{D} \subseteq \mathbf{C}$ is regular with boundary sweeping function $\Phi_{\mathcal{D}} : I \to \partial\mathcal{D}$ and $p(s)$ is a $\mathcal{D}$-stable polynomial, we first write

$$p(s) = K \prod_{i=1}^{n}(s - z_i)$$

with $K \in \mathbf{R}$ and $z_i \in \mathcal{D}$. Now, if we evaluate the rate of change of the angle $\theta_p(\delta)$ along $\partial\mathcal{D}$, we obtain

$$\theta'_p(\delta) = \sum_{i=1}^{n} \theta'_i(\delta),$$

where

$$\theta'_i(\delta) = \frac{d\ \angle(\Phi_{\mathcal{D}}(\delta) - z_i)}{d\delta}.$$

With the aid of Figure 13.6.1, it is easy to see that each angle contribution $\theta'_i(\delta)$ above is positive for the case when $\mathcal{D}$ is convex. For nonconvex $\mathcal{D}$, we also see from the figure that it is possible to have $\theta'_i(\delta) < 0$ for some values of δ. The proof of the lemma below (which we omit) amounts to a formalization of these ideas.

LEMMA 13.6.6 (Monotonic Angle Property):   *Suppose that $\mathcal{D} \subset \mathbf{C}$ is regular and convex with boundary sweeping function $\Phi_{\mathcal{D}} : I \to \partial\mathcal{D}$. Then, given any $\mathcal{D}$-stable polynomial $p(s)$, it follows that the angle $\theta_p(\delta) = \angle p(\Phi_{\mathcal{D}}(\delta))$ is a strictly increasing function of $\delta \in I$. That is, $\theta'_p(\delta) > 0$ for all $\delta \in I$.*

EXERCISE 13.6.7 (Converse):   Consider the converse of the lemma above; i.e., suppose that $\mathcal{D} \subseteq \mathbf{C}$ is regular with an associated boundary sweeping function $\Phi_{\mathcal{D}} : I \to \partial\mathcal{D}$ and has the property that $\theta_p(\delta)$

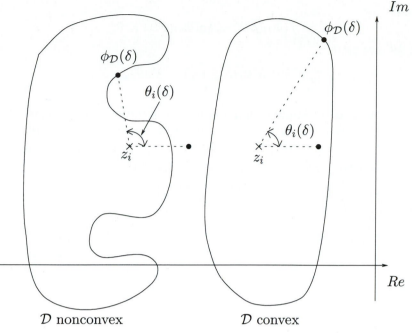

$$\text{FIGURE 13.6.1} \quad \theta_i(\delta) \text{ Increases When } \mathcal{D} \text{ Is Convex}$$

is a strictly increasing function of $\delta \in I$ for every $\mathcal{D}$-stable polynomial $p(s)$. Does it follow that $\mathcal{D}$ is convex? In this regard, is there a difference between real and complex coefficient polynomials?

EXERCISE 13.6.8 (A Mimic: From Stability to $\mathcal{D}$-Stability):   Mimic the proofs of Lemmas 12.4.2 and 12.4.3 to establish the following results under the assumption that $\mathcal{D} \subseteq \mathbf{C}$ is regular with an associated boundary sweeping function $\Phi_{\mathcal{D}} : I \to \partial \mathcal{D}$.
(a) Given any polynomial $p(s)$ and a generalized frequency $\delta \in I$ such that $p(\Phi_{\mathcal{D}}(\delta)) \neq 0$, it follows that

$$\theta_p(\delta) = Im \left[\frac{p'(\Phi_{\mathcal{D}}(\delta))}{p(\Phi_{\mathcal{D}}(\delta))} \Phi'_{\mathcal{D}}(\delta) \right].$$

(b) Given two polynomials $f(s)$ and $g(s)$, a scalar $\lambda^* \in (0,1)$ and a generalized frequency $\delta^* \in I$ such that

$$f(\Phi_{\mathcal{D}}(\delta^*)) \neq 0,$$

$$f(\Phi_{\mathcal{D}}(\delta^*)) + g(\Phi_{\mathcal{D}}(\delta^*)) \neq 0$$

and
$$f(\Phi_\mathcal{D}(\delta^*)) + \lambda g(\Phi_\mathcal{D}(\delta^*)) = 0,$$

it follows that
$$\theta'_g(\delta^*) = \lambda^* \theta'_f(\delta^*) + (1 - \lambda^*)\theta'_{f+g}(\delta^*).$$

LEMMA 13.6.9 (Angle Condition for the Reciprocal): *Suppose that* $\mathcal{D} \subset \mathbf{C}$ *is regular with an associated boundary sweeping function* $\Phi_\mathcal{D} : I \to \partial\mathcal{D}$ *and* $0 \notin \mathcal{D}$. *If* $1/\mathcal{D}$ *is convex, then given any n-th order* $\mathcal{D}$-*stable polynomial* $p(s)$, *it follows that*
$$\theta'_p(\delta) > n\Phi'_\mathcal{D}(\delta)$$

for all $\delta \in I$.

PROOF: We first prove the lemma for all $\mathcal{D}$-stable polynomials of the form $p(s) = s + z$ with $z \in \mathbf{C}$. Indeed, using the boundary sweeping function
$$\tilde{\Phi}_\mathcal{D}(\delta) = \frac{1}{\Phi_\mathcal{D}(\delta)}$$

for $1/\mathcal{D}$, since
$$\tilde{p} = s + \frac{1}{z}$$

is $1/\mathcal{D}$-stable, Lemma 13.6.6 guarantees that for
$$\theta'_{\tilde{p}}(\delta) > 0$$

for all $\delta \in I$. Using this inequality and the expression for $\theta'_{\tilde{p}(\delta)}$ in Exercise 13.6.8, we obtain the chain of inequalities

$$0 < Im \left[\frac{\tilde{p}'(\tilde{\Phi}_\mathcal{D}(\delta))}{\tilde{p}(\tilde{\Phi}_\mathcal{D}(\delta))} \tilde{\Phi}'_\mathcal{D}(\delta) \right]$$

$$= Im \left[\frac{1}{\tilde{\Phi}_\mathcal{D}(\delta) + \frac{1}{z}} \tilde{\Phi}'_\mathcal{D}(\delta) \right]$$

$$= Im \left[\frac{z\Phi_\mathcal{D}(\delta)}{\Phi_\mathcal{D}(\delta) + z} \times \left(-\frac{\Phi'_\mathcal{D}(\delta)}{\Phi^2_\mathcal{D}(\delta)} \right) \right]$$

$$= Im \left[\frac{-z\Phi'_\mathcal{D}(\delta)}{\Phi_\mathcal{D}(\delta)(\Phi_\mathcal{D}(\delta) + \alpha)} \right]$$

$$= Im \left[\frac{\Phi'_\mathcal{D}(\delta)}{\Phi_\mathcal{D}(\delta) + z} \right] - Im \left[\frac{\Phi'_\mathcal{D}(\delta)}{\Phi_\mathcal{D}(\delta)} \right]$$

$$= \theta'_p(\delta) - \Phi'_\mathcal{D}(\delta).$$

Hence,

$$\theta_p'(\delta) > \Phi_{\mathcal{D}}'(\delta).$$

This completes the proof for the case $n = 1$.

Now, to establish the desired inequality for polynomials of arbitrary degree $n > 1$, we need only sum the contributions of each root of $p(s)$. That is, if

$$p(s) = K(s - z_1)(s - z_2) \cdots (s - z_n)$$

for $K \in \mathbf{R}$ and $z_i \in \mathcal{D}$ for $i = 1, 2, \ldots, n$, then

$$\theta_p'(\delta) = \sum_{i=1}^{n} \frac{d \angle (\Phi_{\mathcal{D}}(\delta) - z_i)}{d\delta} > n\Phi_{\mathcal{D}}'(\delta). \quad \blacksquare$$

13.7   Proof of the Theorem

To simplify the proof of Theorem 13.5.2, we assume that $\partial \mathcal{D}$ is C^2 rather than piecewise C^2 and note that the arguments to follow can be readily modified using left and right derivatives whenever appropriate. Proceeding by contradiction, we assume that both $\mathcal{D}$ and $1/\mathcal{D}$ are convex but $\mathcal{D}$ is not a weak Kharitonov region. By the Edge Theorem (see Section 9.4), there exist a pair of $\mathcal{D}$-stable n-th order polynomials $p_1(s)$ and $p_2(s)$, corresponding to the extremes of some n-th order interval polynomial family $\mathcal{P}$, such that

$$p_1(s) - p_2(s) = Ks^k$$

for some $K \in \mathbf{R}$ and some nonnegative integer $k \leq n$. Moreover, the polynomial defined by

$$p(s, \lambda) = p_1(s) + \lambda[p_2(s) - p_1(s)]$$

is not $\mathcal{D}$-stable for some $\lambda \in (0, 1)$. By the Zero Exclusion Condition (see Theorem 7.3.3), there must exist some $\lambda^* \in (0, 1)$ and $\delta^* \in I$ such that

$$p(\Phi_{\mathcal{D}}(\delta^*), \lambda^*) = 0.$$

That is,

$$p_1(\Phi_{\mathcal{D}}(\delta^*)) + \lambda^*[p_2(\Phi_{\mathcal{D}}(\delta^*)) - p_1(\Phi_{\mathcal{D}}(\delta^*))] = 0.$$

Now, using the result of Exercise 13.6.8 and the known form for $p_2(s) - p_1(s)$, it follows that

$$\lambda^* \theta'_{p_1}(\delta^*)\Big|_{\delta=\delta_*} + (1-\lambda^*)\theta_{p_2}(\delta^*)\big|_{\delta=\delta^*} = \frac{d\, K\Phi^k_{\mathcal{D}}(\delta)}{d\,\delta}\bigg|_{\delta=\delta^*} = k\Phi'_{\mathcal{D}}(\delta^*).$$

In view of the convex combination on the left-hand side above, there are two possibilities: Either

$$\theta'_{p_1}(\delta^*) \leq k\Phi'_{\mathcal{D}}(\delta^*)$$

or

$$\theta'_{p_2}(\delta^*) \leq k\Phi'_{\mathcal{D}}(\delta^*).$$

Without loss of generality, we assume that the first of the two inequalities above holds for $p_1(s)$. Now, since $p_1(s)$ is $\mathcal{D}$-stable, we see that the inequality $\theta'_{p_1}(\delta^*) \leq k\Phi'_{\mathcal{D}}(\delta^*)$ contradicts the requirement of Lemma 13.6.9 that

$$\theta_p(\delta) > n\Phi'_{\mathcal{D}}(\delta)$$

for every $\mathcal{D}$-stable polynomial $p(s)$. ∎

13.8   Conclusion

This chapter and its seven predecessors can be viewed as a sequential development of results emanating directly from Kharitonov's Theorem in Chapter 5. In all cases, we heavily exploited the affine linear and independent uncertainty structures in order to obtain either extreme point results or edge results. In the next chapter, where more complicated multilinear uncertainty structures are considered, we see a marked departure from the framework of this chapter. Nevertheless, results developed thus far prove to be useful in a certain "convex hull" context. That is, if Q is a box and $\mathcal{P} = \{p(\cdot, q) : q \in Q\}$ is a family of polynomials whose coefficients depend multilinearly on q, we obtain a simple description of the convex hull of $\mathcal{P}$; i.e., if $a(q)$ is the coefficient vector and $\{q^i\}$ is the set of extreme points of Q, we see that

$$\text{conv } a(Q) = \text{conv}\{a(q^i)\}.$$

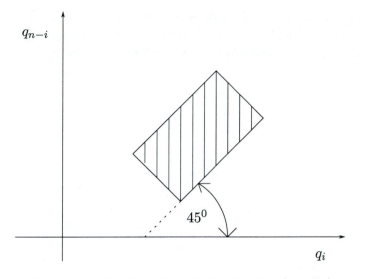

q_{n-i}

45^0

q_i

FIGURE 13.8.1   Bounding Set for the Pair (q_i, q_{n-i})

Notes and Related Literature

NRL 13.1 For more details on extreme point results for robust Schur stability of lower order polynomials, see Cieslik (1987).

NRL 13.2 The ideas introduced in Section 6.2 are further embellished in Kraus, Anderson, Jury and Mansour (1988). For example, for $n = 3$, not all eight extremes need to be tested. Attention can be restricted to a subset of extreme points corresponding to "critical" constraints.

NRL 13.3 In Kraus, Anderson and Mansour (1988), a new uncertainty model is considered in a robust Schur stability context. These authors begin with an uncertain polynomial $p(z, q) = \sum_{i=0}^{n} q_i z^i$ having independent uncertainty structure but dispose with the box bound Q for q. Instead, they consider an uncertainty bound of the sort shown Figure 13.8.1 for pairs of coefficients (q_{n-i}, q_i). To illustrate, for the case when n is even and $i = n/2$, an interval $[q_{\frac{n}{2}}-, q_{\frac{n}{2}}+]$ is taken as a bound for $q_{\frac{n}{2}}$. When the uncertain parameters are bounded in the manner above, we say that Q is a *product of 45° rotated rectangles* and let $\{q^i\}$ denote the extreme points of Q. With this new setup, the resulting family of polynomials $\mathcal{P}$ is robustly Schur stable if and only if the polynomial $p(z, q^i)$ is Schur stable for each extreme point q^i of Q.

NRL 13.4 In a follow-up paper by Mansour, Kraus and Anderson (1988), even stronger results than those described in the note above are established. Under the same hypothesis, they show that a "small" subset of the extreme polynomials need only be considered. Furthermore, working with Chebyshev and Jacobi polynomials, they develop a recipe for selection of these distinguished extremes. The reduction in the number of extremes can be quite dramatic.

NRL 13.5 In the paper by Perez, Decampo and Abdallah (1992), a novel technical method involving barycentric coordinates leads to extreme point results for classes of rotated rectangles with angles other than 45^0. For example, results are given for angular rotations $\phi \in [\pi/4, 3\pi/4]$.

NRL 13.6 A condition for robust aperiodicity of a discrete-time interval polynomial is given in Soh (1986); i.e., we say that a discrete-time interval polynomial family $\mathcal{P} = \{p(\cdot, q) : q \in Q\}$ is *robustly Schur aperiodic* if the following condition holds: Given any $q \in Q$, all roots of $p(z, q)$ are real and positive and lie in $[0, 1)$. Subsequently, it is shown that robust Schur aperiodicity can be ascertained by testing only two distinguished extreme polynomials.

NRL 13.7 For the robust Schur stability problem, a number of authors opt for sufficient conditions in lieu of the combinatorics associated with the Edge Theorem; e.g., see Bose, Jury and Zeheb (1986) where a set of inequalities is given and Vaidyanathan (1990) and Bartlett and Hollot (1988) where transformations of the original polynomial are used.

Chapter 14

Multilinear Uncertainty Structures

Synopsis

The focal point of this chapter is robust $\mathcal{D}$-stability of systems with multilinear uncertainty structures; i.e., we consider uncertain polynomials whose coefficients depend multilinearly on the vector of uncertain parameters q. Using the Mapping Theorem, we can often obtain the tightest possible polytopic overbound for the value sets and coefficient sets of interest.

14.1 Introduction

The basic motivation for this chapter is the following fact: The uncertainty structures which arise in typical applications are more complicated than those which we have analyzed in the polytopic framework of Chapters 8–13. It is easy to describe applications involving highly nonlinear dependence of various system coefficients on the vector q of uncertain parameters. For example, recall the case study involving the Fiat Dedra engine in Chapter 3.

When dealing with complicated uncertainty structures in a robustness context, there are various avenues of attack. In some cases, the number of uncertain parameters is small and no formal theory is required—a practical solution is obtainable via some sort of gridding of the uncertainty bounding set Q. For more formal robustness analyses, we mention some alternatives; to some extent, the discussion below amounts to a review of points raised in Section 1.7.

14.2   More Complicated Uncertainty Structures

In some cases, a family of polynomials or rational functions having a complicated uncertainty structure can be overbounded by a family having a simpler structure for which analytical tools are readily available. For example, in Sections 5.2 and 8.2, it was demonstrated that families of polynomials with complicated dependence on q can often be overbounded by interval polynomials or polytopes of polynomials. Of course, the overbounding process introduces conservatism which may or may not be tolerable. The issue of overbounding prevails in much of the robustness literature; it is not specific to the new tools in this book.

A second approach to dealing with nonlinear uncertainty structures involves reformulation of robustness problems within the framework of mathematical programming. In other words, one massages a robustness problem into an equivalent optimization problem whose solution can then be obtained using a wide variety of software tools. In some cases, the feedback control configuration induces special properties on the resulting mathematical program which can be exploited in a computational algorithm. On the positive side, reformulation of a robustness problem as a mathematical program makes it possible to deal with rather general situations. Complicated uncertainty structures and sophisticated performance specifications can be handled. On the negative side, one must deal with a host of issues involving local versus global minima and computational complexity. Another negative is that a mathematical programming method generally tells us very little about how the solution depends on adjustable parameters such as the compensator gains.

Given any uncertain polynomial or rational function whose coefficients depend nonlinearly on uncertain parameters, the attainment of analytical results for robust stability is generally accomplished at the expense of restricting the class of nonlinearities under consideration. This is the line of attack taken in this chapter. In the next section, we define both multilinear and polynomic uncertainty structures. Subsequently, we describe a transformation relating the polynomic case to the multilinear case.

14.3   Multilinear and Polynomic Uncertainty

In this section, our objectives are twofold. First, we formally define a multilinear uncertainty structure. Second, we demonstrate how multilinear uncertainty structures arise via examples and exercises.

DEFINITION 14.3.1 (Multilinear and Polynomic Uncertainty Structures): An uncertain polynomial $p(s,q) = \sum_{i=0}^{n} a_i(q)s^i$ is said to have a *multilinear uncertainty structure* if each of the coefficient functions $a_i(q)$ is multilinear. That is, if all but one component of the vector q is fixed, then $a_i(q)$ is affine linear in the remaining component of q. More generally, $p(s,q)$ is said to have a *polynomic uncertainty structure* if each of the coefficient functions $a_i(q)$ is a multivariable polynomial in the components of q.

REMARKS 14.3.2 (Multilinear Functions): Although we defined multilinearity in the context of uncertain polynomials above, in the sequel, we have occasion to work with a more general multilinear function $f : \mathbf{R}^n \to \mathbf{R}^k$. When we call f multilinear, the understanding is that each *component function* $f_i : \mathbf{R}^n \to \mathbf{R}$ is multilinear in the sense of the definition above.

EXAMPLE 14.3.3 (Multilinear Uncertainty Structure): The uncertain polynomial

$$p(s,q) = s^3 + (6q_1q_2q_3 + 4q_2q_3 - 5q_1 + 4)s^2$$
$$+ (4q_1q_3 - 6q_1q_2 + q_3)s + (5q_1 - q_2 + 5)$$

has a multilinear uncertainty structure. If the coefficient of s is changed to $a_1(q) = 4q_1 - 6q_1q_2 - q_3^2$, then $p(s,q)$ has a polynomic uncertainty structure.

REMARKS 14.3.4 (Hierarchy): Thus far, we have defined four different types of uncertainty structures for polynomials. Letting $\mathcal{P}_{indep}$, $\mathcal{P}_{aff}$, $\mathcal{P}_{multilin}$ and $\mathcal{P}_{poly}$ denote the set of uncertain polynomials $p(s,q)$ with independent, affine linear, multilinear, and polynomic uncertainty structures, respectively, we draw attention to the obvious inclusion

$$\mathcal{P}_{indep} \subset \mathcal{P}_{aff} \subset \mathcal{P}_{multilin} \subset \mathcal{P}_{poly}.$$

EXERCISE 14.3.5 (Matrix with Independent Uncertainties): Suppose that $A(q)$ is an uncertain $n \times n$ matrix which can be expressed in the form

$$A(q) = \sum_{i=1}^{\ell} A_i q_i,$$

where each A_i is a fixed $n \times n$ matrix having one nonzero entry. Furthermore, assume that if $i_1 \neq i_2$, then the nonzero entries of A_{i_1}

and A_{i_2} are located in different positions; e.g., if the $(2,2)$ entry of A_1 is nonzero, then the $(2,2)$ entry of A_2 must be zero. This type of uncertainty structure is synonymous with the well-known class of interval matrices. Prove that the uncertain characteristic polynomial

$$p(s,q) = \det(sI - A(q))$$

has a multilinear uncertainty structure.

EXAMPLE 14.3.6 (Property of Multilinear Uncertainty Structure): Suppose that an uncertain polynomial has coefficient of s^2 given by $a_2(q) = 6q_1q_2q_3 + 3q_1q_2 + 4q_2q_3 + 5q_1 + 5q_2 + 4$. If q_2 and q_3 are fixed, we can isolate an affine linear function of q_1; i.e.,

$$a_2(q) = q_1 f_{2,1}(q_2, q_3) + g_{2,1}(q_2, q_3),$$

where $f_{2,1}(q_2, q_3) = 6q_2q_3 + 3q_2 + 5$ and $g_{2,1}(q_2, q_3) = 4q_2q_3 + 5q_2 + 4$. Similarly, if q_1 and q_3 are fixed, we can isolate q_2; i.e.,

$$a_2(q) = q_2 f_{2,2}(q_1, q_3) + g_{2,2}(q_1, q_3),$$

where $f_{2,2}(q_1, q_3) = 6q_1q_3 + 3q_1 + 4q_3 + 5$ and $g_{2,2}(q_1, q_3) = 5q_1 + 4$. Finally, for q_1 and q_2 fixed, we can isolate q_3; i.e.,

$$a_2(q) = q_3 f_{2,3}(q_1, q_2) + g_{2,3}(q_1, q_2)$$

with $f_{2,3}(q_1, q_2) = 6q_1q_2 + 4q_2$; $g_{2,3}(q_1, q_2) = 3q_1q_2 + 5q_1 + 5q_2 + 4$.

EXERCISE 14.3.7 (Generalization):   Let $p(s,q) = \sum_{i=0}^{n} a_i(q)s^i$ be an uncertain polynomial with multilinear uncertainty structure. For each component q_k of q, define $q^{\neq k}$ to be the vector q with component q_k deleted. Now, prove that for each coefficient function $a_i(q)$, there exist multilinear functions $f_{i,1}(q^{\neq k})$ and $f_{i,2}(q^{\neq k})$ such that

$$a_i(q) = q_k f_{i,1}(q^{\neq k}) + f_{i,2}(q^{\neq k}).$$

REMARKS 14.3.8 (Transformation):   Further motivation for the study of multilinear uncertainty structures is provided by the lemma below. We see that a rather general class of problems with polynomic uncertainty structure and polytopic uncertainty bounds can be transformed into problems with multilinear uncertainty structure and polytopic uncertainty bounds.

LEMMA 14.3.9 (Sideris and Sanchez Pena (1989)): *Consider the family of polynomials* $\mathcal{P} = \{p(\cdot, q) : q \in Q\}$ *with* $p(s, q)$ *having polynomic uncertainty structure and uncertainty bounding set* Q *which is a polytope. Then there exists a second family of polynomials* $\tilde{\mathcal{P}} = \{\tilde{p}(\cdot, \tilde{q}) : \tilde{q} \in \tilde{Q}\}$ *such that* $\tilde{p}(s, \tilde{q})$ *has multilinear uncertainty structure,* $\tilde{Q}$ *is a polytope and*

$$\tilde{\mathcal{P}} = \mathcal{P}.$$

PROOF: Let $q_i^{k_i}$ denote the highest power of q_i which appears in the coefficient functions. Then we make the substitution

$$q_i^k \mapsto \prod_{j=1}^{k} \tilde{q}_{i,j}$$

in $p(s, q)$ where $\tilde{q}_{i,1}, \tilde{q}_{i,2}, \ldots, \tilde{q}_{i,k_i}$ are new variables which comprise the new set of uncertain parameters

$$\tilde{q} = (q_{1,1}, q_{1,2}, \ldots, q_{1,k_1}, q_{2,1}, q_{2,2}, \ldots, q_{2,k_2}, \ldots, q_{\ell,k_\ell})$$

that replaces the original $q \in \mathbf{R}^\ell$ in $p(s, q)$. Now, we take $\tilde{p}(s, \tilde{q})$ to be the new uncertain polynomial obtained via the substitutions above. Notice that $\tilde{q} \in \mathbf{R}^N$, where

$$N = \sum_{i=0}^{\ell} k_i$$

and, moreover, $\tilde{p}(s, \tilde{q})$ has a multilinear uncertainty structure.

To complete the proof, we now define the uncertainty bounding set $\tilde{Q}$ for $\tilde{q}$. Indeed, for $i \leq \ell$, let

$$\tilde{Q}_i = \{\tilde{q} \in \mathbf{R}^N : q_{i,1} = q_{i,2} = \cdots = q_{i,k_i}\}$$

and take

$$\tilde{Q}_0 = \{\tilde{q} \in \mathbf{R}^N : (q_{1,1}, q_{2,1}, \ldots, q_{\ell,1}) \in Q\}.$$

Now, with

$$\tilde{Q} = \bigcap_{i=0}^{\ell} \tilde{Q}_i,$$

we obtain the desired family $\tilde{\mathcal{P}} = \{\tilde{p}(\cdot, \tilde{q}) : \tilde{q} \in \tilde{Q}\}$. By construction, it follows that $\tilde{\mathcal{P}} = \mathcal{P}$. Furthermore, since $\tilde{Q}_0$ is a polytope and each $\tilde{Q}_i$ is a linear variety, it follows that $\tilde{Q}$ is a polytope; i.e., the intersection of a polytope with a finite collection of linear varieties is a polytope. ∎

EXAMPLE 14.3.10 (From Polynomic to Multilinear):   To illustrate the transformation associated with the lemma above, we begin with the uncertain polynomial

$$p(s, q) = s^3 + (3q_1^3 + q_1^2 q_2 + q_1 q_2 + 3q_1 + 10)s^2$$
$$+ (4q_1^2 + q_2^2 + 15)s + (6q_1 q_2 + 17)$$

with uncertainty bounds $-1 \leq q_1 \leq 1$ and $-2 \leq q_2 \leq 2$. We now "expand" the uncertainty space by defining new variables according to the substitutions $q_1^3 \mapsto \tilde{q}_1 \tilde{q}_2 \tilde{q}_3$ and $q_2^2 \mapsto \tilde{q}_4 \tilde{q}_5$. This leads to a new family of polynomials $\tilde{\mathcal{P}}$ described by $\tilde{q} \in \mathbf{R}^5$, uncertain polynomial with multilinear uncertainty structure given by

$$\tilde{p}(s, \tilde{q}) = s^3 + (3\tilde{q}_1 \tilde{q}_2 \tilde{q}_3 + \tilde{q}_1 \tilde{q}_2 \tilde{q}_4 + \tilde{q}_1 \tilde{q}_4 + 3\tilde{q}_1 + 10)s^2$$
$$+ (4\tilde{q}_1 \tilde{q}_2 + \tilde{q}_4 \tilde{q}_5 + 15)s + (6\tilde{q}_1 \tilde{q}_4 + 17)$$

and polytopic uncertainty bounding set $\tilde{Q}$ described by $-1 \leq \tilde{q}_1 \leq 1$, $\tilde{q}_1 = \tilde{q}_2 = \tilde{q}_3$, $-2 \leq \tilde{q}_4 \leq 2$ and $\tilde{q}_4 = \tilde{q}_5$.

14.4   Interval Matrix Family

Robust stability analysis in a state space setting provides strong motivation for this section. We concentrate on an $n \times n$ uncertain matrix $A(q)$ with independent uncertainty structure (in the sense of matrices) and uncertainty bounding set Q which is a box; the associated characteristic polynomial is

$$p(s, q) = \det (sI - A(q)).$$

DEFINITION 14.4.1 (Interval Matrix Family):   A family of $n \times n$ matrices $\mathcal{A} = \{A(q) : q \in Q\}$ is said to be an *interval matrix family* if Q is a box, the entries $a_{ij}(q)$ depend continuously on q and each component q_i of q enters into only one entry $a_{ij}(q)$ of $A(q)$.

NOTATION 14.4.2 (Lumping and Extreme Matrices):   Analogous to the study of interval polynomials in Chapter 5, we can lump uncertainties within any entry of $A(q)$. It is convenient to view q as a vector in a euclidean space of dimension n^2 or less and create a correspondence between components of q and entries $a_{ij}(q)$ of $A(q)$; i.e., if $a_{ij}(q)$ is nonconstant with respect to q, we write

$$a_{ij}(q) = q_{ij}$$

and consider uncertainty bounds

$$q_{ij}^- \leq q_{ij} \leq q_{ij}^+.$$

Analogous to the case of interval polynomials, we adopt a shorthand notation: We use a matrix whose entries are intervals to describe $\mathcal{A}$. For example, if $n = 2$, we write

$$A(q) = \begin{bmatrix} [q_{11}^-, q_{11}^+] & [q_{12}^-, q_{12}^+] \\ [q_{21}^-, q_{21}^+] & [q_{22}^-, q_{22}^+] \end{bmatrix}.$$

In this context, we call $A(q)$ an *interval matrix*. Since the uncertainty bounding set Q is a box, we can view the family of matrices $\mathcal{A}$ as a box in the euclidean space of at most dimension n^2. The i-th extreme point q^i of Q induces an *extreme matrix* $A(q^i)$ for $\mathcal{A}$.

EXAMPLE 14.4.3 (Some Entries Fixed):   Note that the setup above permits a subset of the entries of $A(q)$ to be fixed. For example, we can consider the interval matrix

$$A(q) = \begin{bmatrix} 1 & [2,3] \\ [4,5] & 6 \end{bmatrix}$$

with $q = (q_{11}, q_{22}) \in \mathbf{R}^2$.

DEFINITION 14.4.4 (Robust $\mathcal{D}$-Stability):   If $\mathcal{D} \subseteq \mathbf{C}$ is open, the family of $n \times n$ matrices $\mathcal{A} = \{A(q) : q \in Q\}$ is said to be *robustly $\mathcal{D}$-stable* if the associated polynomial family induced via the relationship

$$p(s, q) = \det (sI - A(q))$$

is robustly $\mathcal{D}$-stable. Analogous to the case of polynomials, the term *robust stability* is reserved for the case when $\mathcal{D}$ is the strict left half plane and the term *robust Schur stability* is reserved for the case when $\mathcal{D}$ is the interior of the unit disc.

EXERCISE 14.4.5 (Multilinear Uncertainty Structure):   Specialize the result of Exercise 14.3.5 to the interval matrix case; i.e., argue that the uncertain characteristic polynomials associated with an interval matrix family has a multilinear uncertainty structure.

EXERCISE 14.4.6 (Homogeneity Property):   If $p(s,q)$ is the uncertain characteristic polynomial associated with an $n \times n$ interval matrix family, show that each (multilinear) product of the q_{ij} has at most degree n. For example, in the 3×3 case, a fourth order term such as $q_{11}q_{12}q_{22}q_{23}$ cannot arise in the coefficients because this term involves a product of four q_{ij}. For the case when all entries of $A(q)$ are uncertain, argue that each term entering the coefficients involves exactly n uncertainties forming a multilinear product.

14.5   Lack of Extreme Point and Edge Results

In view of the emphasis on extreme point and edge results in the preceding chapters, it is natural to ask whether similar results are available for more general multilinear uncertainty structures. We see below that the answer is no. In order to obtain such results, rather strong additional conditions must be imposed on the multilinearity.

EXERCISE 14.5.1 (Lack of Extreme Point and Edge Results):   We consider the family of polynomials of Barmish, Fu and Saleh (1988) described by

$$
\begin{aligned}
p(s,q) = s^4 &+ (q_1 + q_2 + 2.56)s^3 \\
&+ (q_1q_2 + 2.06q_1 + 1.561q_2 + 2.871)s^2 \\
&+ (1.06q_1q_2 + 4.841q_1 + 1.561q_2 + 3.164)s \\
&+ (4.032q_1q_2 + 3.773q_1 + 1.985q_2 + 1.853)
\end{aligned}
$$

and uncertainty bounds $0 \le q_1 \le 1$ and $0 \le q_2 \le 3$. Verify that the four edges of the uncertainty bounding set Q (take $q_1 = 0$ and $0 \le q_2 \le 3$; $q_1 = 1$ and $0 \le q_2 \le 3$; $0 \le q_1 \le 1$ and $q_2 = 0$; $0 \le q_1 \le 1$ and $q_2 = 3$) all lead to stable polynomials $p(s,q)$. However, for $q_1 = 0.5$ and $q_2 = 1.0$, $p(s,q)$ is unstable. Hence, robust stability cannot be guaranteed by checking extreme points or edges.

REMARKS 14.5.2 (The Realizability Issue):   The next issue which we address is whether examples of the sort above can be ruled out via imposition of a generating mechanism for $p(s,q)$. For example, it is of interest to know whether $p(s,q)$ is realizable as the characteristic polynomial of some state space system. In the exercise below, we see that this is indeed the case. By starting with a rather simple state space description $\dot{x}(t) = A(q)x(t)$, the characteristic polynomial $p(s, q) = \det(sI - A(q))$ turns out to be the "nasty" polynomial $p(s, q)$ in Exercise 14.5.1 above.

EXERCISE 14.5.3 (Realization by an Interval Matrix):   Verify that the characteristic polynomial $p(s, q)$ for the interval matrix

$$A(q) = \begin{bmatrix} [-1.5, -0.5] & -12.06 & -0.06 & 0 \\ -0.25 & -0.03 & 1.0 & 0.5 \\ 0.25 & -4.0 & -1.03 & 0 \\ 0 & 0.5 & 0 & [-4.0, -1.0] \end{bmatrix}$$

is the same as the one given in Exercise 14.5.1 above. Hence, even under the strengthened hypothesis that the multilinear uncertainty structure is *induced* by an interval matrix family, we still cannot guarantee robust stability by restricting our attention to extreme points or edges.

REMARKS 14.5.4 (Lower Order Polynomials):   Since the two exercises above involve fourth order polynomials, it is natural to ask whether some sort of extreme point or edge result is possible for the lower order cases. Via the pair of exercises below, we see that an extreme point result emerges for $n = 2$. In preparation for the second order analysis, we provide a lemma which is fundamental to mathematical programming.

LEMMA 14.5.5 (Multilinear Function on a Box):   *Suppose Q is a box in $\mathbf{R}^\ell$ with set of extreme points $\{q^i\}$ and $f : Q \to \mathbf{R}$ is multilinear. Then both the maximum and minimum of $f(q)$ are attained at extreme points. That is,*

$$\max_{q \in Q} f(q) = \max_i f(q^i)$$

and

$$\min_{q \in Q} f(q) = \min_i f(q^i).$$

PROOF: We establish the maximization result and simply note that a nearly identical proof can be used for minimization. Indeed, suppose that bounds for the components q_i of q are given by $q_i^- \leq q_i \leq q_i^+$ and let $q \in Q$ be arbitrarily selected. Assume that for some component q_k of q, the strict inequality $q_k^- < q_k < q_k^+$ holds.

To complete the proof, it suffices to show that there is a vector $q^* \in Q$ with $q_i^* = q_k$ for $i \neq k$, $q_k^* = q_k^-$ or q_k^+ and $f(q^*) \geq f(q)$. In this way, we can eliminate all nonextreme components of q without decreasing $f(q)$. In order to establish the existence of q^*, we use a factorization as in Exercise 14.3.7; i.e., we express $f(q)$ as

$$f(q) = q_k f_1(q^{\neq k}) + f_2(q^{\neq k}),$$

where $f_1(q^{\neq k})$ and $f_2(q^{\neq k})$ are multilinear functions on $\mathbf{R}^{\ell-1}$. We now consider three possibilities. First, if $f_1(q^{\neq k}) = 0$, it does not matter if we take $q_k^* = q_k^-$ or $q_k^* = q_k^+$. In either event, $f(q^*) = f(q)$. The second possibility is that $f_1(q^{\neq k}) > 0$. Now, by taking $q_k^* = q_k^+$, it follows that $f(q^*) > f(q)$. Finally, if $f_1(q^{\neq k}) < 0$, by taking $q_k^* = q_k^-$, we again conclude that $f(q^*) > f(q)$. ∎

EXERCISE 14.5.6 (Extreme Point Result for Robust Nonsingularity): Let $\mathcal{A} = \{A(q) : q \in Q\}$ be an interval matrix family. Taking q^i to be the i-th extreme point of Q, prove that $\mathcal{A}$ is robustly nonsingular (that is, $A(q)$ is nonsingular for all $q \in Q$) if and only if each of the determinants $A(q^i)$ has the same sign. *Hint*: Apply Lemma 14.5.5 to $f(q) = \det A(q)$.

EXERCISE 14.5.7 (Second Order Case): Consider the uncertain polynomial $p(s, q) = s^2 + a_1(q)s + a_0(q)$ with $a_0(q)$ and $a_1(q)$ having a multilinear uncertainty structure.
(a) Prove that the family of polynomials $\mathcal{P} = \{p(\cdot, q) : q \in Q\}$ is robustly stable if and only if the set of extremes $\{p(\cdot, q^i)\}$ is stable. *Hint*: Consider the result in Lemma 14.5.5 in conjunction with the fact that in the second order case, coefficient positivity is equivalent to stability.
(b) Using the result in (a), prove that a 2×2 interval matrix is robustly stable if and only if each member of the set of extreme matrices $\{A(q^i)\}$ is stable.

14.6   The Mapping Theorem

For uncertain polynomials with multilinear uncertainty structures, robust stability analysis can be quite complicated. However, with the aid of the Mapping Theorem below, we can often establish robust stability using a "special" overbounding family of polynomials. The power of the Mapping Theorem is derived from the fact that this overbounding family turns out to be the convex hull of the original family. Moreover, this convex hull family is seen to be a polytope of polynomials. Hence, it becomes possible to apply many of the results covered in the last several chapters while recognizing the fact that we are working in a sufficiency context. In contrast to other overbounding methods, the fact that we are working with the convex hull lends credibility to the "tightness" of our approximation. After stating the theorem below, we move immediately to interpretation and illustrative examples. The proof of the theorem below is relegated to Sections 14.9 and 14.10.

THEOREM 14.6.1 (The Mapping Theorem): *Suppose $Q \subset \mathbf{R}^\ell$ is a box with extreme points $\{q^i\}$ and $f : Q \to \mathbf{R}^k$ is multilinear. Let*

$$f(Q) = \{f(q) : q \in Q\}$$

denote the range of f. Then it follows that

$$conv\ f(Q) = conv\{f(q^i)\}.$$

14.7   Geometric Interpretation

The geometry associated with the Mapping Theorem is illustrated in Figure 14.7.1 for $k = 2$. Notice that we obtain the tightest pos-

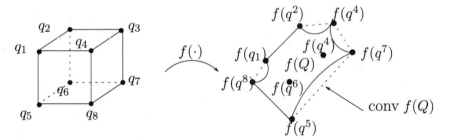

FIGURE 14.7.1   Geometry Associated with the Mapping Theorem

sible polygon bounding the range set $f(Q)$. In view of the Mapping Theorem, we can rule out a number of possible shapes for the range $f(Q)$. Roughly speaking, any $f(Q)$ which "curves outward" rather than "inward" is not realizable. For example, since taking the convex hull of the $f(q^i)$ in Figure 14.7.2 does not yield the convex hull of $f(Q)$, it follows that $f(Q)$ cannot be the range of some multilinear function on a box. In particular, the arc joining $f(q^1)$ and $f(q^2)$ is inconsistent with the requirement of inward curvature.

There are also other inconsistencies in the figure. For example, since the straight line joining q^2 and q^3 defines an edge of Q, it must

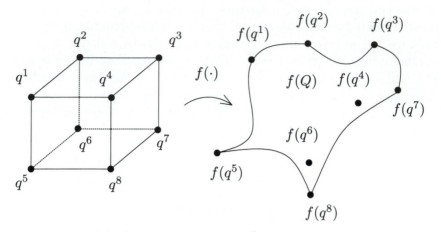

FIGURE 14.7.2   A Geometry Which Is Not Realizable

be the case that $f(Q)$ includes the straight line joining $f(q^2)$ and $f(q^3)$; notice that this line is missing. A similar comment applies to the pairs (q^7, q^8), (q^5, q^8) and (q^1, q^5).

EXERCISE 14.7.1 (The Range of $f(Q)$): Suppose that Q is the unit square in $\mathbf{R}^2$ and $f(q)$ is the two-dimensional multilinear function with components $f_1(q) = q_1 q_2$ and $f_2(q) = q_1 + q_2$. Describe the range $f(Q)$ and sketch it. Compare $f(Q)$ and conv $f(Q)$.

REMARKS 14.7.2 (Coefficient Interpretation): Suppose that $p(s, q)$ is an uncertain polynomial with coefficient vector $a(q)$ depending multilinearly on q. Then, if Q is a box with set of extremes $\{q^i\}$, the Mapping Theorem provides a simple description of the convex hull of the coefficient set; i.e.,

$$\text{conv } a(Q) = \text{conv}\{a(q^i)\}.$$

14.8   Value Set Interpretation

The ideas in the preceding section have interesting and useful interpretations in a value set context. Indeed, suppose $Q \subset \mathbf{R}^\ell$ is a box with extreme points $\{q^i\}$ and $p(s, q)$ is an uncertain polynomial having a multilinear uncertainty structure. Then, given any $z \in \mathbf{C}$, we consider the mapping $f : Q \to \mathbf{R}^2$ described by

$$q \mapsto (Re\ p(z, q), Im\ p(z, q)).$$

Since $Re\ p(z,q)$ and $Im\ p(z,q)$ are multilinear with respect to q, the Mapping Theorem tells us that

$$\text{conv } p(z,Q) = \text{conv}\{p(z,q^i)\}.$$

Having this convex hull description available, it is now easy to state a sufficient condition for robust $\mathcal{D}$-stability. The lemma below is an immediate consequence of the Zero Exclusion Condition (Theorem 7.4.2) in conjunction with the Mapping Theorem above.

LEMMA 14.8.1 (Robust $\mathcal{D}$-Stability Criterion):   *Consider a family of polynomials* $\mathcal{P} = \{p(\cdot,q) : q \in Q\}$ *with invariant degree, multilinear uncertainty structure and at least one $\mathcal{D}$-stable member $p(s,q^0)$. In addition, assume that Q is a box with extreme points $\{q^i\}$ and the desired root location region $\mathcal{D}$ is open. Then $\mathcal{P}$ is robustly $\mathcal{D}$-stable if the Zero Exclusion Condition*

$$0 \notin \text{conv}\{p(z,q^i)\}$$

is satisfied for all $z \in \partial\mathcal{D}$.

REMARKS 14.8.2 (Conservatism):   In view of the lemma above, we now provide a value set interpretation for the fact that the Mapping Theorem only leads to a sufficient condition for robust stability. Suppose $Q \subseteq \mathbf{R}^\ell$ is a box and the uncertain polynomial $p(s,q)$ has a multilinear uncertainty structure. Then, in a robust $\mathcal{D}$-stability analysis, it is possible that at some frequency $\omega^* > 0$, we have $0 \in \text{conv } p(j\omega^*,Q)$ but $0 \notin p(j\omega^*,Q)$. This situation is depicted in Figure 14.8.1 Notice that we must deem the robust stability test inconclusive. Said another way, if we are applying the polytope stability theory to the overbounding family obtained via the Mapping Theorem, we do not know if stability is lost when $z = 0$ penetrates the set conv $p(j\omega,Q)$. Roughly speaking, the true value set $p(j\omega,Q)$ is unobservable through the eyes of the Mapping Theorem. In the example below, we illustrate this undesirable phenomenon.

EXAMPLE 14.8.3 (Nasty Value Set):   For the family of polynomials described in Exercise 14.5.1, we begin with the set of extreme points $q^1 = (0,0)$, $q^2 = (0,3)$, $q^3 = (1,0)$ and $q^4 = (1,3)$ and generate

$$p(s,q^1) = s^4 + 2.56s^3 + 2.871s^2 + 3.164s + 1.853;$$
$$p(s,q^2) = s^4 + 5.56s^3 + 7.544s^2 + 7.847s + 7.808;$$
$$p(s,q^3) = s^4 + 3.56s^3 + 5.931s^2 + 8.065s + 5.885;$$
$$p(s,q^4) = s^4 + 6.56s^3 + 12.614s^2 + 15.868s + 23.677.$$

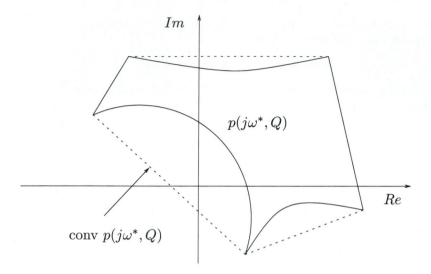

FIGURE 14.8.1   Inconclusive Robust Stability Test

Now, motivated by the remarks above, we examine both the "true" value set $p(j\omega, Q)$ and its convex hull $\text{conv}\{p_i(j\omega)\}$ at the frequency $\omega = 1.5$; see Figure 14.8.2. Notice that $z = 0$ lies inside conv $p(j\omega, Q)$ but does not lie inside $p(j\omega, Q)$ itself. In other words, the nasty value set geometry discussed in Remarks 14.8.2 above is realizable.

14.9   Machinery for Proof of the Mapping Theorem

In this section, we develop some basic machinery for the proof of the Mapping Theorem. To this end, we provide a quick primer on some basics from convex analysis; the reader is also referred to Section 8.3 where more elementary concepts from convex analysis are covered.

If $X \subseteq \mathbf{R}^n$, then the *support function* $h : \mathbf{R}^n \to \mathbf{R}$ on X is defined by

$$h(y) = \sup_{x \in X} y^T x,$$

where x and y are viewed as column vectors above and y^T denotes the transpose of y. The first point to note is that the support function on X can be identified with hyperplanes supporting X. In other words, for a fixed $y \in \mathbf{R}^n$, X is contained in the closed halfspace described by

$$y^T x \le h(y).$$

This situation is depicted in Figure 14.9.1. The second point to note

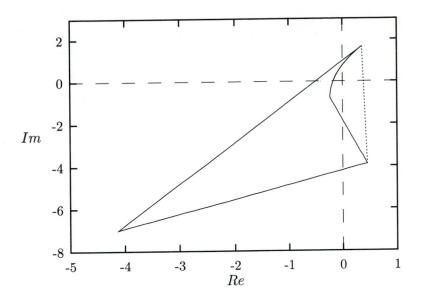

FIGURE 14.8.2   Value Set and Its Convex Hull for Example 14.8.3

is that if X is closed, then the convex hull of X, denoted conv X, is obtained by intersecting all possible closed halfspaces as indicated above with y ranging over $\mathbf{R}^n$; that is,

$$\text{conv } X = \bigcap_{y \in \mathbf{R}^n} \mathcal{H}_y,$$

where

$$\mathcal{H}_y = \{x : y^T x \le h(y)\}.$$

The formation of conv X in this manner is depicted in Figure 14.9.2. On the left side of Figure 14.9.2, an approximation to conv X is shown using finitely many supporting hyperplanes. On the right side of Figure 14.9.2, the true convex hull is shown.

A fundamental difference between general convex sets and polytopes is the nature of their support functions. For the polytope

$$X = \text{conv}\{x^1, x^2, \ldots, x^p\},$$

it is well known that for any given $y \in \mathbf{R}^n$, we can compute the support function for X by using only the extreme points of X. Since the generating set $\{x^1, x^2, \ldots, x^p\}$ contains all extreme points, it follows that

$$h(y) = \max_{i \le p} y^T x^i.$$

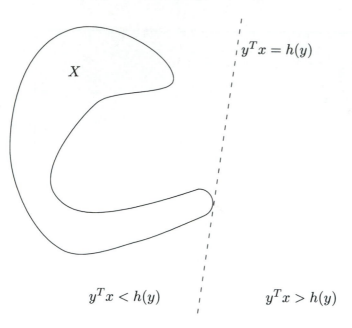

FIGURE 14.9.1   Supporting Hyperplane for X

This fact is basic to the theory of linear programming. It is really just another way of saying that a linear function on a polytope achieves its maximum (and minimum) at an extreme point. The converse is also true. That is, if $\{x^1, x^2, \ldots, x^p\}$ is a set of points for which equality holds in the equation above for all $y \in \mathbf{R}^n$, then it follows that X is a polytope with $\{x^1, x^2, \ldots, x^p\}$ as a generating set. In the language of convex analysis, we say that the support function is *finitely generated*. In the proof of the Mapping Theorem in the section to follow, we make use of the lemma below.

LEMMA 14.9.1 (Equal Convex Hulls):   *If X_1 and X_2 are two closed and bounded sets in $\mathbf{R}^n$ having the same support function $h(y)$, then*

$$conv\, X_1 = conv\, X_2.$$

14.10   Proof of the Mapping Theorem

In view of Lemma 14.9.1 above, it suffices to prove that $f(Q)$ and the finite point set

$$\mathcal{F} = \{f(q^i)\}$$

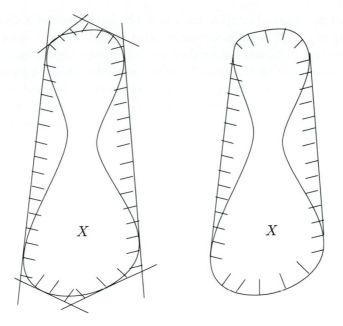

FIGURE 14.9.2   Finite Approximation to conv X and the True conv X

have the same support function. To this end, let $y \in \mathbf{R}^k$ be arbitrarily fixed and let $h_f(y)$ and $h_{\mathcal{F}}(y)$ denote the support functions on $f(Q)$ and $\mathcal{F}$, respectively. Now, by definition of the support function, we know that

$$h_f(y) = \sup_{q \in Q} \; y^T f(q).$$

Since $y \in \mathbf{R}^k$ is fixed, we can view the computation of $h_f(y)$ as a problem of maximizing the multilinear function

$$J(q) = y^T f(q)$$

on the box Q. Now, applying Lemma 14.5.5, we conclude that

$$h_f(y) = \max_i \; J(q^i) = \max_i \; y^T f(q^i) = h_{\mathcal{F}}(y). \quad \blacksquare$$

14.11   Conclusion

As shown in this chapter, robust stability analysis based on the Mapping Theorem leads to sufficient but not necessary conditions. However, the fact that we use a "tight" overbound (the convex hull of the original family) raises an interesting question: Are there large classes of polynomial families for which the Zero Exclusion Condition $0 \notin \operatorname{conv} p(j\omega, Q)$ is both necessary and sufficient for robust stability? Another direction for further work involves algorithm development. Essentially, one reformulates the robust stability problem in a mathematical programming context and proceeds toward a solution via some iteration process which exploits the Mapping Theorem; for further discussion see Sections 1.7, 1.8, 14.1 and the notes to follow.

Notes and Related Literature

NRL 14.1 The issue of overbounding arises in many places in the robustness literature. For example, in the line of research on robustness margins (see Doyle (1982) and Safonov (1982)), the quality of the bounds obtained for the structured singular value are of paramount importance. A similar issue arises in H^∞ theory; e.g., if one wants to use the Riccati equations in Doyle, Glover, Khargonekar and Francis (1989) for systems with structured real uncertainty, one approach involves overbounding of real parametric uncertainty by discs in the complex plane.

NRL 14.2 In the recent robustness literature, a line of research popularizing the mathematical programming approach begins with Kiendl (1985), Kiendl (1987), de Gaston and Safonov (1988), Sideris and Sanchez Pena (1989), Chang and Ekdal (1989), Ossadnik and Kiendl (1990) and Vicino, Tesi and Milanese (1990). In the first six of these papers, various schemes are proposed for partitioning of the uncertainty bounding set Q. Subsequently, stability (or instability) is established on local subdomains and the issue of covering Q is addressed. For a nice exposition of the branch and bound techniques associated with such an approach, see the textbook by Boyd and Barratt (1990).

NRL 14.3 In view of the transformation from polynomic to multilinear uncertainty structures in Section 14.3, an important question arises: Does the Mapping Theorem hold if Q is a polytope instead of a box? Since the answer to this question is no, a future breakthrough for multilinear uncertainty structures may have limited applicability to more general polynomic uncertainty structures.

NRL 14.4 A number of authors obtain results for special cases of the robust stability problem by imposing stronger assumptions on the multilinear uncertainty structure. For example, in Kharitonov (1979), no uncertainty entering even order coefficients is allowed to enter into odd order coefficients and vice versa. Subsequently, an extreme point result for robust stability is attained; see Panier, Fan and Tits (1989) for further extensions. Slight generalizations of this type of even–odd decoupling result are given in papers by Djaferis (1988) and Djaferis and Hollot (1989b).

NRL 14.5 Another special class of the multilinear uncertainty structures is studied in Barmish and Shi (1990). The uncertain polynomial $p(s,q)$ is assumed to have the form $XY+UV$ where X, Y, U and V, correspond to interval polynomial families. Subsequently, it is shown that the satisfaction of a "covering condition" is both necessary and sufficient for robust stability.

NRL 14.6 In the interesting Ph.D. dissertation of Zong (1990), an "inner intersection exclusion" condition for robust stability is given in the context of multilinear uncertainty structures. The use of this condition is roughly analogous to the way the Mapping Theorem is exploited to establish a sufficient condition for robust stability. The dissertation also includes interesting results for the 3×3 interval matrix problem.

NRL 14.7 In Wei and Yedavalli (1989), nonlinear uncertainty structures are treated by transforming real and imaginary parts of $p(s, q)$. However, no systematic method for constructing the desired transformation is given.

NRL 14.8 In the control literature, the Mapping Theorem appears at least as early as 1963 in the book by Zadeh and Desoer (1963). The paper by Saeki (1986) is credited with revival of these ideas. Since the Mapping Theorem leads to conservative results, the papers by de Gaston and Safonov (1988) and Sideris and Sanchez Pena (1989) are relevant. In both cases, a domain-splitting algorithm for robust stability is proposed. Motivated by computational inefficiencies associated with the frequency sweep used in these two papers, Sideris and Sanchez Pena (1989) provide a different algorithm which is based on the Routh table.

NRL 14.9 With the goal of eliminating conservatism associated with application of the Mapping Theorem, a number of authors have concentrated on the characterization of polynomials $p(s, q)$ having multilinear uncertainty structure, polytopic uncertainty bound Q and satisfying the following condition: At each frequency $\omega \geq 0$, condition $p(j\omega, Q) = \text{conv } p(j\omega, Q)$. For such cases, the value set is a polytope and a complete solution of the robust stability problem is straightforward. This line of research begins with the paper by Hollot and Xu (1989) where a conjecture is given involving the image of a polytope under a multilinear function; see also Polyak (1992) and Tsing and Tits (1992) for further work in this direction.

Part IV

The Spherical Theory

Chapter 15

Spherical Polynomial Families

Synopsis

In this chapter, The ℓ^2 analogue of Kharitonov's problem is considered. We work with a family of polynomials $\mathcal{P}$ with independent uncertainty structure and uncertainty bounding set Q which is a sphere. The highlight of the chapter is the Soh–Berger–Dabke Theorem. This theorem provides a simple method for robust stability testing using a frequency dependent scalar function.

15.1 Introduction

To motivate the technical exposition in this chapter, we begin with a family of polynomials $\mathcal{P}$ described by

$$p(s, q) = \sum_{i=0}^{n} a_i(q) s^i$$

and $q \in Q$. We pose two questions which are important to answer when working in an applications context: First, what uncertainty structure is being assumed for the uncertain coefficient functions $a_i(q)$? This question has already occupied much of our attention in the earlier chapters. Second, what type of uncertainty bounding set Q is being assumed? Given that most results in the robustness literature involve sets Q which are either boxes or spheres, some comments are in order.

The point of view in this text is that in most applications, it is not worth agonizing whether to use a box or a spherical representation for Q. Imprecision in the engineering problem formulation enables us to use either model. We elaborate on this point. For illustrative purposes, suppose that we are dealing with two uncertain parameters q_1 and q_2 (mass and coefficient of friction, respectively) and we ask the engineer what bounds should be assumed. A typical answer might be: The coefficient of friction q_1 can experience variations up to 20 or 30 percent about its nominal value $q_1^0 = 0.4$, and the mass q_2 can vary up to 10 or 20 percent about its nominal value $q_2^0 = 56.8$ kilograms. To further embellish this scenario, suppose that we ask the engineer whether we should assume a sphere or a box for the uncertain parameter vector q. The answer we receive is: Of course, we must assume a box because the mass and friction variations are independent; it does not make sense to use spheres.

15.2   Boxes Versus Spheres

Continuing with the hypothetical scenario above, suppose that we go into our mathematical toolbox and find that the only theoretical tool available for the problem at hand requires an assumption that Q is a sphere. Or more precisely, application of the available tool requires Q to be an ellipsoid which we view as a sphere using an appropriately weighted norm. Should we ignore the engineer's advice and apply the available spherical theory?

To decide whether to ignore the engineer, we sketch the uncertainty bounding set Q as shown in Figure 15.2.1. In view of the imprecise description of the uncertainty bounds, we know that the "true" bounding set Q lies between Q_{min} and Q_{max}; that is, $Q_{min} \subseteq Q \subseteq Q_{max}$. After a few moments' reflection, we conclude that we can take advantage of this latitude in the description of Q. By appropriate choice of weights $w_1 > 0$ and $w_2 > 0$ and a radius of uncertainty $r > 0$, we can approximately represent Q via the inequality

$$w_1^2(q_1 - q_1^0)^2 + w_2^2(q_2 - q_2^0)^2 \le r^2$$

and then go ahead and apply our spherically based theory to the problem at hand. In conclusion, an engineer's insistence on the use of boxes versus spheres is not really justifiable when the uncertainty bounds are not tightly specified.

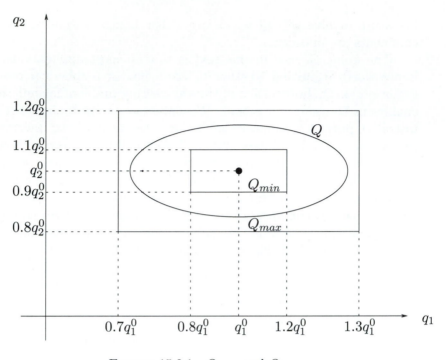

FIGURE 15.2.1 Q_{max} and Q_{min}

The choice of sphere versus box should be dictated by the available tools for solution of the problem at hand. This point serves as our launching point for this chapter and the next. By working with a spherical uncertainty structure, we obtain some powerful new tools to facilitate robustness analysis.

15.3 Spherical Polynomial Families

In this section, we provide the formal definition of a *spherical polynomial family*. If we associate an interval polynomial family with the ℓ^∞ norm (Q is a box), then it is natural to think of the theory in this chapter as the ℓ^2 analogue of the theory in Chapter 5. To this end, we now provide a slightly more general definition of the ℓ^2 norm than that given in Section 2.4.

DEFINITION 15.3.1 (Weighted ℓ^2 Norm and Ellipsoid): Given a $k \times k$ positive-definite symmetric matrix W and $x \in \mathbf{R}^k$, the *weighted euclidean norm* of x is given by

$$\|x\|_{2,W} = (x^T W x)^{\frac{1}{2}}.$$

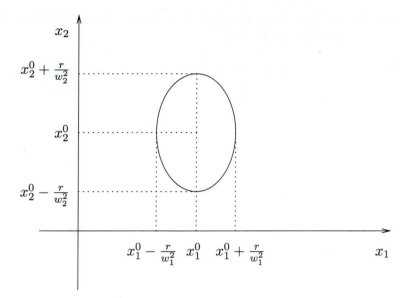

FIGURE 15.3.1   Ellipsoidal Geometry for Weighted Euclidean Norm

We call W the *weighting matrix*. Furthermore, given $r \geq 0$ and $x^0 \in \mathbf{R}^k$, we define an *ellipsoid* in $\mathbf{R}^k$ *centered at* x^0 by the inequality

$$(x - x^0)^T W (x - x^0) \leq r^2.$$

Equivalently,

$$\|x - x^0\|_{2,W} \leq r.$$

REMARKS 15.3.2 (Weighted Euclidean Norms):   The standard definitions above can be easily visualized. For example, if $k = 2$ and $W = \text{diag}\{\text{w}_1^2, \text{w}_2^2\}$, we obtain the ellipsoid depicted in Figure 15.3.1.

DEFINITION 15.3.3 (Spherical Polynomial Family):   A family of polynomials $\mathcal{P} = \{p(\cdot, q) : q \in Q\}$ is said to be a *spherical polynomial family* if $p(s, q)$ has an independent uncertainty structure and Q is an ellipsoid.

DEFINITION 15.3.4 (Spherical Plant Family):   A set of rational functions $\mathcal{P} = \{P(s, q, r) : q \in Q; r \in R\}$ is called a *spherical plant family* if $P(s, q, r)$ can be expressed as the quotient of uncertain polynomials,

$$P(s, q, r) = \frac{N(s, q)}{D(s, r)},$$

with Q and R being ellipsoids and $N(s,q)$ and $D(s,r)$ having independent uncertainty structures.

REMARKS 15.3.5 (Alternative Definition):   In some cases (for example, see the latter part of Chapter 16), it is convenient to define
a spherical plant family using a joint bound for (q,r) rather than
individual bounds for q and r. If W_1 and W_2 are square weighting
matrices with dimension $n_1 = \dim q$ and $n_2 = \dim r$, respectively,
we can work with the weighted norm

$$\|(q,r)\|_W = \sqrt{\|q\|_{W_1}^2 + \|r\|_{W_2}^2}$$

and bounding set

$$(Q,R) = \{(q,r) : \|(q,r)\|_W \le 1\}$$

when describing the unit sphere.

EXAMPLE 15.3.6 (Centering):   We consider a spherical polynomial
family $\mathcal{P}$ described by

$$p(s,q) = (4 + q_3)s^3 + (2 + q_2)s^2 + (1 + q_1)s + (0.5 + q_0)$$

with ellipsoidal uncertainty bound $\|q\|_{2,\mathrm{w}} \le 1$ and weighting matrix
$W = \mathrm{diag}\{2,5,3,1\}$ for $q = (q_0, q_1, q_2, q_3)$. Notice that by centering
this family on the vector $\tilde{q}^0 = (0.5, 1, 2, 4)$, we obtain an equivalent
description of $\mathcal{P}$. That is, we can work with the uncertain polynomial
$\tilde{p}(s, \tilde{q}) = \tilde{q}_0 + \tilde{q}_1 s + \tilde{q}_2 s^2 + \tilde{q}_3 s^3$ and uncertainty bounding set $\tilde{Q}$
described by $\|\tilde{q} - \tilde{q}^0\|_{2,\mathrm{w}} \le 1$.

REMARKS 15.3.7 (Representations):   In view of the fact that a
spherical polynomial family can be centered, we often begin with

$$p(s,q) = \sum_{i=0}^{n} q_i s^i$$

and uncertainty bound

$$\|q - q^0\|_{2,\mathrm{w}} \le r$$

with W being a positive-definite symmetric matrix, q^0 representing
the *nominal* and $r \ge 0$ being the radius of uncertainty. An equivalent representation is obtained by extracting the nominal polynomial

$p_0(s) = p(s, q^0)$ and using the representation

$$p(s, q) = p_0(s) + \sum_{i=0}^{n} q_i s^i$$

with $\|q\|_{2,\mathrm{w}} \leq r$. With this representation, we can adopt the point of view that the uncertainty bounding set Q is centered at zero.

EXAMPLE 15.3.8 (Not All Coefficients Uncertain): Analogous to the case of interval polynomials, in some cases we might only have a strict subset of the coefficients being uncertain. For example, consider $p(s, q) = s^3 + (2 + q_2)s^2 + 4s + (3 + q_0)$.

15.4   Lumping

To further extend the analogy between spherical and interval polynomial families, we now consider the issue of lumping; recall Exercise 5.3.5. For the case of spherical polynomial families, however, the lumping process is somewhat more subtle.

EXAMPLE 15.4.1 (Motivation): Consider the uncertain polynomial

$$p(s, q) = (3 + q_2)s^2 + (4 + 3q_1 - 5q_3)s + (2 + 2q_0 + 6q_4)$$

with uncertainty bound $\|q\|_2 \leq 2$. Our intuition tells us that various uncertainties can be combined; i.e., lump q_0 and q_4 together, lump q_1 and q_3 together. To carry out this lumping process precisely, we need some machinery. A standard result from matrix algebra is stated as a lemma below.

LEMMA 15.4.2 (Minimum Norm): *Consider* $\mathbf{R}^n$ *with* $\| \cdot \|_{2,W}$ *and let A be a fixed real $n \times m$ matrix having rank n. Then, given any $b \in \mathbf{R}^n$, the minimum norm solution of*

$$Ax = b$$

is given by

$$x^{MN} = W^{-1}A^T(AW^{-1}A^T)^{-1}b.$$

NOTATION 15.4.3 (Subvectors and Matrices): Given the fact that we want to allow for the possibility that only a subset of the coefficients of $p(s, q)$ are uncertain, it is convenient to introduce the notion of *subvectors*. Indeed, suppose $x = (x_0, x_1, \ldots, x_n)$ and index set $I \subseteq \{0, 1, 2, \ldots, n-1, n\}$ is nonempty. Then we let x^I denote the subvector of x obtained by retaining components x_i for $i \in I$ and deleting components x_i for $i \notin I$. To avoid permutations among the components of x^I, it is always assumed that if $I = \{i_1, i_2, \ldots, i_r\}$, then $i_1 < i_2 < \cdots < i_{r-1} < i_r$. This convention makes x^I uniquely defined. To illustrate, if $x = (x_0, x_1, x_2, x_3, x_4)$ and $I = \{2, 4\}$, then we obtain $x^I = (x_2, x_4)$.

REMARKS 15.4.4 (Interpretation for Uncertain Polynomials):   We now interpret the subvector notation above in terms of uncertain polynomials. If $p(s, q) = \sum_{i=0}^{n} a_i(q) s^i$ is an uncertain polynomial whose *coefficient vector* is $a(q)$, then $a^I(q)$ is the subvector of $a(q)$ generated using the components $a_i(q)$ with $i \in I$. Note that for a spherical polynomial family with $q \in \mathbf{R}^{\ell}$, the independent uncertainty structure enables us to express any subvector $a^I(q)$ as

$$a^I(q) = A_I q + b^I,$$

where A_I is a matrix having $\dim a^I(q)$ rows and ℓ columns and b^I is a column vector having $\dim a^I(q)$ entries. Finally, if we allow some of the coefficients $a_i(q)$ to be fixed, say $a_i(q) = a_i$ for $i \notin I$, it is often convenient to write

$$p(s, q) = \sum_{i \in I} a_i(q) s^i + \sum_{i \notin I} a_i s^i.$$

In this way, it is easy to emphasize which coefficients are fixed and which coefficients are uncertain.

EXAMPLE 15.4.5 (Subvectors and Matrix Representation):   For the uncertain polynomial

$$p(s, q) = (2 + q_4) s^4 + (3 + q_3 + 2q_5) s^3 + (2 - q_2 + 4q_6) s^2 + (6 + 2q_1) s + (q_0 + 4)$$

with $I = \{0, 2, 4\}$, we have

$$a^I(q) = \begin{bmatrix} q_0 + 4 \\ 2 - q_2 + 4q_6 \\ 2 + q_4 \end{bmatrix}.$$

Furthermore, we can write $a^I(q) = A_I q + b^I$, where

$$
A_I = \begin{bmatrix} 1 & 0 & 0 & 0 & 0 & 0 & 0 \\ 0 & 0 & -1 & 0 & 0 & 0 & 4 \\ 0 & 0 & 0 & 0 & 1 & 0 & 0 \end{bmatrix} ; \qquad b^I = \begin{bmatrix} 4 \\ 2 \\ 2 \end{bmatrix} .
$$

THEOREM 15.4.6 (Lumping for a Spherical Polynomial Family):
Suppose that $\mathcal{P} = \{p(\cdot, q) : q \in Q\}$ is a spherical polynomial family described by

$$
p(s, q) = \sum_{i \in I} a_i(q) s^i + \sum_{i \notin I} a_i s^i
$$

and

$$
Q = \{q \in \mathbf{R}^\ell : \|q - q^0\|_{2,W} \le r\}.
$$

Using the representation

$$
a^I(q) = A_I q + b
$$

for the n_I-dimensional subvector of $a(q)$ whose rows are not constant with respect to q, define the new spherical polynomial family $\tilde{\mathcal{P}}$ by

$$
\tilde{p}(s, \tilde{q}) = \sum_{i \in I} \tilde{q}_i s^i + \sum_{i \notin I} a_i s^i
$$

and $\tilde{q} \in \tilde{Q}$, where

$$
\tilde{Q} = \{\tilde{q} \in \mathbf{R}^{n_I} : \|\tilde{q} - \tilde{q}^0\|_{2,\tilde{W}} \le r\},
$$

$$
\tilde{q}^0 = A_I q^0 + b^I
$$

and

$$
\tilde{W} = (A_I W^{-1} A_I^T)^{-1}.
$$

Then it follows that

$$
\tilde{\mathcal{P}} = \mathcal{P}.
$$

PROOF: Denoting the range of $a^I(q)$ by

$$
a^I(Q) = \{A_I q + b : q \in Q\},
$$

it suffices to show that $a^I(Q) = \tilde{Q}$. Indeed, notice that $\tilde{q} \in a^I(Q)$ if and only if there exists some $q \in Q$ such that

$$
A_I q + b = \tilde{q}.
$$

Adding $A_I q^0$ on each side above, it follows that $\tilde{q} \in a^I(Q)$ if and only if there exists $\hat{q} \in \mathbf{R}^\ell$ such that $\|\hat{q}\|_{2,W} \leq r$ and $A_I \hat{q} = \tilde{q} - \tilde{q}^0$. Now, an important observation to make is that the independent uncertainty structure (each q_i enters only one coefficient) guarantees that each column of A_I has exactly one nonzero entry. Hence, A_I has rank equal to n_I, its number of rows. Therefore, using Lemma 15.4.2, an appropriate $\hat{q}$ exists if and only if the minimum norm solution of the equation $A_I \hat{q} = \tilde{q} - q^0$ has norm less than or equal to r. This minimum norm solution is given by

$$\hat{q}^{MN} = W^{-1} A_I^T (A_I W^{-1} A_I^T)^{-1} (\tilde{q} - \tilde{q}^0).$$

The proof is concluded by observing that the condition

$$\|\hat{q}^{MN}\|_2 \leq r$$

is equivalent to the condition $\tilde{q} \in \tilde{Q}$. ∎

EXAMPLE 15.4.7 (Lumping):   Returning to Example 15.4.1, for the uncertain polynomial

$$p(s, q) = (3 + q_2)s^2 + (4 + 3q_1 - 5q_3)s + (2 + 2q_0 + 6q_4)$$

with uncertainty bound $\|q\|_2 \leq 2$, we now reduce the number of uncertainties from five to three. In accordance with Theorem 15.4.6, we take index set $I = \{0, 1, 2\}$, center $q^0 = 0$, identity weighting matrix $W = I$ and radius $r = 2$. We use the representation

$$a^I(q) = \begin{bmatrix} 2 & 0 & 0 & 0 & 6 \\ 0 & 3 & 0 & -5 & 0 \\ 0 & 0 & 1 & 0 & 0 \end{bmatrix} \begin{bmatrix} q_0 \\ q_1 \\ q_2 \\ q_3 \\ q_4 \end{bmatrix} + \begin{bmatrix} 2 \\ 4 \\ 3 \end{bmatrix} = A_I q + b^I$$

and compute $\tilde{W} = \mathrm{diag}\{1/40, 1/34, 1\}$. Now, we calculate center $\tilde{q}^0 = A_I q^0 + b = [\,2\;\;4\;\;3\,]^T$ and obtain the lumped spherical family of polynomials $\tilde{\mathcal{P}}$ described by $\tilde{p}(s, \tilde{q}) = \tilde{q}_2 s^2 + \tilde{q}_1 s + \tilde{q}_0$ and

$$\frac{1}{40}(\tilde{q}_0 - 2)^2 + \frac{1}{34}(\tilde{q}_1 - 4)^2 + (\tilde{q}_2 - 3)^2 \leq 4.$$

EXERCISE 15.4.8 (Lumping): Apply Theorem 15.4.6 to the spherical family of polynomials described by

$$p(s,q) = s^5 + (4 + q_4 - 3q_5)s^4 + (2 + q_3)s^3 + (6 + 5q_2 - 3q_6)s^2 + 6s + (q_0 + 6q_1)$$

and $\|q - q^0\|_{2,W} \leq 10$. Take nominal $q^0 = (1,1,1,0,0,0,1)$ and weighting matrix $W = \text{diag}\{1,2,6,1,20,5,10\}$.

REMARKS 15.4.9 (Lumping for a Spherical Plant Family): We note that Theorem 15.4.6 is readily adapted to handle a spherical plant family $\mathcal{P} = \{P(\cdot,q,r) : q \in Q; r \in R\}$. Indeed, if we express $P(s,q,r)$ as the quotient of uncertain polynomials, we simply apply Theorem 15.4.6 to the numerator and denominator separately.

15.5   The Soh–Berger–Dabke Theorem

The focal point of this section is the Soh–Berger–Dabke Theorem. This theorem provides a complete solution for the ℓ^2 analogue of Kharitonov's problem. We consider the robust stability problem for a spherical family of polynomials and generate a special scalar function of frequency. This function is used to compute the robustness margin—the largest uncertainty bound r for which robust stability is guaranteed. For simplicity of exposition, we state the theorem below with all coefficients equally weighted. In the exercises following the theorem, we indicate modifications in the theory which are needed to handle the cases where the description of Q includes a weighting matrix W or only a subset of the coefficients are uncertain. The proof of the theorem is relegated to Sections 15.6 and 15.7.

THEOREM 15.5.1 (Soh, Berger and Dabke (1985)): *Consider the spherical family of polynomials $\mathcal{P}$ with invariant degree $n \geq 1$ described by*

$$p(s,q) = p_0(s) + \sum_{i=0}^{n} q_i s^i$$

with nominal

$$p_0(s) = \sum_{i=0}^{n} a_i s^i$$

and uncertainty bounding set $\|q\|_2 \leq r$. For $\omega > 0$, let

$$G_{SBD}(\omega) = \frac{[Re\ p_0(j\omega)]^2}{\sum\limits_{i\ even} \omega^{2i}} + \frac{[Im\ p_0(j\omega)]^2}{\sum\limits_{i\ odd} \omega^{2i}}.$$

*Then $\mathcal{P}$ is robustly stable if and only if $p_0(s)$ is stable, the zero fre-
quency condition*

$$|a_0| > r$$

is satisfied and

$$G_{SBD}(\omega) > r^2$$

for all frequencies $\omega > 0$.

REMARKS 15.5.2 (Robustness Margin): The Soh–Berger–Dabke
theorem suggests a simple method to compute a robustness margin.
Indeed, if we use the notation $Q_r = \{q \in \mathbf{R}^{n+1} : \|q\|_2 \leq r\}$ and $\mathcal{P}_r = \{p(\cdot, q) : q \in Q_r\}$ to emphasize the dependence on the uncertainty
bound $r \geq 0$, the quantity of interest is

$$r_{max} = \sup\{r : \mathcal{P}_r \text{ has invariant degree and is robustly stable}\}.$$

Now, if we define

$$r^+_{max}(\omega) = \sqrt{G_{SBD}(\omega)}$$

and

$$r^+_{max} = \inf_{\omega > 0} r_{max}(\omega),$$

we obtain

$$r_{max} = \min\{|a_0|, |a_n|, r^+_{max}\}$$

as the robustness margin.

EXAMPLE 15.5.3 (Computation of r_{max}): To compute a robustness
margin for the uncertain polynomial

$$p(s, q) = (1 + q_3)s^3 + (1.5 + q_2)s^2 + (1.4 + q_1)s + (2 + q_0),$$

we consider nominal $p_0(s) = s^3 + 1.5s^2 + 1.4s + 2$ and generate

$$G_{SBD} = \frac{(2 - 1.5\omega^2)^2}{1 + \omega^4} + \frac{(1.4\omega - \omega^3)^2}{\omega^2 + \omega^6} = \frac{3.25\omega^4 - 8.8\omega^2 + 5.96}{1 + \omega^4}.$$

From the plot of $r^+_{max}(\omega)$ versus $\omega > 0$ in Figure 15.5.1, the minimum
value is approximately $r^+_{max} \approx 0.0011$. With $a_0 = 2$ and $a_3 = 1$, we
compute $r_{max} = \min\{|a_0|, |a_n|, r^+_{max}\} \approx \min\{2, 1, 0.0011\} = 0.0011$.

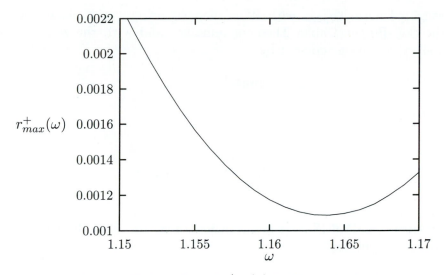

FIGURE 15.5.1   Plot of $r^+_{max}(\omega)$ for Example 15.5.3

EXERCISE 15.5.4 (Ship Steering):   The following state variable model for an automatic ship steering system is given in Dorf (1974):

$$\dot{x}(t) = \begin{bmatrix} -0.05 & -6 & 0 & 0 \\ -0.003 & -0.15 & 0 & 0 \\ 1 & 0 & 0 & 13 \\ 0 & 1 & 0 & 0 \end{bmatrix} x(t) + \begin{bmatrix} -0.2 \\ 0.03 \\ 0 \\ 0 \end{bmatrix} u(t);$$

$$y(t) = x_3(t),$$

where $x_1(t)$ is the transverse velocity, $x_2(t)$ is the angular rate of the ship's coordinate frame relative to its response frame, $x_3(t)$ is the deviation distance on an axis perpendicular to the track and $x_4(t)$ is the deviation angle.

(a) Using a linear feedback control $u(t) = K_1 x_1(t) + K_3 x_3(t)$, determine the set $\mathcal{K}$ of gain pairs (K_1, K_3) for which closed loop stability of the system is guaranteed.

(b) Fix some "centrally located" stabilizing gain pair $(K_1, K_3) \in \mathcal{K}$ and let $p_0(s)$ denote the resulting closed loop polynomial. With these gains fixed, consider a spherical family of polynomials with nominal $p_0(s)$ and compute the associated robustness margin r_{max} using the formula in Remarks 15.5.2 above.

EXERCISE 15.5.5 (Weighted Norm):   For the case when q has weighted norm $\|q\|_{2,w}$ with $W = \text{diag}\{w_0^2, w_1^2, \dots, w_n^2\}$, show that the Soh–Berger–Dabke Theorem remains valid with the zero frequency condition replaced by

$$|w_0 a_0| > r$$

and with modified testing function

$$G_{SBD}(\omega) = \frac{[Re\ p_0(j\omega)]^2}{\displaystyle\sum_{i\ even} w_i^2 \omega^{2i}} + \frac{[Im\ p_0(j\omega)]^2}{\displaystyle\sum_{i\ odd} w_i^2 \omega^{2i}}.$$

EXERCISE 15.5.6 (Subset of Coefficients Fixed): Consider the uncertain polynomial

$$p(s, q) = p_0(s) + \sum_{i \in I} q_i s^i$$

with I denoting the index set associated with those coefficients which are uncertain. Modify the statement of the Soh–Berger–Dabke Theorem to handle this case.

15.6 The Value Set for a Spherical Polynomial Family

This section and the next can be skipped by the reader interested primarily in application of the results. Our objective is to provide a characterization of the value set associated with a spherical polynomial family $\mathcal{P} = \{p(\cdot, q) : q \in Q\}$. The proof of the Soh–Berger–Dabke Theorem, given in the next section, exploits this characterization. To this end, we now show that at each frequency $\omega > 0$, the value set $p(j\omega, Q)$ is an ellipse in the complex plane. Throughout this section, to avoid trivialities, we assume that $\deg p(s, q) \geq 1$. In addition, we work with an uncertain polynomial of the form $p(s, q) = p_0(s) + \sum_{i=0}^{n} q_i s^i$ with uncertainty bound $\|q\|_2 \leq r$.

To begin the analysis, we hold the frequency $\omega > 0$ fixed and observe that $z \in p(j\omega, Q)$ if and only if there exists some $q \in Q$ such that

$$\vec{z} = A(\omega)q + b(\omega),$$

where

$$\vec{z} = \begin{bmatrix} Re\ z \\ Im\ z \end{bmatrix},$$

$$A(\omega) = \begin{bmatrix} 1 & 0 & -\omega^2 & 0 & \omega^4 & 0 & -\omega^6 & \cdots \\ 0 & \omega & 0 & -\omega^3 & 0 & \omega^5 & 0 & \cdots \end{bmatrix}$$

and

$$b(\omega) = \begin{bmatrix} Re\ p_0(j\omega) \\ Im\ p_0(j\omega) \end{bmatrix}.$$

Next, observing that rank $A(\omega) = 2$ when $\omega > 0$, it follows that $z \in p(j\omega, Q)$ if and only if the minimum norm solution $q^{MN}(\omega)$ of the equation $\vec{z} = A(\omega)q + b(\omega)$ has norm less than or equal to r. Using Lemma 15.4.2, we generate

$$q^{MN}(\omega) = A^T(\omega)(A(\omega)A^T(\omega))^{-1}(\vec{z} - b(\omega))$$

and enforce the condition $\|q^{MN}\|_2 \le r$ to arrive at the following conclusion: For $\omega > 0$, the value set $p(j\omega, Q)$ is the ellipse in the complex plane described by

$$(\vec{z} - b(\omega))^T W(\omega)(\vec{z} - b(\omega)) \le r^2,$$

where

$$W(\omega) = (A(\omega)A^T(\omega))^{-1}.$$

To further simplify the description of the ellipse $p(j\omega, Q)$, we now develop a closed form description of the weighting matrix $W(\omega)$. Indeed, using the formulas for $A(\omega)$ and $W(\omega)$ above, a straightforward computation yields

$$W(\omega) = \text{diag} \left\{ \frac{1}{\displaystyle\sum_{i\ even} \omega^{2i}}, \frac{1}{\displaystyle\sum_{i\ odd} \omega^{2i}} \right\}.$$

Now, substituting for $\vec{z}$, $b(\omega)$ and $W(\omega)$ in the elliptical value set description $(\vec{z} - b(\omega))^T W(\omega)(\vec{z} - b(\omega)) \le r^2$, we reach the conclusion that $z \in p(j\omega, Q)$ if and only if

$$\frac{[Re\ z - Re\ p_0(j\omega)]^2}{\displaystyle\sum_{i\ even} \omega^{2i}} + \frac{[Im\ z - Im\ p_0(j\omega)]^2}{\displaystyle\sum_{i\ odd} \omega^{2i}} \le r^2.$$

The set of $z \in \mathbf{C}$ satisfying the inequality above is clearly an ellipse; see Figure 15.6.1. This ellipse is centered at the nominal $p_0(j\omega)$

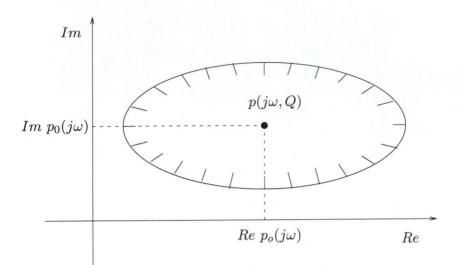

FIGURE 15.6.1   Value Set Ellipse $p(j\omega, Q)$ for $\omega > 0$

with major axis in the real direction having length

$$R_0 = 2r \left(\sum_{i \ even} \omega^{2i} \right)^{1/2}$$

and major axis in the imaginary direction having length

$$I_0 = 2r \left(\sum_{i \ odd} \omega^{2i} \right)^{1/2}$$

For the case when $\omega = 0$, the analysis above does not hold because rank $A(0) = 1$. Nevertheless, for this degenerate case, the value set is easy to describe. Indeed, notice that for $\omega = 0$, the value set is the real interval given by

$$p(j0, Q) = [a_0 - r, a_0 + r].$$

By combining the two analyses for the cases $\omega > 0$ and $\omega = 0$, we have a complete description of the value set.

15.7   Proof of the Soh–Berger–Dabke Theorem

First, it is easy to see that stability of $p_0(s)$ is necessary for robust stability of $\mathcal{P}$ because $q = 0$ is admissible. Also, when $\omega = 0$, in order to avoid the possibility of a root at $s = 0$, it is also necessary to have $|a_0| > r$; equivalently, $0 \notin p(j0, Q)$. Therefore, in the remainder of the proof, we assume that $p_0(s)$ is stable and $|a_0| > r$. We must prove that $G_{SBD}(\omega) > r^2$ for all $\omega > 0$ is both necessary and sufficient for robust stability. Indeed, since all the preconditions for application of the Zero Exclusion Condition (Theorem 7.4.2) are satisfied, it follows that $\mathcal{P}$ is robustly stable if and only if $0 \notin p(j\omega, Q)$ for all $\omega > 0$. To complete the proof, we use the inequality characterizing $z \in p(j\omega, Q)$; i.e., in accordance with the conclusion reached in the preceding section, it follows that $0 \notin p(j\omega, Q)$ if and only if

$$\frac{[Re\ p_0(j\omega)]^2}{\sum_{i\ even} \omega^{2i}} + \frac{[Im\ p_0(j\omega)]^2}{\sum_{i\ odd} \omega^{2i}} > r^2.$$

Hence, robust stability of $\mathcal{P}$ is guaranteed if and only if

$$G_{SBD}(\omega) > r^2$$

for all frequencies $\omega > 0$. ∎

15.8   Overbounding via a Spherical Family

In this section, we see that a family of polynomials which is *non-spherical* can often be overbounded by a spherical family. The reader should recall, however, that overbounding can lead to a conservative result; see the discussion in Section 5.11. To describe the key idea involved in the overbounding process, we begin with a family of polynomials $\mathcal{P}$ described by

$$p(s, q) = p_0(s) + \sum_{i \in I} a_i(q) s^i$$

and $q \in Q$ with $p_0(s)$ being the nominal and $I \subseteq \{0, 1, 2, \ldots, n\}$ denoting the index set describing the coefficients which are uncertain. Suppose that this family does not have an independent uncertainty structure and the uncertainty bounding Q is not necessarily a sphere.

As a first step, we compute a bound for the norm of $a^I(q)$. That is, we find $\bar{r} > 0$ such that

$$\bar{r} \geq \max_{q \in Q} \|a^I(q)\|_2.$$

Subsequently, it follows that $\mathcal{P}$ is a subset of the overbounding family $\bar{\mathcal{P}}$ described by

$$\bar{p}(s, \bar{q}) = p_0(s) + \sum_{i \in I} \bar{q}_i s^i + \sum_{i \notin I} a_i s^i,$$

and $\bar{q} \in \bar{Q}$, where

$$\bar{Q} = \{\bar{q} \in \mathbf{R}^{n_I} : \|\bar{q}\|_2 \leq \bar{r}\}.$$

In view of the set inclusion $\mathcal{P} \subseteq \bar{\mathcal{P}}$, any robustness criterion which is satisfied by the *overbounding family* $\bar{\mathcal{P}}$ is automatically satisfied for the original family $\mathcal{P}$.

EXAMPLE 15.8.1 (Centering):   In some cases, we can improve upon the bound $\bar{\mathcal{P}}$ above by "centering" the $\bar{Q}$ set. We now illustrate this technique via an example which is considered in two different ways. Suppose that $p(s, q) = s^2 + (3 \cos^2 q)s + 5$ and $Q = [0, 5]$. We first overbound without centering $\bar{Q}$. Using the prescription above, we take $I = \{1\}$ and obtain

$$\bar{r} = \max_{q \in Q} a_1(q) = \max_{q \in [0,5]} 3 \cos^2 q = 3.$$

This leads to $\bar{p}(s, \bar{q}) = s^2 + \bar{q}_1 s + 5$ and $\bar{Q} = [-3, 3]$. We now provide a second solution which involves centering. After writing $p(s, q) = s^2 + (3 \cos^2 q - 1.5)s + 1.5s + 5$, we take the nominal to be $p_0(s) = s^2 - 1.5s + 5$ and obtain

$$\bar{r} = \max_{q \in [0,5]} |3 \cos^2 q - 1.5| = 1.5,$$

$\bar{p}(s, \bar{q}) = s^2 + (1.5 + \bar{q}_1)s + 5$ and $\bar{Q} = [-1.5, 1.5]$. Observe that this family of polynomials is a strict subset of the one obtained without centering. In conclusion, centering of $\bar{Q}$ can be effective in tightening the overbound $\bar{\mathcal{P}}$. In many cases, there are other commonsense considerations which can be used to improve the bounding process.

EXERCISE 15.8.2 (Maximum Eigenvalue Overbound):   Consider the family of polynomials $\mathcal{P}$ described by $p(s, q) = p_0(s) + \sum_{i \in I} a_i(q)s^i$ and $\|q\|_2 \leq r$. Assume that $a^I(q)$ is linear with respect to q; i.e., one can write $a^I(q) = A_I q$, where A_I is an $n_I \times \ell$ matrix. Now, if rank $A_I = \ell$, show that an overbounding spherical family of polynomials $\bar{\mathcal{P}}$ is described by $\bar{p}(s, \bar{q}) = \sum_{i \in I} \bar{q}_i s^i$ and $\|\bar{q}\|_2 \leq \bar{r}$ where,

$$\bar{r} = r \lambda_{max}^{\frac{1}{2}}(A_I^T A_I)$$

and $\lambda_{max}(A_I^T A_I)$ denotes the largest eigenvalue of $A_I^T A_I$.

EXAMPLE 15.8.3 (Conservatism in Overbounding):   To see that the overbounding procedure (using the maximum eigenvalue) above

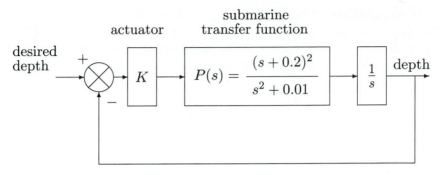

FIGURE 15.8.1   Depth Control System for Example 15.8.4

can be conservative, consider the uncertain polynomial

$$p(s, q) = s^3 + (q + 3)s^2 + 4s + q + 2$$

with uncertainty bounding set $Q = [-1.5, 1.5]$. First, we generate the Routh table:

$$
\begin{array}{lll}
s^3 & 1 & 4 \\
s^2 & q + 3 & q + 2 \\
s^1 & \frac{3q+10}{q+3} & \\
s^0 & q + 2 &
\end{array}
$$

Since there are no sign changes in the first column of the table for all $q \in Q$, this family of polynomials is robustly stable. On the other hand, if we overbound as prescribed in Exercise 15.8.2, we use $I = \{0, 2\}$, $A_I = [1\ 1]^T$ and $r = 1.5$ and obtain an overbounding family $\bar{\mathcal{P}}$ described by

$$\bar{p}(s, \bar{q}) = s^3 + (\bar{q}_2 + 3)s^2 + 4s + (\bar{q}_0 + 2)$$

and $\|\bar{q}\|_2 \leq 1.5\sqrt{2}$. It is now easy to see that $\bar{\mathcal{P}}$ is not robustly stable. For example, to induce instability, it suffices to take $\bar{q}_2 = 0$ for any $\bar{q}_0$ satisfying $-1.5\sqrt{2} \leq \bar{q}_0 \leq -2$.

EXAMPLE 15.8.4 (Depth Control System):   To demonstrate the overbounding process in the context of rational functions, we consider the model of a submarine depth control system in Dorf (1974); see Figure 15.8.1. A pressure transducer is used to measure the depth, and the gain of the stem plane actuator is set at $K = 0.2$.

It is straightforward to verify that the closed loop system is stable. We now examine the effect on stability for uncertainty up to 20% in the pole and zero locations in the approximate submarine transfer function. Hence, in lieu of $P(s)$, we take

$$P(s, q, r) = \frac{(s + 0.2 + q_0)^2}{s^2 + 0.01 + r_0}$$

with $-0.04 \leq q_0 \leq 0.04$ and $-0.002 \leq r_0 \leq 0.002$. Notice that the denominator of $P(s, q, r)$ is a spherical family of polynomials but the numerator

$$N(s, q) = s^2 + (0.4 + 2q_0)s + (q_0^2 + 0.4q_0 + 0.04)$$

is not. To obtain a spherical overbound for the numerator, we take nominal $N_0(s) = s^2 + 0.4s + 0.04$, index set $I = \{0, 1\}$ and compute

$$\max_{|q_0| \leq 0.04} \|a^I(q)\| = \max_{|q_0| \leq 0.04} \sqrt{4q_0^2 + (q_0^2 + 0.4q_0)^2} \approx 0.082.$$

Hence, we end up with a numerator overbounding family which is described by

$$\bar{N}(s, \bar{q}) = s^2 + (0.4 + \bar{q}_1)s + (0.04 + \bar{q}_0)$$

and $\|\bar{q}\| \leq 0.082$. Now, to study robustness of closed loop stability, we can use the overbounding polynomial

$$\begin{aligned}\bar{p}(s, \bar{q}, r) &= K\bar{N}(s, \bar{q}) + sD(s, r) \\ &= s^3 + 0.2s^2 + (0.2\bar{q}_1 + r_0 + 0.09)s + 0.2\bar{q}_0 + 0.008.\end{aligned}$$

EXERCISE 15.8.5 (Robustness Margin):   For the depth control system in the example above, use the formula for r_{max} in Section 15.5 to compute a robustness margin for the overbounding family of polynomials which was obtained.

15.9   Conclusion

In this chapter, the ℓ^2 analogue of Kharitonov's interval polynomial problem was studied. The highlight was the Soh–Berger–Dabke Theorem. Modulo some minor technical assumptions about $\omega = 0$ and avoidance of degree dropping, the theorem tells us that robust stability of a spherical polynomial family can be checked by generating the frequency function $G_{SBD}(\omega)$ and finding its infimum.

Notes and Related Literature

NRL 15.1 The important paper by Fam and Meditch (1978) appears to have motivated the work of Soh, Berger and Dabke (1985). In a sense, it is reasonable to say that Soh, Berger and Dabke were the first to realize that the key ideas of Fam and Meditch had important ramifications in robustness theory.

NRL 15.2 For spherical polynomial families, the development of robust stability concepts in a feedback control context are attributable to Biernacki, Hwang and Bhattacharryya (1987). In the next chapter, we describe their control theoretic extension of the Soh–Berger–Dabke Theorem. In Hinrichsen and Pritchard (1988) and Keel, Bhattacharryya and Howze (1988), additional results are given.

NRL 15.3 The ℓ^2 version of the result of Tsypkin and Polyak (1991) is directly related to the result given in the Soh–Berger–Dabke Theorem. Namely, the complex frequency function

$$G_{TP}(\omega) = \frac{Re\ p_0(j\omega)}{\sqrt{\sum_{i\ even} \omega^{2i}}} + j\frac{Im\ p_0(j\omega)}{\sqrt{\sum_{i\ odd} \omega^{2i}}}$$

of Tsypkin and Polyak (see Section 6.7) is simply related to the scalar function of Soh, Berger and Dabke; i.e.,

$$G_{SBD}(\omega) = |G_{TP}(\omega)|^2.$$

Chapter 16

Embellishments for Spherical Families

Synopsis

In this chapter, we continue to concentrate on spherical families. In the first part of the chapter, we provide the spectral set characterization of Barmish and Tempo. In the second part of the chapter, a spherical plant family and a compensator are connected in a feedback loop. With this control system setup, the results of Biernacki, Hwang and Bhattacharyya facilitate robust stability analysis.

16.1 Introduction

Central to the technical development of the last chapter is the minimum norm solution to a set of linear equations $Ax = b$. In the first part of this chapter, we use the same set of technical ideas to develop a characterization of the spectral set for a spherical polynomial family. With the spectral set in hand, we have complete knowledge of all possible $\mathcal{D}$ regions for which robust stability is guaranteed. In the second part of this chapter, we concentrate on robust stability of feedback systems involving a spherical plant family and some given compensator. Recalling the analysis of interval plants in Chapter 10, the reader is reminded that the independent uncertainty structure which is present in the plant is no longer present in the closed loop polynomial $p(s, q)$. Instead, $p(s, q)$ has an affine linear uncertainty structure; e.g., see Lemma 8.2.3. For this reason, the theory in

Chapter 15 (which applies to independent uncertainty structures) is extended later in the chapter.

16.2   The Spectral Set

For a family of polynomials $\mathcal{P} = \{p(\cdot, q) : q \in Q\}$, in Section 9.6 we briefly introduced the spectral set

$$\sigma[\mathcal{P}] = \{z \in \mathbf{C} : p(z, q) = 0 \text{ for some } q \in Q\}$$

in the context of the root version of the Edge Theorem. We now study the spectral set in greater detail. For the case of spherical polynomial families, our goal is to obtain a useful characterization of the boundary of $\sigma[\mathcal{P}]$. One obvious motivation for the study of $\sigma[\mathcal{P}]$ is a desire to know something more about the distribution of the roots of $\mathcal{P}$ (rather than a simple yes or no answer as to whether all roots lie in a given region $\mathcal{D}$). From a control systems point of view, knowing the root distribution is often useful in robust performance analysis. For example, simply knowing that a system has strict left half plane poles does not indicate whether the damping is acceptable. More detailed information about the distribution of the system's poles is important from a performance point of view.

To illustrate the usefulness of a spectral set characterization in comparison to a yes–no solution by robust $\mathcal{D}$-stability analysis, suppose that we seek a description of the "minimal" damping cone which contains all the roots of $p(s, q)$ for $q \in Q$. Notice that a candidate cone is parameterized by an angle θ; e.g., for $0 < \theta < \pi/2$, let

$$\mathcal{D}_\theta = \{z \in \mathcal{C} : \pi - \theta < \sphericalangle z < \pi + \theta\}.$$

Now, by gradually decreasing θ from $\pi/2$ to 0 and applying a robust $\mathcal{D}$-stability test for each θ, we obtain a practical solution to the problem. However, this type of solution method involves some sort of recursive scheme requiring iterative adjustment of θ. We arrive at the minimal damping cone but gain little insight into the root distribution. For more complicated root location regions, a similar iterative scheme can also be developed. Once the shape of the $\mathcal{D}$ region is specified, we can carry out iterations involving expansion and contraction of $\mathcal{D}$.

In the next section, our main objective is to avoid $\mathcal{D}$ region iteration completely. To this end, we develop formulas which describe the boundary of $\sigma[\mathcal{P}]$. With these formulas, we can display $\sigma[\mathcal{P}]$ graphically and immediately know the solution to the robust $\mathcal{D}$-stability problem for every possible $\mathcal{D}$ region. That is, $\mathcal{P}$ is robustly $\mathcal{D}$-stable if and only if $\sigma[\mathcal{P}] \subset \mathcal{D}$. This situation is depicted in Figure 16.2.1.

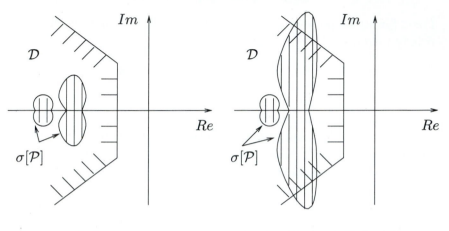

Robust $\mathcal{D}$-Stability   Lack of Robust $\mathcal{D}$-Stability

FIGURE 16.2.1   Relationship Between $\sigma(\mathcal{P})$ and $\mathcal{D}$

16.3   Formula and Theorem of Barmish and Tempo

Given a spherical polynomial family $\mathcal{P}$ having invariant degree, we now develop a formula which characterizes the boundary of $\sigma[\mathcal{P}]$. To this end, we construct a function $\Phi(z)$ having the property that $z \in \sigma[\mathcal{P}]$ if and only if $\Phi(z) \leq 0$. The justification of the construction below is relegated to the proof of Theorem 16.3.4.

DEFINITION 16.3.1 (The Spectral Set Weighting Matrix):   Given a spherical polynomial family $\mathcal{P} = \{p(\cdot, q) : q \in Q\}$ with invariant degree $n \geq 1$, the *spectral set weighting matrix* is defined by

$$W(z) = \left(A(z) A^T(z) \right)^{-1},$$

where

$$A(z) = \begin{bmatrix} 1 & Re\ z & \cdots & Re\ z^n \\ 0 & Im\ z & \cdots & Im\ z^n \end{bmatrix}.$$

EXERCISE 16.3.2 (Closed Form Description of $W(z)$):   Show that a closed form for the spectral set weighting matrix above is

$$W(z) = \frac{1}{\Delta(z)} \begin{bmatrix} \sum_{i=1}^{n}(Im\ z^i)^2 & -\sum_{i=1}^{n}(Re\ z^i)(Im\ z^i) \\ -\sum_{i=1}^{n}(Re\ z^i)(Im\ z^i) & \sum_{i=0}^{n}(Re\ z^i)^2 \end{bmatrix},$$

where

$$\Delta(z) = \sum_{i=0}^{n}(Re\ z^i)^2 \cdot \sum_{i=1}^{n}(Im\ z^i)^2 - \left(\sum_{i=1}^{n}(Re\ z^i)(Im\ z^i)\right)^2.$$

DEFINITION 16.3.3 (Spectral Set Boundary Function $\Phi(z)$): Given a spherical polynomial family $\mathcal{P} = \{p(\cdot, q) : q \in Q\}$ with invariant degree $n \geq 1$, uncertainty bound $r \geq 0$ and nominal $p_0(s)$, the *spectral set boundary function* $\Phi : \mathbf{C} \to \mathbf{R}^2$ is defined by

$$\Phi(z) = \begin{cases} \vec{p_0}^T(z) W(z) \vec{p_0}(z) - r^2 & \text{if } Im\ z \neq 0; \\ p_0^2(z) - r^2 \sum_{i=0}^{n} z^{2i} & \text{if } Im\ z = 0. \end{cases}$$

where

$$\vec{p_0}(z) = \begin{bmatrix} Re\ p_0(z) \\ Im\ p_0(z) \end{bmatrix}$$

is the vector representation for $p_0(z)$.

THEOREM 16.3.4 (Barmish and Tempo (1991)): *Given a spherical polynomial family $\mathcal{P} = \{p(\cdot, q) : q \in Q\}$ with invariant degree $n \geq 1$, it follows that $z \in \sigma[\mathcal{P}]$ if and only if*

$$\Phi(z) \leq 0.$$

PROOF: Given a candidate point $z \in \mathbf{C}$, observe that $z \in \sigma[\mathcal{P}]$ if and only if $p(z, q) = 0$ for some $q \in Q$. Now, with $A(z)$ as given in Definition 16.3.1 and $p(s, q) = p_0(s) + \sum_{i=0}^{n} q_i s^i$, we express the condition $p(z, q) = 0$ in matrix form; i.e., $z \in \sigma[\mathcal{P}]$ if and only if

$$A(z)q = -\vec{p_0}(z)$$

for some $q \in Q$. Equivalently, $z \in \sigma[\mathcal{P}]$ if and only if every minimum norm solution $q^{MN}(z)$ of the equation above satisfies

$$\|q^{MN}(z)\|_2 \leq r.$$

We now consider two cases.

Case 1: *Im* $z \neq 0$. Then rank $A(z) = 2$ and in accordance with Lemma 15.4.2, the minimum norm solution is unique and given by

$$q^{MN}(z) = -A^T(z)\left(A(z)A^T(z)\right)^{-1}\vec{p}_0(z) = -A^T(z)W(z)\vec{p}_0(z),$$

where $W(z)$ is the spectral set weighting matrix. Now, by enforcing the condition $\|q^{MN}(z)\|_2 \leq r$, it follows that $z \in \sigma[\mathcal{P}]$ if and only if

$$\vec{p}_0^T(z)W(z)\vec{p}_0(z) \leq r^2.$$

Equivalently, $z \in \sigma[\mathcal{P}]$ if and only if $\Phi(z) \leq 0$.

Case 2: *Im* $z = 0$. In this case, $\vec{p}_0(z)$ is real, *Re* $z^i = z^i$ for $i = 0, 1, 2, \ldots, n$ and *Im* $z^i = 0$ for $i = 1, 2, \ldots, n$. Hence, the spectral set membership condition $A(z)q = -\vec{p}_0(z)$ becomes

$$\begin{bmatrix} 1 & z & z^2 & \cdots & z^n \end{bmatrix} q = -p_0(z).$$

Again invoking Lemma 15.4.2, the minimum norm solution for this equation is uniquely given by

$$q^{MN}(z) = -\frac{1}{\displaystyle\sum_{i=0}^{n} z^{2i}} \begin{bmatrix} 1 \\ z \\ z^2 \\ \vdots \\ z^n \end{bmatrix} p_0(z).$$

Enforcing the condition $\|q^{MN}(z)\|_2 \leq r$, a straightforward calculation leads to the conclusion that $z \in \sigma[\mathcal{P}]$ if and only if $\Phi(z) \leq 0$. ∎

REMARKS 16.3.5 (Computational Aspects): The characterization of the spectral set via the inequality $\Phi(z) \leq 0$ makes it possible to use a contour plotting routine to display results in an easy-to-understand manner. In this regard, notice that $\Phi(z)$ is rational in z off the real axis and polynomic in z on the real axis. Moreover,

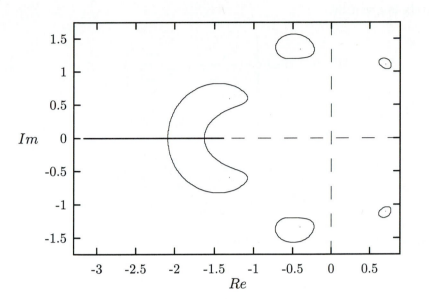

FIGURE 16.3.1   Spectral Set for Example 16.3.6

since z is only two-dimensional, we obtain a graphical display. This is illustrated via an example below.

EXAMPLE 16.3.6 (Spectral Set Generation):   For the spherical polynomial family $\mathcal{P}$ with nominal $p_0(s) = s^6 + 2s^5 + 3s^4 + 4s^3 + 5s^2 + 6s + 7$ and uncertainty bound $r = 0.55$, a contour plot for the function $\Phi(z)$ was generated. The resulting spectral set $\sigma[\mathcal{P}]$, characterized by $\Phi(z) \leq 0$, is displayed in Figure 16.3.1.

EXERCISE 16.3.7 (Coalescence of Root Clouds):   For the example above, generate $\sigma[\mathcal{P}]$ for smaller values of the uncertainty bound r and verify that the six distinct root clouds are obtained when the uncertainty bound r is small enough.

EXERCISE 16.3.8 (Recovery of Soh–Berger–Dabke Theorem):   The objective of this exercise is to develop a concrete connection between the robust stability problem and the spectral set generation problem. To this end, suppose that $\mathcal{P}$ is a spherical polynomial family with invariant degree $n \geq 1$ and stable nominal $p_0(s)$. Argue that robust stability is guaranteed if and only if the spectral set does not cross the imaginary axis. Now, restricting attention to the imaginary axis, take $z = j\omega$ and show that for $\omega > 0$, the spectral set weighting matrix is given by

$$W(\omega) = \text{diag} \left\{ \frac{1}{\displaystyle\sum_{i\ even} \omega^{2i}} \, , \, \frac{1}{\displaystyle\sum_{i\ odd} \omega^{2i}} \right\}.$$

Subsequently, by invoking Theorem 16.3.4, conclude that $j\omega \in \sigma[\mathcal{P}]$ if and only if

$$\frac{(Re\ p_0(j\omega))^2}{\displaystyle\sum_{i\ even} \omega^{2i}} + \frac{(Im\ p_0(j\omega))^2}{\displaystyle\sum_{i\ odd} \omega^{2i}} \leq r^2.$$

Equivalently, $j\omega \notin \sigma[\mathcal{P}]$ if and only if

$$G_{SBD}(\omega) > r^2,$$

where $G_{SBD}(\omega)$ is the Soh–Berger–Dabke function given in the statement of Theorem 15.5.1.

16.4   Affine Linear Uncertainty Structures

Our objective in this section is to refine the analysis of Chapter 15 to account for affine linear uncertainty structures which arise from feedback interconnections involving a spherical plant family and a compensator; see Section 8.2 where this issue is first discussed. In this regard, note that the Soh–Berger–Dabke Theorem (see Section 15.5) does not apply because the closed loop polynomial does not inherit the independent uncertainty structure of the plant.

The analysis begins with the spherical plant family $\mathcal{P}$ described by the quotient of uncertain polynomials,

$$P(s, q, r) = \frac{N(s, q)}{D(s, r)},$$

and uncertainty bound (Q, R) which is a sphere of radius $\rho \geq 0$; recalling the notational convention in Section 15.3, $(q, r) \in (Q, R)$ if and only if $\|(q, r)\|_2 \leq \rho$. To make the uncertainty structure more explicit, we write

$$N(s, q) = N_0(s) + \sum_{i=0}^{m} q_i s^i$$

and

$$D(s,r) = D_0(s) + \sum_{i=0}^{n} r_i s^i$$

with $N_0(s)$ and $D_0(s)$ representing the *nominal* numerator and denominator, respectively. Finally, we also assume that a compensator $C(s)$ is given. Expressing the compensator as the quotient of coprime polynomials

$$C(s) = \frac{N_C(s)}{D_C(s)},$$

the resulting closed loop polynomial is

$$p(s, q, r) = N(s, q)N_C(s) + D(s, r)D_C(s).$$

16.5   The Testing Function for Robust Stability

With the setup above, our first objective is to describe the construction of the robust stability testing function $G_{BHB}(\omega)$ of Biernacki, Hwang and Bhattacharyya (1987). Analogous to the function given in the Soh–Berger–Dabke Theorem (see Section 15.5), the infimum of $G_{BHB}(\omega)$ is central to the robust stability test.

There are two basic ingredients involved in the recipe for the function $G_{BHB}(\omega)$. We first introduce the vector notation

$$\vec{p}(j\omega, q, r) = \begin{bmatrix} Re\ p(j\omega, q, r) \\ Im\ p(j\omega, q, r) \end{bmatrix}$$

to represent the closed loop polynomial $p(s, q, r)$ at frequency $\omega \geq 0$. Particularizing below to $q = 0$ and $r = 0$, the first ingredient in the $G_{BHB}(\omega)$ formula is

$$\vec{p}(j\omega, 0, 0) = \begin{bmatrix} Re(N_0(j\omega)N_C(j\omega) + D_0(j\omega)D_C(j\omega)) \\ Im(N_0(j\omega)N_C(j\omega) + D_0(j\omega)D_C(j\omega)) \end{bmatrix}.$$

The second ingredient in the $G_{BHB}(\omega)$ formula is a 2×2 symmetric weighting matrix $W(\omega)$ whose inverse $W^{-1}(\omega)$ has entries

$$[W^{-1}(\omega)]_{1,1} = (Re\ N_C(j\omega))^2 \sum_{\substack{i\ even}}^{m} \omega^{2i} + (Re\ D_C(j\omega))^2 \sum_{\substack{i\ even}}^{n} \omega^{2i}$$

$$+ (Im\ N_C(j\omega))^2 \sum_{\substack{i\ odd}}^{m} \omega^{2i} + (Im\ D_C(j\omega))^2 \sum_{\substack{i\ odd}}^{n} \omega^{2i},$$

$$[W^{-1}(\omega)]_{1,2} = [W^{-1}(\omega)]_{2,1}$$

$$= (Re\ N_C(j\omega))(Im\ N_C(j\omega))\left(\sum_{\substack{i \\ even}}^{m} \omega^{2i} - \sum_{i\ odd}^{m} \omega^{2i}\right)$$

$$+ (Re\ D_C(j\omega))(Im\ D_C(j\omega))\left(\sum_{\substack{i \\ even}}^{n} \omega^{2i} - \sum_{i\ odd}^{n} \omega^{2i}\right)$$

and

$$[W^{-1}(\omega)]_{2,2} = (Im\ N_C(j\omega))^2 \sum_{i\ even}^{m} \omega^{2i} + (Re\ N_C(j\omega))^2 \sum_{i\ even}^{m} \omega^{2i}$$

$$+ (Im\ D_C(j\omega))^2 \sum_{i\ odd}^{n} \omega^{2i} + (Re\ D_C(j\omega))^2 \sum_{i\ odd}^{n} \omega^{2i}.$$

Using the notation above, we now define

$$G_{BHB}(\omega) = \vec{p}^T(j\omega, 0, 0)W(\omega)\vec{p}(j\omega, 0, 0).$$

We are now prepared to state the robustness criterion. To avoid degenerate cases, we assume below that the plant numerator and denominator are at least first order; the degenerate zero order cases are analyzed in the exercises of Section 16.8. The proof of theorem below is relegated to Sections 16.6 and 16.7.

THEOREM 16.5.1 (Biernacki, Hwang and Bhattacharyya (1987)):
Consider a spherical family of plants which is described by $\mathcal{P} = \{P(\cdot, q, r) : (q, r) \in (Q, R)\}$ with uncertainty bound $\|(q, r)\|_2 \le \rho$ and with compensator $C(s) = N_C(s)/D_C(s)$ with $N_C(s)$ and $D_C(s)$ coprime. Assume that the associated family of closed loop polynomials $\mathcal{P}_{CL} = \{p(\cdot, q, r) : (q, r) \in (Q, R)\}$ has invariant degree and the nominal plant numerator $N_0(s)$ and denominator $D_0(s)$ are polynomials of order one or more. Then $\mathcal{P}_{CL}$ is robustly stable if and only if

$$p(s, 0, 0) = N_0(s)N_C(s) + D_0(s)D_C(s)$$

is stable, the zero frequency condition

$$\rho < \left|\frac{p(j0, 0, 0)}{C(j0)}\right|$$

is satisfied and

$$G_{BHB}(\omega) > \rho^2$$

for all frequencies $\omega > 0$.

REMARKS 16.5.2 (Associated Robustness Margin):   Before proving the theorem, we complete our generalization of the theory in Chapter 15 by developing robustness margin formulas. The standing assumptions are those given in Theorem 16.5.1 above. To emphasize the dependence on the uncertainty bound ρ, we use the notation $(Q, R)_\rho$ instead of (Q, R) and take

$$\mathcal{P}_{CL,\rho} = \{p(\cdot, q, r) : (q, r) \in (Q, R)_\rho\}$$

to be the associated family of closed loop polynomials. Hence, the *robustness margin* of interest is

$$\rho_{max} = \sup\{\rho : \mathcal{P}_{CL,\rho} \text{ has invariant degree and is robustly stable}\}.$$

Analogous to Section 15.5, we are going to describe the robustness margin as the minimum of three quantities $|\rho_0|$, $|\rho_n|$ and ρ_{max}^+. Beginning with the enforcement of the zero frequency condition in Theorem 16.5.1, we obtain

$$\rho_0 = \left| \frac{p(j0, 0, 0)}{C(j0)} \right| = \left| \frac{N_0(j0)N_C(0) + D_0(j0)D_C(j0)}{C(j0)} \right|.$$

Next, for frequencies $\omega > 0$, we define

$$\rho_{max}^+(\omega) = G_{BHB}^{\frac{1}{2}}(\omega)$$

and obtain

$$\rho_{max}^+ = \inf_\omega \rho_{max}^+(\omega).$$

Finally, to enforce the invariant degree requirement, we consider two cases. In the analysis below, we use the notation a_m, b_n, c_{m_c} and d_{n_c} to denote the highest order coefficients of $N_0(s)$, $D_0(s)$, $N_C(s)$ and $D_C(s)$, respectively.

Case 1: If $P(s, q, r)C(s)$ is strictly proper for all $(q, r) \in (Q, R)$, then the family $\mathcal{P}_{CL}$ has invariant degree if and only if the plant denominator has invariant degree; i.e., $|r_n| < b_n$. Hence, in this case, we take

$$\rho_n = |b_n|.$$

Case 2: If $P(s, q, r)C(s)$ is proper for all $(q, r) \in (Q, R)$ (but not strictly proper), a straightforward calculation leads to the invariant degree condition $|a_m c_{m_c} + b_n d_{n_c}| > q_m c_{m_c} + r_n d_{n_c}$. Hence, by maximizing the right-hand side above with respect to all pairs (q_m, r_n)

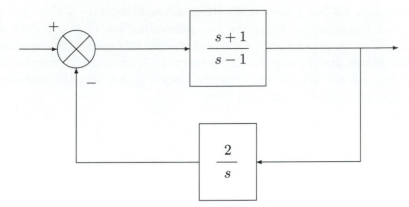

FIGURE 16.5.1   Nominal System for Example 16.5.3

satisfying $\|(q,r)\|_2 \leq \rho$, we conclude that invariant degree of the family $\mathcal{P}_{CL,\rho}$ is guaranteed if and only if $\rho < \rho_n$, where

$$\rho_n = \frac{a_m c_{m_c} + b_n d_{n_c}}{\sqrt{c_{m_c}^2 + d_{n_c}^2}}.$$

In summary, from the analysis above, it follows that

$$\rho_{max} = \min\{\rho_0, \rho_n, \rho_{max}^+\}.$$

EXAMPLE 16.5.3 (Computation of Robustness Margin):   For the first order unstable system controlled via an integrator in Figure 16.5.1, the nominal closed loop polynomial $p(s) = s^2 + s + 2$ is stable. We now compute the robustness margin ρ_{max} for the uncertain plant

$$P(s,q) = \frac{(q_1 + 1)s + (q_0 + 1)}{(r_1 + 1)s + (r_0 - 1)}.$$

Using the formulas above, we obtain $\rho_0 = 1$ and $\rho_n = 1$ by inspection. Now, to generate $\rho_{max}^+(\omega)$, we require the weighting matrix $W(\omega)$. Using the formulas in Section 16.5, we compute the entries $\left[W^{-1}(\omega)\right]_{1,1} = \omega^4 + 4$; $\left[W^{-1}(\omega)\right]_{1,2} = \left[W^{-1}(\omega)\right]_{2,1} = 0$; $\left[W^{-1}(\omega)\right]_{2,2} = \omega^4 + 4$. Next, we calculate

$$\vec{p}(j\omega, 0, 0) = \begin{bmatrix} 2 - \omega^2 \\ \omega \end{bmatrix}$$

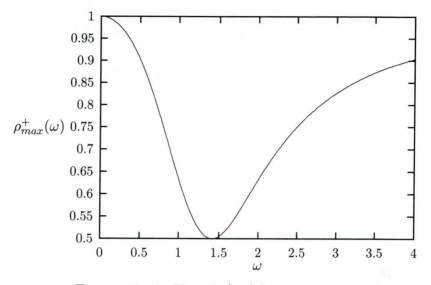

FIGURE 16.5.2   Plot of $\rho_{max}^+(\omega)$ for Example 16.5.3

and

$$G_{BHB}(\omega) = \bar{p}^T(j\omega, 0, 0)W(\omega)\bar{p}(j\omega, 0, 0) = \frac{\omega^4 - 3\omega^2 + 4}{\omega^4 + 4},$$

which leads to

$$\rho_{max}^+(\omega) = G_{BHB}^{\frac{1}{2}}(\omega) = \sqrt{\frac{\omega^4 - 3\omega^2 + 4}{\omega^4 + 4}}.$$

The plot of $\rho_{max}^+(\omega)$ over the relevant frequency range is indicated in Figure 16.5.2. From the graph, we see that the minimum occurs at $\omega^* \approx 1.41$ with associated value $\rho_{max}^+ \approx 0.50$. Hence, we conclude that $\rho_{max} = \min\{|\rho_0|, |\rho_n|, \rho_{max}^+\} \approx \min\{1, 1, 0.50\} = 0.50$.

16.6   Machinery for Proof of the Theorem

We now develop some technical machinery associated with the proof of Theorem 16.5.1. Paralleling the development in Section 15.6, we proceed to characterize the value set

$$p(j\omega, Q, R) = \{p(j\omega, q, r) : (q, r) \in (Q, R)\}$$

for the family of closed loop polynomials. Throughout this section, we work with the feedback system described in Section 16.4 and the standing assumptions in Theorem 16.5.1. Indeed, for fixed $\omega > 0$ and $z \in \mathbf{C}$, we begin by noting that $z \in p(j\omega, Q, R)$ if and only if

$$
z = p(j\omega, 0, 0) + \sum_{i=0}^{m} q_i \omega^i N_C(j\omega) + \sum_{i=0}^{n} r_i \omega^i D_C(j\omega)
$$

for some $(q, r) \in (Q, R)$. Equivalently, using vector notation, it follows that $z \in p(j\omega, Q, R)$ if and only if

$$
\vec{z} = \begin{bmatrix} Re\ N_C(j\omega) & -\omega Im\ N_C(j\omega) & -\omega^2 Re\ N_C(j\omega) & \omega^3 Im\ N_C(j\omega) & \cdots \\ Im\ N_C(j\omega) & \omega Re\ N_C(j\omega) & -\omega^2 Im\ N_C(j\omega) & -\omega^3 Re\ N_C(j\omega) & \cdots \end{bmatrix} q
$$

$$
+ \begin{bmatrix} Re\ D_C(j\omega) & -\omega Im\ D_C(j\omega) & -\omega^2 Re\ D_C(j\omega) & \omega^3 Im\ D_C(j\omega) & \cdots \\ Im\ D_C(j\omega) & \omega Re\ D_C(j\omega) & -\omega^2 Im\ D_C(j\omega) & -\omega^3 Re\ D_C(j\omega) & \cdots \end{bmatrix} r
$$

$$
+ \vec{p}(j\omega, 0, 0)
$$

for some $(q, r) \in (Q, R)$. Factoring the expressions above, it follows that $z \in p(j\omega, Q, R)$ if and only if

$$
\vec{z} = \begin{bmatrix} Re\ N_C(j\omega) & -Im\ N_C(j\omega) \\ Im\ N_C(j\omega) & Re\ N_C(j\omega) \end{bmatrix} \begin{bmatrix} 1 & 0 & -\omega^2 & 0 & \cdots \\ 0 & \omega & 0 & -\omega^3 & \cdots \end{bmatrix} q
$$

$$
+ \begin{bmatrix} Re\ D_C(j\omega) & -Im\ D_C(j\omega) \\ Im\ D_C(j\omega) & Re\ D_C(j\omega) \end{bmatrix} \begin{bmatrix} 1 & 0 & -\omega^2 & 0 & \cdots \\ 0 & \omega & 0 & -\omega^3 & \cdots \end{bmatrix} r
$$

$$
+ \vec{p}(j\omega, 0, 0)
$$

for some $(q, r) \in (Q, R)$. We define the 2×2 *compensator matrices*

$$
\mathcal{N}_C(\omega) = \begin{bmatrix} Re\ N_C(j\omega) & -Im\ N_C(j\omega) \\ Im\ N_C(j\omega) & Re\ N_C(j\omega) \end{bmatrix},
$$

$$
\mathcal{D}_C(\omega) = \begin{bmatrix} Re\ D_C(j\omega) & -Im\ D_C(j\omega) \\ Im\ D_C(j\omega) & Re\ D_C(j\omega) \end{bmatrix},
$$

the $2 \times (m + 1)$ matrix

$$
A_N(\omega) = \begin{bmatrix} 1 & 0 & -\omega^2 & 0 & \omega^4 & 0 & -\omega^6 & \cdots \\ 0 & \omega & 0 & -\omega^3 & 0 & \omega^5 & 0 & \cdots \end{bmatrix}
$$

for the numerator and the $2 \times (n+1)$ matrix $A_D(\omega)$ for the denominator having the same form as $A_N(\omega)$. Now, with

$$\Omega_C(\omega) = [\mathcal{N}_C(\omega) \ \ \mathcal{D}_C(\omega)]$$

and

$$A(\omega) = \begin{bmatrix} A_N(\omega) & 0 \\ 0 & A_D(\omega) \end{bmatrix},$$

it follows that $z \in p(j\omega, Q, R)$ if and only if

$$\vec{z} = \Omega_C(\omega)A(\omega) \begin{bmatrix} q \\ r \end{bmatrix} + \vec{p}(j\omega, 0, 0)$$

for some $(q, r) \in (Q, R)$.

The arguments to follow parallel those given in Section 15.6. We first observe that the condition $z \in p(j\omega, Q, R)$ is equivalent to the following: Any minimum norm solution $(q^{MN}(\omega), r^{MN}(\omega))$ of the $\vec{z}$ equation above satisfies

$$\|(q^{MN}(\omega), r^{MN}(\omega))\|_2 \leq \rho.$$

We now claim that the minimum norm solution is unique. To prove this claim, we recall Lemma 15.4.2; i.e., it suffices to show that the standing assumptions guarantee

$$\text{rank } \Omega_C(\omega)A(\omega) = 2.$$

Indeed, notice that the presence of the diagonal subblocks $\text{diag}\{1, \omega\}$ in both $A_N(\omega)$ and $A_D(\omega)$ assure that rank $A(\omega) = 4$. Furthermore, exploiting coprimeness of $N_C(s)$ and $D_C(s)$, it follows that either $\det \mathcal{N}_C(\omega) \neq 0$ or $\det \mathcal{D}_C(\omega) \neq 0$. Hence, we are sure that rank $\Omega_C(\omega) = 2$, which, when combined with rank $A(\omega) = 4$, guarantees that rank $\Omega_C(\omega)A(\omega) = 2$.

Having established the rank condition above, we now obtain the unique minimum norm solution. To this end we call

$$W(\omega) = [\Omega_C(\omega)A(\omega)A^T(\omega)\Omega_C^T(\omega)]^{-1}$$

the *value set weighting matrix* and generate

$$[q^{MN}(\omega) \ \ r^{MN}(\omega)] = A^T(\omega)\Omega_C^T(\omega)W(\omega)[\vec{z} - \vec{p}(j\omega, 0, 0)].$$

Furthermore, a straightforward calculation using the expression for $(q^{MN}(\omega), r^{MN}(\omega))$ above leads to

$$\|(q^{MN}(\omega), r^{MN}(\omega))\|_2^2 = [\vec{z} - \vec{p}(j\omega, 0, 0)]^T W(\omega)[\vec{z} - \vec{p}(j\omega, 0, 0)] .$$

In summary, given any frequency $\omega > 0$ and $z \in \mathbf{C}$, it follows that $z \in p(j\omega, Q, R)$ if and only if $(q^{MN}(\omega), r^{MN}(\omega))$ has at most norm ρ. Equivalently, z lies in the ellipse described by

$$[\vec{z} - \vec{p}(j\omega, 0, 0)]^T W(\omega)[\vec{z} - \vec{p}(j\omega, 0, 0)] \le \rho^2.$$

EXERCISE 16.6.1 (The Weighting Matrix):   After expressing the inverse of the *value set weighting matrix* as

$$W^{-1}(\omega) = [\mathcal{N}_C(\omega)A_N(\omega)A_N^T(\omega)\mathcal{N}_C^T(\omega) + \mathcal{D}_C(\omega)A_D(\omega)A_D^T(\omega)\mathcal{D}_C^T(\omega)]^{-1},$$

verify that for $\omega > 0$, the entries of 2×2 inverse $W^{-1}(\omega)$ are those given in Section 16.5.

EXERCISE 16.6.2 (Zero Frequency):   For frequency $\omega = 0$, show that the value set is the real interval

$$p(j0, Q, R) = [p(j0, 0, 0) - \rho|C(j0)|, p(j0, 0, 0) + \rho|C(j0)|].$$

16.7   Proof of the Theorem

To prove Theorem 16.5.1, we first note that stability of $p(s, 0, 0)$ is necessary for robust stability of $\mathcal{P}_{CL}$ because $(q, r) = (0, 0)$ is admissible. Also, at $\omega = 0$, using the fact that the value set is the interval given in Exercise 16.6.2, it follows that satisfaction of the zero frequency condition

$$\rho < \left|\frac{p(j0, 0, 0)}{C(j0)}\right|$$

is also necessary for robust stability of $\mathcal{P}_{CL}$. Therefore, in the remainder of the proof, without loss of generality, we assume that $p(s, 0, 0)$ is stable and the zero frequency condition is satisfied. It remains to prove that for $\omega > 0$, the condition $G_{BHB}(\omega) > \rho^2$ is both necessary and sufficient for robust stability of $\mathcal{P}_{CL}$.

Since all the preconditions for application of the Zero Exclusion Condition (Theorem 7.4.2) are satisfied, it follows that $\mathcal{P}_{CL}$ is robustly stable if and only if $0 \notin p(j\omega, Q, R)$ for all $\omega > 0$. Now, using the description of the value set ellipse given in Section 16.6 above, we see immediately that $0 \notin p(j\omega, Q, R)$ if and only if

$$\vec{p}^T(j\omega, 0, 0)W(\omega)\vec{p}(j\omega, 0, 0) > \rho^2.$$

Using the expressions for the entries of the weighting matrix found in Exercise 16.6.1, we recognize that the inequality above is equivalent to $G_{SBD}(\omega) > \rho^2$. ∎

16.8   Some Refinements

In the preceding analysis, we made a number of assumptions for pedagogical purposes. The objective in this section is provide some exercises aimed at various special cases which we excluded for simplicity. Throughout this section, $\mathcal{P} = \{P(\cdot, q, r) : (q, r) \in (Q, R)\}$ is taken to be a spherical plant family with compensator $C(s)$ as described in Section 16.4 and Theorem 16.5.1.

EXERCISE 16.8.1 (Zero Order Numerator and Denominator):   One degenerate case which we omitted in Theorem 16.5.1 is characterized by $m = n = 0$. If the family of closed loop polynomials $\mathcal{P}_{CL}$ has invariant degree and $p(s, 0, 0)$ is stable, verify that robust stability of $\mathcal{P}_{CL}$ is equivalent to satisfaction of the Zero Exclusion Condition

$$0 \notin [p(j\omega, 0, 0) - \rho|C(j\omega)|, p(j\omega, 0, 0) + \rho|C(j\omega)|]$$

at all frequencies $\omega \geq 0$.

EXERCISE 16.8.2 (Other (m, n) Combinations):   With $m = 0$ and $n \geq 1$, assume that $\mathcal{P}_{CL}$ has invariant degree and $p(s, 0, 0)$ is stable. Now, refine the $G_{BHB}(\omega)$ formula in Theorem 16.5.1 and observe that the zero frequency condition remains unchanged.

EXERCISE 16.8.3 (Only a Subset of Coefficients Uncertain):   Letting $I_N \subseteq \{0, 1, 2, \ldots, m\}$ denote an index set for the plant numerator, we take

$$N(s, q) = N_0(s) + \sum_{i \in I_N} q_i s^i.$$

Similarly, letting $I_D \subseteq \{0, 1, 2, \ldots, m\}$ denote an index set for the plant denominator, we take

$$D(s, r) = D_0(s) + \sum_{i \in I_D} r_i s^i.$$

For the nondegenerate cases when I_N and I_D each have cardinality two or more, provide a refinement of Theorem 16.5.1.

EXERCISE 16.8.4 (Weighted Norm): Provide a refinement of Theorem 16.5.1 for the case when a weighted euclidean norm $\|(q,r)\|_{2,W}$ is used in lieu of the standard euclidean norm $\|(q,r)\|_2$.

16.9   Conclusion

This chapter completes Part IV of this text. The remaining chapters provide a sampling of results which may be of interest to the more advanced reader. Some of the technical developments are slightly more abstract than those given in the earlier chapters and some of the results are probably of more interest to the researcher than the practitioner. In particular, Chapter 19 includes detailed descriptions of a number of open problems.

Notes and Related Literature

NRL 16.1 The results given on the spectral set also have an obvious root locus interpretation. Namely, one can introduce a gain K into $p(s, q)$ and generate spectral sets for different values of K; see the paper by Barmish and Tempo (1990) for further details.

NRL 16.2 As early as the fifties, authors have formulated multivariable root locus problems which are similar to the spectral set generation problem; e.g., see the textbook by Truxal (1955). In the later work by Zeheb and Walach (1977), a two-parameter root locus problem is considered and rather specific assumptions (motivated by circuit theory) are made about the uncertainty structure. A number of papers following Zeheb and Walach's work deal with the so-called *zero set concept*; e.g., see Zeheb and Walach (1981) and Fruchter, Srebro and Zeheb (1987). It is seen that the zero set provides a rather general framework for dealing with multivariable root loci. In practice, however, the computational complexity associated with this approach is high.

NRL 16.3 An extension of the spectral set theory of Sections 16.2 and 16.3 is pursued in a paper by Monov (1992). The author develops a characterization of the spectral set for a family of matrices $\mathcal{A}$. This set is described by an uncertain $n \times n$ matrix $A(q) = A_0 + \sum_{i=0}^{\ell} q_i A_i$ with $A_i \in \mathbf{R}^{n \times n}$ fixed for $i = 0, 1, 2, \dots, \ell$ and a spherical uncertainty bounding set Q for q. In addition, it is assumed that the A_i can be simultaneously upper triangularized; i.e., there exists a fixed matrix V such that $V^{-1} A_i V$ is upper triangular for $i = 0, 1, 2, \dots, \ell$. An example of such a set of A_i is a commutative family; i.e., if $A_{i_1} A_{i_2} = A_{i_2} A_{i_1}$ for all $i_1, i_2 \in \{0, 1, 2, \dots, \ell\}$, then V can be taken to be unitary; e.g., see Horn and Johnson (1988) for a more complete description of sets of matrices which can be simultaneously upper triangularized.

NRL 16.4 Our analysis in this chapter applies to the class of affine linear uncertainty structures generatable from the feedback system setup described in Section 16.4. A number of authors (for example, see Hinrichsen and Prichard (1989) and Tsypkin and Polyak (1991)) begin at the more basic level of polynomials and derive similar results. To this end, these authors work with an uncertain polynomial of the form $p(s, q) = p_0(s) + \sum_{i=0}^{\ell} q_i p_i(s)$, where $p_0(s), p_1(s), \dots, p_\ell(s)$ are fixed polynomials. Now, with uncertainty bounding set taken to be a sphere, robust stability criteria are developed. With this setup, however, a characterization of the spectral set has not been given.

NRL 16.5 Some of the robustness margin problems of Chapters 15 and 16 also have matrix versions. For example, if A is a stable real $n \times n$ matrix, it is of interest to find $r_{max} = \sup\{r : A + \Delta A \text{ is stable for all } \|\Delta A\| \leq r\}$, where admissible perturbations are real $n \times n$ matrices with $\|\Delta A\| = \bar{\sigma}(\Delta A)$ and $\bar{\sigma}(\Delta A)$ denotes the largest singular value of ΔA. Note that no "clean" formula for r_{max} has been given in the literature to date.

NRL 16.6 For the matrix problem in the note above, the lower bound of Qiu and Davison (1992) is interesting because it may turn out to be sharp. They show that

$$r_{max} \geq \min\{\underline{\sigma}(A), \beta(A), \}$$

where $\underline{\sigma}(A)$ is the smallest singular value of A,

$$\beta(A) = \inf_{\omega > 0} \; \sup_{0 < \gamma \leq 1} \; \sigma_{2n-1} \begin{bmatrix} A & \gamma \omega I \\ -\frac{\omega}{\gamma} I & A \end{bmatrix}$$

and $\sigma_k(M)$ denotes the k-th largest singular value of a matrix M.

Part V

Some Happenings at the Frontier

Chapter 17

An Introduction to Guardian Maps

Synopsis

In Chapter 4, we developed eigenvalue criteria for robust stability problems with one uncertain parameter entering affine linearly into the coefficients of the polynomial of interest. This chapter concentrates on the more general framework of Saydy, Tits and Abed. Using the guardian map concept, it becomes possible to deal with a general root location region $\mathcal{D}$ and classes of uncertainty structures which are not necessarily affine linear.

17.1   Introduction

To motivate the technical exposition of this chapter, we review the Bialas criterion given in Section 4.11. Namely, given a family of polynomials $\mathcal{P}$ with invariant degree described by $\lambda \in [0, 1]$ and

$$p(s, \lambda) = \lambda p_0(s) + (1 - \lambda)p_1(s)$$

with $p_0(s)$ stable, with positive coefficients and $\deg p_0(s) > \deg p_1(s)$, robust stability is assured if and only if the matrix $H^{-1}(p_0)H(p_1)$ has no purely real nonpositive eigenvalues; recall that $H(p_i)$ is the Hurwitz matrix for $p_i(s)$.

The criterion of Bialas above has some attractive features. First, it is quite general in the sense that no stringent assumptions on $p_0(s)$ and $p_1(s)$ are required. In contrast to the theory in Chapter 12, we are not insisting on an extreme point result. Recalling Chapter 12, the class of polynomials for which extreme point results apply is limited by the requirement that Rantzer's Growth Condition is satisfied. The second attractive feature of the Bialas criterion is more conceptual in nature. Since it is based on coefficient manipulation rather than frequency sweeping, we gain a degree of insight about the computational complexity associated with robust stability testing. The existence of a nonpositive eigenvalue of $H^{-1}(p_0)H(p_1)$ is decidable via a finite number of arithmetic operations; this issue is further pursued in the notes at the end of this chapter.

The third attractive feature of the Bialas criterion which we mention serves as the takeoff point for this chapter—its generalizability. We see in this chapter that eigenvalue criteria for robust stability can be derived for rather general one-parameter uncertainty structures. In fact, many of the ideas in the sections to follow admit generalizations involving more than one uncertain parameter.

17.2 Overview

Since this chapter is somewhat more technical and abstract than many of its predecessors, we provide a brief overview of the exposition to follow: Given a desired root location region $\mathcal{D}$, we plan to construct a *guardian map* ν. When we want to solve stability problems involving polynomials, the domain of ν is the set of n-th order polynomials; when we want to solve matrix stability problems, the domain of ν is the set of $n \times n$ matrices. In either event, the range of ν is the reals and the construction of ν involves the desired root location region $\mathcal{D}$ and the declared order n of the polynomial.

Once ν is determined, we generate a *representation* with respect to the family of polynomials or matrices under consideration. For example, for a family of matrices $\mathcal{A} = \{A(\lambda) : \lambda \in [0, 1]\}$, we construct a polynomial matrix function

$$f(\lambda) = \sum_{i=0}^{N} F_i \lambda^i$$

which represents the action of ν on $A(\lambda)$ in the sense that

$$\nu(A(\lambda)) = \det\, f(\lambda).$$

The final step of the analysis involves construction of a matrix M (depending on $F_0, F_1, \ldots, F_N$) whose eigenvalues tell us whether the family $\mathcal{A}$ is robustly $\mathcal{D}$-stable.

17.3   Topological Preliminaries for Guardian Maps

The technical exposition to follow makes use of a number of basic topological concepts. We now consolidate this critical material.

DEFINITION 17.3.1 (Open Neighborhood and Open Sets):   Let $\mathcal{P}^n$ denote the set of n-th order polynomials. To each n-th order polynomial $p(s) = \sum_{i=0}^{n} a_i s^i$, we take $a = (a_0, a_1, \ldots, a_n)$ to be its coefficient representation and note that $a_n \neq 0$. In addition, we assume that some norm $\| \cdot \|$ is declared on $\mathbf{R}^{n+1}$. Now, given any $\epsilon > 0$ and some $p \in \mathcal{P}^n$, we call

$$B_\epsilon(p) = \{p' : p'(s) = \sum_{i=0}^{n} a_i' s^i \text{ and } \|a' - a\| \leq \epsilon\}$$

the *ball of radius ϵ centered at $p(s)$* and also refer to the interior of $B_\epsilon(p)$ as an *open neighborhood* of $p(s)$. We say that a set $\mathcal{P} \subseteq \mathcal{P}^n$ is *open* if the following property holds: Given any $p \in \mathcal{P}$ with associated coefficient vector $a \in \mathbf{R}^{n+1}$, there exists an $\epsilon > 0$ satisfying the following condition: If $a' \in \mathbf{R}^{n+1}$ is any other vector in $\mathbf{R}^{n+1}$ defining an n-th order polynomial p' and $\|a - a'\| < \epsilon$, then $p' \in \mathcal{P}$. In other words, every polynomial $p \in \mathcal{P}$ admits an open neighborhood which is wholly contained in $\mathcal{P}$.

REMARKS 17.3.2 (Characterizations):   Note that an open neighborhood $\mathcal{P}$ in $\mathcal{P}^n$ can be associated with an open neighborhood in $\mathbf{R}^{n+1}$. Furthermore, such a neighborhood has the following property: If $\{p_k\}_{k=1}^{\infty}$ is a sequence of polynomials converging to p (in the sense that their coefficient representations $\{a_k\}_{k=1}^{\infty}$ converge to a), then there exists a positive integer N such that $p_k \in \mathcal{P}$ for all $k \geq N$.

EXERCISE 17.3.3 (Association with Polynomial Coefficients):   Argue that every open set $\mathcal{A} \subseteq \mathbf{R}^{n+1}$ which does not contain points of the form $a = (a_0, a_1, \ldots, a_{n-1}, 0)$ can be associated with an open set $\mathcal{P}$ in the space of n-th order polynomials $\mathcal{P}^n$.

EXERCISE 17.3.4 (Open Sets in the Complex Plane):   Let $\mathcal{D}$ be an open subset of the complex plane. Using continuous dependence of roots on coefficients (Lemma 4.8.2), argue that the set of n-th order polynomials $\mathcal{P}$ having its roots in $\mathcal{D}$ is open.

DEFINITION 17.3.5 (Closure):   Let $\mathcal{P}$ be a set of n-th order polynomials. Then a polynomial $p^* \in \mathcal{P}^n$ is said to be a *point of closure* of $\mathcal{P}$ if the following condition is satisfied: Every open neighborhood of p^* contains points in $\mathcal{P}$. We use

$$\operatorname{cl} \mathcal{P} = \{p \in \mathcal{P}^n : p \text{ is a point of closure of } \mathcal{P}\}$$

to denote the *closure* of $\mathcal{P}$.

REMARKS 17.3.6 (Alternative Definition):   The closure of $\mathcal{P}$ can also be defined via sequences. Indeed, if the family $\mathcal{P}$ has coefficient set $\mathcal{A} \subseteq \mathbf{R}^{n+1}$, then a polynomial $p^*(s) = \sum_{i=0}^{n} a_i^* s^i$ is a point of closure of $\mathcal{P}$ if $a_n^* \neq 0$ and there exists a sequence $\{a_k\}_{k=1}^{\infty}$ in $\mathcal{A}$ converging to $a^* = (a_0^*, a_1^*, \ldots, a_n^*)$. That is, $\|a_k - a^*\| \to 0$ as $k \to \infty$.

EXERCISE 17.3.7 (Closure):   Let $\mathcal{P}$ be a set of n-th order polynomials which are all stable. Using continuous dependence of roots on coefficients (Lemma 4.8.2), argue that $\operatorname{cl} \mathcal{P}$ consists of all n-th order polynomials having all its roots in the closed left half plane. This set includes polynomials with roots on the imaginary axis.

DEFINITION 17.3.8 (Boundary of a Set of Polynomials):   Let $\mathcal{P}$ be a set of n-th order polynomials and take $p^*(s)$ to be an n-th order polynomial (not necessarily in $\mathcal{P}$) having associated coefficient vector given by a^*. Then we say that p^* *lies on the boundary of* $\mathcal{P}$ if every neighborhood of p^* contains n-th order polynomials both in $\mathcal{P}$ and in $\mathcal{P}^c$, the complement of $\mathcal{P}$. Equivalently,

$$\partial \mathcal{P} = \{p^* \in \mathcal{P}^n : B_\epsilon(p^*) \cap \mathcal{P} \neq \phi \text{ and } B_\epsilon(p^*) \cap \mathcal{P}^c \neq \phi \text{ for all } \epsilon > 0\}$$

denotes the set of boundary points of $\mathcal{P}$.

EXERCISE 17.3.9 (Boundary):   If $\mathcal{P}$ is the set of stable n-th order polynomials, argue that $\partial \mathcal{P}$ consists of all n-th order polynomials having one or more roots on the imaginary axis. *Hint*: First express the polynomial as $p(s) = K \prod_{i=1}^{n} (s + z_i)$.

REMARKS 17.3.10 (Topological Considerations for Matrices):   We now proceed to develop matrix analogues for some of the definitions above.   To this end, we view an $n \times m$ matrix A as an element in $\mathbf{R}^{m \times n}$ and define the relevant topological concepts in a manner entirely analogous to the polynomial case. For example, if $\| \cdot \|$ is a declared norm on the space of $n \times m$ matrices and $\epsilon > 0$ is given, the *ball of radius ϵ* centered at A is given by

$$B_\epsilon(A) = \{A' \in \mathbf{R}^{m \times n} : \|A' - A\| \leq \epsilon\}.$$

Similarly, a set of matrices $\mathcal{A} \subseteq \mathbf{R}^{m \times n}$ is said to be open if given any $A \in \mathcal{A}$, there exists an open neighborhood of A which is wholly contained in $\mathcal{A}$. Finally, if $\mathcal{A}$ is a set of $m \times n$ matrices, we define points of closure of $\mathcal{A}$ and the set cl $\mathcal{A}$ exactly as in the polynomial case. That is, A is a point of closure of $\mathcal{A}$ if every ball $B_\epsilon(A)$ about A contains points in both $\mathcal{A}$ and cl $\mathcal{A}$. As in the polynomial case, the boundary of $\mathcal{A}$, denoted by $\partial \mathcal{A}$, consists of all $m \times n$ matrices A having the property that every open neighborhood of A meets both $\mathcal{A}$ and $\mathcal{A}^c$, the complement of $\mathcal{A}$. Finally, we note that all of the topological concepts above can be interpreted in terms of sequences as in the polynomial case.

EXERCISE 17.3.11 (Topological Concepts for Matrices):   If $\mathcal{A}$ is the set of nonsingular $n \times n$ matrices, argue that cl $\mathcal{A} = \mathbf{R}^{n \times n}$ and $\partial \mathcal{A}$ consists of all $n \times n$ singular matrices.

17.4   The Guardian Map

In this section, we provide the formal definition of the guardian map and some simple exercises illustrating its construction. In order to construct other useful guardian maps, we introduce some additional mathematical machinery in the next section.

DEFINITION 17.4.1 (Guardian Map):   Let $\nu : \mathcal{P}^n \to \mathbf{R}$ be a given mapping. Then ν is said to *guard* an open set of n-th order polynomials $\mathcal{P}$ if $\nu(p) \neq 0$ for $p \in \mathcal{P}$ and $\nu(p) = 0$ for $p \in \partial \mathcal{P}$. For such a case, we call ν a *guardian map* for $\mathcal{P}$. For matrices, a similar definition applies. That is, if $\nu : \mathbf{R}^{n \times n} \to \mathbf{R}$, then ν is said to guard an open set of $n \times n$ matrices $\mathcal{A}$ if $\nu(A) \neq 0$ for $A \in \mathcal{A}$ and $\nu(A) = 0$ for $A \in \partial \mathcal{A}$. For such cases, we call ν a *guardian map* for $\mathcal{A}$.

EXERCISE 17.4.2 (Guardian Map for Nonsingularity):   Argue that the map defined by

$$\nu(A) = \det A$$

guards the set of $n \times n$ nonsingular matrices.

EXERCISE 17.4.3 (Polynomials Nonvanishing at Zero):   Argue that the map defined by

$$\nu(p) = p(0)$$

guards the set of n-th order polynomials $p(s)$ which do not vanish at the point $z = 0$.

17.5 Some Useful Guardian Maps

The objective of this section is to construct and catalog a number of guardian maps associated with robust $\mathcal{D}$-stability problems. To construct the relevant maps, we exploit Kronecker products and Kronecker sums. These notions have already been covered in Section 4.12 in the context of affine linear uncertainty structures. If A is an $n \times n$ matrix, we see below that the relationship between the eigenvalues of A and $A \oplus A$ proves to be quite useful.

EXAMPLE 17.5.1 (Guardian Map for Stable Matrices):   We claim that the map $\nu : \mathbf{R}^{n \times n} \to \mathbf{R}$ described by

$$\nu(A) = \det A \oplus A$$

for $A \in \mathbf{R}^{n \times n}$ is a guardian map for the set $\mathcal{A}$ of stable $n \times n$ matrices. To justify this claim, we first observe that if $A \in \partial \mathcal{A}$, then, using the result in Exercise 17.3.9, it is easy to see that A has one or more eigenvalues on the imaginary axis. Now, we must show that $\nu(A) = 0$. Indeed, let $\lambda_{i_1}(A)$ be any eigenvalue on the imaginary axis and let $\lambda_{i_2}(A)$ be an eigenvalue of A which is the complex conjugate of $\lambda_{i_1}(A)$; if $\lambda_{i_1}(A) = 0$, then take $\lambda_{i_2}(A) = \lambda_{i_1}(A)$. Now, in view of Remarks 4.12.4, $\lambda_{i_1}(A) + \lambda_{i_2}(A) = 0$ is an eigenvalue of $A \oplus A$. Hence, $A \oplus A$ is singular and it follows that $\nu(A) = 0$.

To complete the justification, we now assume that $A \in \mathcal{A}$ is such that $\nu(A) = 0$; we must show that $A \in \partial \mathcal{A}$. Indeed, since $\lambda = 0$ is an eigenvalue of $A \oplus A$, using a similar argument to the one above, A must have an eigenvalue on the imaginary axis; Hence, it follows that $A \in \partial \mathcal{A}$.

EXERCISE 17.5.2 (Guardian Map for Stable Polynomials):   As given in Section 4.7, take $H(p)$ to be the Hurwitz matrix associated with a polynomial $p(s) = \sum_{i=0}^{n} a_i s^i$ with $a_n > 0$. Show that

$$\nu(p) = \det H(p)$$

guards the set of n-th order stable polynomials with $a_n > 0$.

EXAMPLE 17.5.3 (Guardian Map for Schur Stability of Matrices):   Notice that if A is an $n \times n$ matrix and λ_i is one of its eigenvalues, then $A \otimes A - I \otimes I$ has $|\lambda_i|^2 - 1$ as an eigenvalue; see Remarks 4.12.4. From this observation, it is easy to see that the map defined by

$$\nu(A) = \det(A \otimes A - I \otimes I)$$

guards the set of $n \times n$ Schur stable matrices; i.e., the set of $n \times n$ matrices having eigenvalues in the interior of the unit disc.

EXAMPLE 17.5.4 (Guardian Map for Schur Stability of Polynomials): To construct a guardian map associated with the Schur stability problem for polynomials, we begin with a polynomial

$$p(s) = \sum_{i=0}^{n} a_i s^i$$

and form the $(n-1) \times (n-1)$ matrix

$$S(p) = \begin{bmatrix} a_n & a_{n-1} & a_{n-2} & \cdots & a_3 & a_2 - a_0 \\ 0 & a_n & a_{n-1} & \cdots & a_4 - a_0 & a_3 - a_1 \\ \vdots & \vdots & \vdots & & \vdots & \vdots \\ 0 & -a_0 & -a_1 & \cdots & a_n - a_{n-4} & a_{n-1} - a_{n-3} \\ -a_0 & -a_1 & -a_2 & \cdots & -a_{n-3} & a_n - a_{n-2} \end{bmatrix}$$

of Jury and Pavlidis (1963). The proposed guardian ν is now given by the formula

$$\nu(p) = p(1)p(0)\det S(p).$$

Verification that $\nu(p)$ satisfies the requirements for guarding the set of Schur stable polynomials is relegated to the exercise below.

EXERCISE 17.5.5 (Verification): With setup as in Example 17.5.4, let $z_1, z_2, \ldots, z_n$ denote the roots of $p(s)$. Using the fact that

$$\det S(p) = a_n^{n-1} \prod_{i_2 > i_1 \geq 1}^{n} (1 - z_{i_1} z_{i_2}),$$

prove that $\nu(p)$ is indeed an appropriate guardian map for the set of Schur stable polynomials.

EXERCISE 17.5.6 (Guardian Map for a Damping Region): Given any angle $\theta \in [\pi/2, \pi)$, prove that the map defined by

$$\nu(A) = \det(e^{j\theta} A \ominus e^{-j\theta} A)$$

guards the set of $n \times n$ matrices having eigenvalues which lie in the damping cone given by

$$C = \{z \in \mathbf{C} : \theta < \angle z < 2\pi - \theta\}.$$

Now construct a guardian map for the set of polynomials having all roots in this cone.

EXERCISE 17.5.7 (Guardian Map for a Strip):   Given any $\beta > 0$, prove that the map defined by

$$\nu(A) = \det[(A + j\beta I) \ominus (A - j\beta I)]$$

guards the set of $n \times n$ matrices with eigenvalues in the strip

$$\mathcal{S} = \{z \in \mathbf{C} : |Im\ z| < \beta\}.$$

Now construct a guardian map for the set of polynomials having all roots in this strip.

17.6   Families with One Uncertain Parameter

We now proceed toward a generalization of the result of Bialas; see Section 4.11. Instead of working with the strict left half plane, we allow for rather general $\mathcal{D}$ regions. In addition, we work in a more general setting allowing polynomic dependence on the uncertain parameter. For comparison purposes, note that in Chapter 4, affine linear dependence was assumed.

For the remainder of this chapter, we work with either an uncertain polynomial of the form

$$p(s, \lambda) = \sum_{i=0}^{\ell} \lambda^i p_i(s)$$

or an uncertain matrix of the form

$$A(\lambda) = \sum_{i=0}^{\ell} \lambda^i A_i,$$

where the set of polynomials $p_0(s)$, $p_1(s)$,...,$p_\ell(s)$ and matrices A_0, A_1,...,A_ℓ above are fixed. Note that for $i = 0$, we obtain the nominals $p(s, 0) = p_0(s)$ and $A(0) = A_0$. Without loss of generality, we take the uncertainty bounding set to be $\Lambda = [0, 1]$ and consider the family of polynomials $\mathcal{P} = \{p(\cdot, \lambda) : \lambda \in \Lambda\}$ and the family of matrices $\mathcal{A} = \{A(\lambda) : \lambda \in \Lambda\}$.

Taking note of the natural embedding of the coefficients of a polynomial $p(s) = s^n + \sum_{i=0}^{n-1} a_i s^i$ into the companion form matrix

$$A_c(p) = \begin{bmatrix} 0 & 1 & 0 & 0 & \cdots & 0 \\ 0 & 0 & 1 & 0 & \cdots & 0 \\ \vdots & \vdots & & & & \vdots \\ 0 & 0 & 0 & 0 & \cdots & 1 \\ -a_0 & -a_1 & -a_2 & -a_3 & \cdots & -a_{n-1} \end{bmatrix},$$

we henceforth restrict our attention to the family of matrices $\mathcal{A}$. Once the matrix result is given, an interpretation for the polynomial case is readily available.

17.7 Polynomic Determinants

In preparation for the main result of this chapter, we need one more technical concept. To motivate the definition to follow, notice that in the examples of the preceding section, each of the guardian map evaluations $\nu(A)$ is generated by taking the determinant of some matrix function $f(A)$. Furthermore, observe that $f(A)$ depends polynomially on the entries of A.

DEFINITION 17.7.1 (Polynomial Determinant Representation): Let $\mathcal{A}$ be an open set of $n \times n$ matrices with guardian map ν. We say that ν admits a *polynomial determinant representation* if there exists a matrix function $f : \mathbf{R}^{n \times n} \to \mathbf{R}^{k \times k}$ such that $f(A)$ depends polynomially on the entries of A and

$$\nu(A) = \det f(A)$$

for all $A \in \mathbf{R}^{n \times n}$.

DEFINITION 17.7.2 (Matrix Representation): Suppose that

$$A(\lambda) = \sum_{i=0}^{\ell} \lambda^i A_i$$

is an uncertain $n \times n$ matrix and $f : \mathbf{R}^{n \times n} \to \mathbf{R}^{k \times k}$ is matrix function such that $f(A)$ depends polynomially on the entries of A. Then, excluding the trivial mapping $F(A) \equiv 0$, the unique set of matrices

$F_0, F_1,\ldots,F_N$ such that $F_N \neq 0$ and

$$f(A(\lambda)) = \sum_{i=0}^{N} \lambda^i F_i$$

for all λ is called the *matrix representation* of $f(A(\lambda))$.

EXERCISE 17.7.3 (Existence and Uniqueness):   Under the conditions of the definition above, argue that the matrices F_i exist and are unique. *Hint*: To establish uniqueness, notice that two polynomials with the same values must have the same coefficients.

EXAMPLE 17.7.4 (Derivation of Matrix Representation):   To illustrate the concepts associated with the definition above, consider $A(\lambda) = A_0 + \lambda A_1$ with

$$A_0 = \begin{bmatrix} 1 & 2 \\ -1 & 3 \end{bmatrix}; \qquad A_1 = \begin{bmatrix} 1 & 2 \\ 4 & 0 \end{bmatrix}.$$

Now, with $f(A) = A^2$, a straightforward computation yields

$$f(A(\lambda)) = \begin{bmatrix} 9\lambda^2 + 8\lambda - 1 & 2\lambda^2 + 10\lambda + 8 \\ 4\lambda^2 + 15\lambda - 4 & 8\lambda^2 + 6\lambda + 7 \end{bmatrix}$$

$$= \begin{bmatrix} -1 & 8 \\ -4 & 7 \end{bmatrix} + \lambda \begin{bmatrix} 8 & 10 \\ 15 & 6 \end{bmatrix} + \lambda^2 \begin{bmatrix} 9 & 2 \\ 4 & 8 \end{bmatrix}.$$

Hence, the matrix representation of $f(A(\lambda))$ is described by $N = 2$ and

$$F_0 = \begin{bmatrix} -1 & 8 \\ -4 & 7 \end{bmatrix}; \qquad F_1 = \begin{bmatrix} 8 & 10 \\ 15 & 6 \end{bmatrix}; \qquad F_2 = \begin{bmatrix} 9 & 2 \\ 4 & 8 \end{bmatrix}.$$

From the expansion above, it is obvious that F_0, F_1 and F_2 provide a unique representation.

17.8   The Theorem of Saydy, Tits and Abed

We are now prepared to present the main result of this chapter. The theorem below applies to a large class of robust $\mathcal{D}$-stability problems with one uncertain parameter.

THEOREM 17.8.1 (Saydy, Tits and Abed (1990)): *Let $\mathcal{D} \subseteq \mathbf{C}$ be open and let $\mathcal{A} = \{A(\lambda) : \lambda \in \Lambda\}$ be a family of $n \times n$ matrices described by*

$$A(\lambda) = \sum_{i=0}^{\ell} \lambda^i A_i$$

and $\Lambda = [0, 1]$. In addition, suppose that the set of $n \times n$ matrices having all roots in $\mathcal{D}$ admits a polynomial determinant representation $\nu(A) = \det f(A)$ with matrices $F_0, F_1, \ldots, F_N$ representing $f(A(\lambda))$. Then, for $N = 1$, the family $\mathcal{A}$ is robustly $\mathcal{D}$-stable if and only if A_0 is $\mathcal{D}$-stable and the matrix

$$M(\mathcal{A}, \mathcal{D}) = -F_0^{-1} F_1$$

has no real eigenvalues in the interval $[1, +\infty)$. More generally, for $N > 1$, the same result holds with

$$M(\mathcal{A}, \mathcal{D}) = \begin{bmatrix} 0 & I & 0 & 0 & \cdots & 0 \\ 0 & 0 & I & 0 & \cdots & 0 \\ \vdots & \vdots & \vdots & \ddots & & \vdots \\ 0 & 0 & 0 & 0 & \cdots & I \\ -F_0^{-1} F_N & -F_0^{-1} F_{N-1} & -F_0^{-1} F_{N-2} & -F_0^{-1} F_{N-3} & \cdots & -F_0^{-1} F_1 \end{bmatrix}$$

PROOF: The proof for $N = 1$ involves mimicking the more general case with the slight change in the definition of $M(\mathcal{A}, \mathcal{D})$ as indicated above. Hence, we concentrate on $N > 1$ and claim that $\mathcal{A}$ is robustly $\mathcal{D}$-stable if and only if $\nu(A(\lambda)) \neq 0$ for all $\lambda \in \Lambda$. To prove this claim, we first note that robust $\mathcal{D}$-stability of $\mathcal{A}$ in conjunction with the guarding property of ν implies that $\nu(A(\lambda)) \neq 0$ for all $\lambda \in \Lambda$. Now, to prove that $\nu(A(\lambda)) \neq 0$ for all $\lambda \in \Lambda$ implies robust $\mathcal{D}$-stability, we proceed by contradiction; i.e., we assume that $\nu(A(\lambda)) \neq 0$ for all $\lambda \in \Lambda$ but there exists some $\lambda^* \in \Lambda$ such that $A(\lambda^*)$ is not $\mathcal{D}$-stable. Letting $\lambda_\alpha = \alpha \lambda^*$ for $\alpha \in [0, 1]$, it is apparent that $A(\lambda_0) = A_0$ and $A(\lambda_1) = A(\lambda^*)$. Noting that $A(\lambda_0)$ is $\mathcal{D}$-stable and $A(\lambda_1)$ is not $\mathcal{D}$-stable, continuous dependence of the eigenvalues of $A(\lambda_\alpha)$ on λ (Lemma 4.8.2) implies that there exists some $\alpha^* \in [0, 1]$ such that $A(\lambda_{\alpha^*})$ has an eigenvalue in the boundary $\partial \mathcal{D}$ of $\mathcal{D}$. This implies that $\nu(A_{\lambda_{\alpha^*}}) = 0$, which is the contradiction we seek.

Continuing with the proof of the theorem, our objective is to establish that $\nu(A(\lambda)) \neq 0$ for all $\lambda \in \Lambda$ is equivalent to $M(\mathcal{A}, \mathcal{D})$

not having any real eigenvalues in the interval $[1, +\infty)$. Since A_0 is $\mathcal{D}$-stable, we need only establish this condition for $\lambda \in (0, 1]$. Now, since the guarding property of ν implies that $\nu(A_0) = \det F_0$ does not vanish, we use the matrix representation for $f(A(\lambda))$ and write

$$f(A(\lambda)) = F_0 \left(I + \sum_{i=1}^{N} \lambda^i F_0^{-1} F_i \right).$$

Hence,

$$\nu(A(\lambda)) = \det F_0 \, \det \left(I + \sum_{i=1}^{N} \lambda^i F_0^{-1} F_i \right),$$

and it is apparent that $\nu(A(\lambda)) \neq 0$ for all $\lambda \in (0, 1]$ if and only if

$$\det \left(I + \sum_{i=1}^{N} \lambda^i F_0^{-1} F_i \right) \neq 0$$

for all $\lambda \in (0, 1]$.

To complete the proof, we interpret the nonvanishing determinant condition above in terms of the characteristic polynomial

$$\det(\lambda I - M(\mathcal{A}, \mathcal{D})) = \det \left(\lambda^N I + \sum_{i=0}^{N-1} \lambda^i F_0^{-1} F_{i-N} \right).$$

Noting that the nonvanishing property of this determinant is invariant to the change of variable $\tilde{\lambda} = 1/\lambda$, it follows that $\nu(A(\lambda)) \neq 0$ for $\lambda \in (0, 1]$ if and only if

$$\det \left(\tilde{\lambda} I - M(\mathcal{A}, \mathcal{D}) \right) \neq 0$$

for all $\tilde{\lambda} \in [1, \infty)$. That is, $M(\mathcal{A}, \mathcal{D})$ has no real eigenvalues in the real interval $[1, +\infty)$. ∎

EXERCISE 17.8.2 (General Uncertainty Bounds):   Modify the statement of Theorem 17.8.1 to handle the more general interval of the form $\Lambda = [\lambda^-, \lambda^+]$ instead of $\Lambda = [0, 1]$.

EXAMPLE 17.8.3 (Application of the Theorem):   We consider the family of matrices $\mathcal{A}$ described by

$$A(\lambda) = A_0 + \lambda A_1 + \lambda^2 A_2$$

with

$$A_0 = \begin{bmatrix} -5 & 6 \\ 2 & -4 \end{bmatrix}, \quad A_1 = \begin{bmatrix} 1 & 1 \\ 1 & 1 \end{bmatrix}, \quad A_2 = \begin{bmatrix} -1 & -1 \\ -1 & -1 \end{bmatrix},$$

$\lambda \in [0, 1]$ and $\mathcal{D}$ region which is the strict left half plane. To create input for Theorem 17.8.1, we use the guardian map $\nu(A) = \det A \oplus A$ of Example 17.5.1 and form

$A(\lambda) \oplus A(\lambda)$

$$= \begin{bmatrix} -10 + 2\lambda - 2\lambda^2 & 6 + \lambda - \lambda^2 & 6 + \lambda - \lambda^2 & 0 \\ 2 + \lambda - \lambda^2 & -9 + 2\lambda - 2\lambda^2 & 0 & 6 + \lambda - \lambda^2 \\ 2 + \lambda - \lambda^2 & 0 & -9 + 2\lambda - 2\lambda^2 & 6 + \lambda - \lambda^2 \\ 0 & 2 + \lambda - \lambda^2 & 2 + \lambda - \lambda^2 & -8 + 2\lambda - 2\lambda^2 \end{bmatrix}.$$

This leads to polynomial determinant representation

$$f(A(\lambda)) = \det \nu(A(\lambda)) = F_0 + \lambda F_1 + \lambda^2 F_2,$$

where the F_i are obtained by inspection of the Kronecker sum above; we obtain

$$F_0 = \begin{bmatrix} -10 & 6 & 6 & 0 \\ 2 & -9 & 0 & 6 \\ 2 & 0 & -9 & 6 \\ 0 & 2 & 2 & -8 \end{bmatrix}; F_1 = \begin{bmatrix} 2 & 1 & 1 & 0 \\ 1 & 2 & 0 & 1 \\ 1 & 0 & 2 & 1 \\ 0 & 1 & 1 & 2 \end{bmatrix}; F_2 = \begin{bmatrix} -2 & -1 & -1 & 0 \\ -1 & -2 & 0 & -1 \\ -1 & 0 & -2 & -1 \\ 0 & -1 & -1 & -2 \end{bmatrix}.$$

To apply the theorem, we verify that A_0 is stable and we need to form the 8×8 matrix

$$M(\mathcal{A}, \mathcal{D}) = \begin{bmatrix} 0 & I \\ -F_0^{-1} F_2 & -F_0^{-1} F_1 \end{bmatrix}.$$

A straightforward calculation yields

$$F_0^{-1} F_1 \approx \begin{bmatrix} -0.6667 & -0.75 & -0.75 & -0.8333 \\ -0.3889 & -0.6528 & -0.4306 & -0.6944 \\ -0.3889 & -0.4306 & -0.6528 & -0.6944 \\ -0.1944 & -0.3958 & -0.3958 & -0.5972 \end{bmatrix}$$

and

$$F_0^{-1}F_2 \approx \begin{bmatrix} 0.6667 & 0.75 & 0.75 & 0.8333 \\ 0.3889 & 0.6528 & 0.4306 & 0.6944 \\ 0.3889 & 0.4306 & 0.6528 & 0.6944 \\ 0.1944 & 0.3958 & 0.3958 & 0.5972 \end{bmatrix}.$$

An eigenvalue computation now leads to $\lambda_{1,2}(M) \approx 1.06 \pm j0.99$, $\lambda_{3,4}(M) \approx 0.11 \pm j0.46$, $\lambda_{5,6}(M) \approx 0.11 \pm j0.46$ and $\lambda_{7,8}(M) \approx 0$. Since none of the eigenvalues above are approaching any purely real $\lambda \in [1, \infty)$, we deem $\mathcal{A}$ to be robustly stable.

EXERCISE 17.8.4 (Alternative Method): For the family of matrices $\mathcal{A}$ in the example above, show that the characteristic polynomial has the form $p(s, \lambda) = s^2 + a_1(\lambda)s + a_0(\lambda)$. Find expressions for $a_0(\lambda)$ and $a_1(\lambda)$. By examining the roots of the equations $a_0(\lambda) = 0$ and $a_1(\lambda) = 0$, conclude that $\mathcal{A}$ is robustly stable.

17.9 Schur Stability

To demonstrate the generality of Theorem 17.8.1, we see below that we can recover a known result for robust Schur stability. Subsequently, we deal with a more general robust Schur stability problem involving quadratic dependence on λ.

EXERCISE 17.9.1 (The Result of Ackermann and Barmish (1988)): Consider the family of uncertain polynomials $\mathcal{P} = \{p(\cdot, \lambda) : \lambda \in \Lambda\}$ described by

$$p(s, \lambda) = \lambda p_0(s) + (1 - \lambda)p_1(s),$$

where $p_0(s)$ and $p_1(s)$ are fixed polynomials and $p_0(s)$ is assumed to be Schur stable with $\deg p_0(s) > \deg p_1(s)$. Using the guardian map $\nu(p) = p(1)p(0)\det S(p)$ given in Exercise 17.5.5, argue that $\mathcal{P}$ is robustly Schur stable if and only if $S^{-1}(p_0)S(p_1)$ has no real eigenvalues in $(-\infty, 0]$. In the expression above, $p_1(s)$ is viewed as a polynomial with the same degree as $p_0(s)$ for conformability of matrix multiplication.

EXERCISE 17.9.2 (Quadratic Dependence): Consider the uncertain family of matrices $\mathcal{A}$ described by

$$A(\lambda) = A_0 + \lambda A_1 + \lambda^2 A_2^2$$

with $A_0, A_1, A_2 \in \mathbf{R}^{n \times n}$, A_0 Schur stable and $\lambda \in [0, 1]$. Letting $\bar{A}_1 = A_1 - A_0$ and taking F_0, F_1 and F_2 to be a matrix representation for

$$f(A(\lambda)) = A_0 \otimes A_0 - I \otimes I + \lambda[A_0 \otimes \bar{A}_1 + \bar{A}_1 \otimes A_0] + \lambda^2 \bar{A}_1 \otimes \bar{A}_1,$$

conclude that with $\mathcal{D}$ being the interior of the unit disc, $\mathcal{A}$ is robustly Schur stable if and only if the matrix

$$M(\mathcal{A}, \mathcal{D}) = \begin{bmatrix} 0 & I \\ -F_0^{-1} F_2 & -F_0^{-1} F_1 \end{bmatrix}$$

has no real eigenvalues in the interval $[1, +\infty)$.

17.10   Conclusion

For robust $\mathcal{D}$-stability problems involving families of polynomials and matrices having a single uncertain parameter, the theory of guardian maps appears to be quite powerful in a robust $\mathcal{D}$-stability context. The ideas in this chapter can be extended to address more general $\mathcal{D}$ regions which admit *semiguardian* maps rather than guardian maps and systems with more than one uncertain parameter; see the notes to follow for further discussion. However, the computational complexity associated with the multiparameter case can be prohibitive.

Notes and Related Literature

NRL 17.1 In Section 17.1, we mentioned that robust stability is often decidable in a finite number of steps. To elaborate, suppose $p(s, q)$ is an uncertain polynomial with coefficients depending polynomially on the components q_i of q and the uncertainty bounding set Q is a box. Since the leading principle minors $\Delta_i(q)$ of the associated Hurwitz matrix also depend polynomially on q, the robust stability problem boils down to a positivity problem; i.e., determine if the set of multivariable polynomials $\{\Delta_i(q)\}$ is positive on the box Q. This problem is solvable in a finite number of steps; e.g., see Bose (1982) for a nice exposition.

NRL 17.2 In the feedback control literature, the issue of decidability was first raised in the context of output stabilization by Anderson, Bose and Jury (1975). At the heart of their theory is a tree-branch-type algorithm motivated by the decision calculus of Tarski (1951). The practical applicability of these ideas is severely inhibited by the fact that the number of tree-branch contingencies grows intolerably fast as a function of data dimension. Although less explicit about the use of Tarski-like ideas, the zero set theory in papers such as Hertz, Jury and Zeheb (1987) and Zeheb and Walach (1981) is limited in the same way.

NRL 17.3 In Theorem 17.8.1, the hypothesis that $\nu(A)$ is a polynomic determinant map can be weakened. If $\nu(A)$ is simply polynomic but not necessarily "realizable" via a determinant operation, a robust $\mathcal{D}$-stability criterion may still be achievable via the theory of Sturm sequences; e.g., see Marden (1966).

NRL 17.4 By using the slightly more technical definition of a *semiguardian map*, an even richer theory results; see Saydy, Tits and Abed (1990). These authors also provide some interesting extensions of the theory for systems with more than one uncertain parameter. For example, for the case of a bivariate polynomial described by $p(s, \lambda_1, \lambda_2) = \sum_{i_1, i_2} \lambda_1^{i_1} \lambda_2^{i_2} p_{i_1, i_2}(s)$ with bounding intervals Λ_1 and Λ_2 for λ_1 and λ_2 respectively, under rather weak hypotheses on $\mathcal{D}$, it is possible to reduce the robust $\mathcal{D}$-stability problem to a finite set of single-variable problems.

NRL 17.5 For a fuller exposition of many fundamental concepts underlying the theory in this chapter, see the paper on root clustering by Gutman and Jury (1981). For a more complete treatise, see the book by Gutman (1990).

NRL 17.6 In some cases, the guardian maps constructed in this chapter are obtained from matrices whose dimension is not necessarily minimal. For example, instead of using the $n^2 \times n^2$ matrix $A \oplus A$, one can work the lower Schlaflian matrix $A_{[2]}$; see Brockett (1973). It is also possible to use the theory of bialternate products, Bezoutians and Sylvester resultants to construct a number of novel guardian and semiguardian maps. This topic is pursued in the paper by Saydy, Tits and Abed (1990).

Chapter 18

The Arc Convexity Theorem

Synopsis

Associated with a polynomial $p(s)$ and an interval $\Omega \subseteq \mathbf{R}$ is a frequency response arc. This arc is obtained by sweeping the frequency ω over Ω and plotting $p(j\omega)$ in the complex plane. In this chapter, we establish the Arc Convexity Theorem of Hamann and Barmish. That is, if $p(s)$ is stable with net phase change of 180 degrees or less, as ω increases over Ω, the associated arc must be convex. The chapter also includes some extensions and ramifications.

18.1 Introduction

Throughout this text, we have seen that the proof of robustness results involving uncertain polynomials often relies on classical properties of fixed polynomials—properties that have been known for decades. For example, in the proof of the Zero Exclusion Condition in Section 7.4, we exploited the well-known fact that the roots of a polynomial depend continuously on its coefficients. A second example is provided by the proof of Kharitonov's Theorem, which involved exploitation of the Monotonic Angle Property; see Lemma 5.7.6. In this chapter, we establish a fundamental convexity property which is stronger than the Monotonic Angle Property. That is, given a stable polynomial $p(s)$, if we plot $p(j\omega)$ for $\omega \geq 0$, we obtain an arc which has a special convexity property; roughly speaking, arc

314

convexity implies a monotonic angle but not conversely. A second objective of this chapter is to raise possibilities for the application of arc convexity in a robustness context. It is felt that further research along these lines would be fruitful.

18.2 Definitions for Frequency Response Arcs

In this section, we provide the basic definitions which are essential for the exposition of the main result of this chapter. In the definition below, the "arcs" which we describe are in fact portions of the well known Mikhailov plot; see Mikhailov (1938).

DEFINITION 18.2.1 (Frequency Response Arc):   Given a polynomial $p(s)$ and an interval $\Omega \subseteq \mathbf{R}$, the plot of $p(j\omega)$ for ω increasing over Ω is called a *frequency response arc* or simply an *arc*.

DEFINITION 18.2.2 (Properness):   Given a polynomial $p(s)$ and an interval $\Omega \subseteq \mathbf{R}$, the associated frequency response arc is said to be *proper* if it does not pass through the origin and the net change in the phase of $p(j\omega)$ is no more than 180 degrees as ω increases over Ω; otherwise, the arc is said to be *improper*.

DEFINITION 18.2.3 (Convexity):   Given a polynomial $p(s)$ and an interval $\Omega \subseteq \mathbf{R}$, the associated frequency response arc is said to be *convex* if the following condition holds: Given any two distinct frequencies $\omega_1, \omega_2 \in \Omega$, the arc does not intersect the interior of the "triangle" with vertices $z_0 = 0$, $z_1 = p(j\omega_1)$ and $z_2 = p(j\omega_2)$. This definition also applies to the degenerate cases where z_0, z_1 and z_2 are collinear. When this condition fails, the arc is said to be *nonconvex*.

REMARKS 18.2.4 (Geometry for Convexity Definition):   Taking $p(s)$, ω_1 and ω_2 as in the definition above, the notion of arc convexity is easily understood with the aid of Figure 18.2.1. We see that arc A is convex; notice that triangle z_0, z_1, z_2 is not penetrated. On the other hand, arc B is nonconvex because there is a frequency range (ω_1, ω^*) for which the arc is interior to triangle z_0, z_1, z_2. Another important point to observe is that arc B can be identified with a polynomial having monotonically increasing angle but which is nonconvex. In the theorem to follow, we see that such behavior is impossible if $p(s)$ is stable. In other words, arc convexity is not implied by monotonicity of the angle; a deeper level of analysis is required.

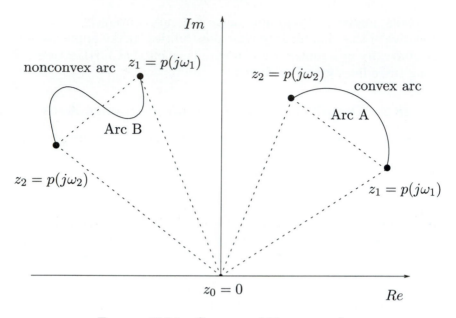

FIGURE 18.2.1   Convex and Nonconvex Arcs

EXAMPLE 18.2.5 (Four Basic Arc Types):   To illustrate the basic ideas above, notice that in Figure 18.2.2, arc A is proper and convex, arc B is proper and nonconvex, arc C is improper and convex and arc D is improper and nonconvex.

18.3   The Arc Convexity Theorem

The proof of the theorem below is relegated to the next two sections.

THEOREM 18.3.1 (Hamann and Barmish (1992)):   *All proper arcs associated with the frequency response of a stable polynomial are convex.*

EXERCISE 18.3.2 (Strict Convexity):   We say that a frequency response arc associated with a polynomial $p(s)$ and an interval $\Omega \subseteq \mathbf{R}$ is *strictly* convex if, given any two distinct pair of frequencies $\omega_1, \omega_2 \in \Omega$, the arc intersects the triangle with vertices $z_0 = 0$, $z_1 = p(j\omega_1)$ and $z_2 = p(j\omega_2)$ at only the vertices z_1 and z_2. For stable polynomials of degree $n \geq 2$, prove that all proper frequency response arcs are strictly convex. *Hint*: Since convexity is established in the theorem above, rule out the existence of linear sections of an arc; i.e., if for some constants $a, b \in \mathbf{R}$ we have $Re\, p(j\omega) = a Im\, p(j\omega) + b$ on

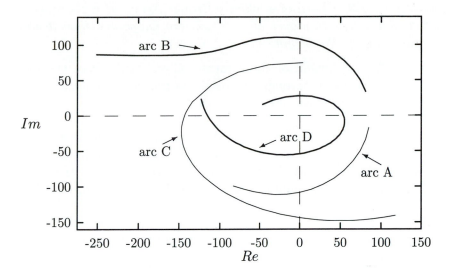

FIGURE 18.2.2   Four Basic Arc Types

an interval of frequency having positive length, argue that analyticity of $p(j\omega)$ demands that $Re\ p(j\omega) = aIm\ p(j\omega) + b$ for all $\omega \in \mathbf{R}$.

18.4   Machinery 1: Chords and Derivatives

This section and the next two are devoted to the development of the technical machinery which is used in the proof of the Arc Convexity Theorem. The reader interested solely in application of the theorem can proceed directly to Section 18.8. We begin the exposition by making a distinction between properness and strict properness of an arc. An obvious property relating strictly proper arcs and chords is also given.

DEFINITION 18.4.1 (Strictly Proper Arc):   Given a polynomial $p(s)$ and a frequency interval $\Omega = [\omega_1, \omega_2]$, we say that the associated arc is *strictly proper* if, as ω increases from ω_1 to ω_2, $p(j\omega) \neq 0$ and the net angle change of $p(j\omega)$ is less than 180 degrees.

DEFINITION 18.4.2 (Chord):   Given a polynomial $p(s)$ and a frequency interval $\Omega = [\omega_1, \omega_2]$, consider the straight line segment joining the *endpoints* $p(j\omega_1)$ and $p(j\omega_2)$ of the associated arc. We call this line the *chord* associated with this arc.

LEMMA 18.4.3 (Chord of a Strictly Proper Arc): *If a chord associated with a strictly proper arc has nonzero endpoints, then it does not pass through the origin.*

PROOF: The conclusion of this lemma follows easily from the fact that a chord passing through the origin would correspond to a net angle change of at least 180 degrees. ∎

REMARKS 18.4.4 (Derivatives): We now provide a lemma which generalizes the fact that stability of a polynomial $p(s)$ implies stability of its derivative $p'(s)$. More generally, we relate the roots of $p(s)$ to those of $p'(s)$. The lemma below is due to Lucas; e.g., see Marden (1966).

LEMMA 18.4.5 (Roots of a Derivative Polynomial): *Given a polynomial $p(s)$, the roots of its derivative*

$$p'(s) = \frac{dp(s)}{ds}$$

lie within the convex hull of the set of roots of $p(s)$.

EXERCISE 18.4.6 (Proof): Prove the lemma above. *Hint:* First write $p(s) = K \prod_{i=1}^{n}(s - s_i)$ and then show that

$$p'(s) = p(s) \sum_{i=1}^{n} \frac{1}{s - s_i}.$$

Subsequently, if z is a root of $p'(s)$, argue that if $z \notin \text{conv}\{s_i\}$, then

$$\sum_{i=1}^{n} \frac{1}{z - s_i} \neq 0.$$

REMARKS 18.4.7 ($\mathcal{D}$-Stable Case): Notice that if $p(s)$ is a $\mathcal{D}$-stable polynomial and $\mathcal{D}$ is convex, it follows that the convex hull of its root set must be a polygon which is wholly contained in $\mathcal{D}$. Hence, $\mathcal{D}$-stability of $p(s)$ implies $\mathcal{D}$-stability of $p'(s)$. In the sequel, we take $\mathcal{D}$ to be the strict left half plane, noting that a similar arc convexity result can also be established for the more general case when $\mathcal{D}$ is convex; see the notes at the end of this chapter.

EXERCISE 18.4.8 (Curvature): Given twice-differentiable functions of frequency $X(\omega)$ and $Y(\omega)$, let $\omega \in \mathbf{R}$ be a frequency at which at least one of these functions has a nonvanishing derivative. For such ω, the curvature of the curve associated with

$$Z(\omega) = X(\omega) + jY(\omega)$$

is given by

$$C_Z(\omega) = \frac{\frac{dX}{d\omega}\frac{d^2Y}{d\omega^2} - \frac{d^2X}{d\omega^2}\frac{dY}{d\omega}}{\left[\left(\frac{dX}{d\omega}\right)^2 + \left(\frac{dY}{d\omega}\right)^2\right]^{3/2}}.$$

(a) Taking $p(s)$ to be a stable polynomial, $X(\omega) = Re\ p(j\omega)$ and $Y(\omega) = Im\ p(j\omega)$, prove that

$$C_Z(\omega) > 0$$

for all $\omega \in \mathbf{R}$. *Hint*: Taking note of Lemma 18.4.5, exploit the fact that stability of $p(s)$ implies stability of the derivative polynomial $p'(s)$. Hence, the angle of $p'(j\omega) = Y'(\omega) - jX'(\omega)$ is increasing.
(b) Provide an example of a polynomial $p(s)$ which has the property that the associated curvature of $p(j\omega)$ is positive over a proper arc which is nonconvex.

18.5   Machinery 2: Flow

For a polynomial $p(s)$, we now develop some ideas involving the trajectory of the frequency response as ω is increased. Associated with each frequency $\omega \in \mathbf{R}$ is the *flow* of the response which one might liken to a velocity of $p(s)$ at the point $s = j\omega$.

DEFINITION 18.5.1 (Flow at $s = j\omega$):   Given a polynomial $p(s)$ and a frequency $\omega \in \mathbf{R}$, the *flow* at $s = j\omega$ is given by

$$F_p(\omega) = \frac{d\,Re\ p(j\omega)}{d\omega} + j\,\frac{d\,Im\ p(j\omega)}{d\omega}.$$

LEMMA 18.5.2 (Monotonic Angle of Flow):   *Given a stable polynomial $p(s)$, its flow $F_p(\omega)$ has a angle which is a continuous, non-decreasing function of the frequency $\omega \in \mathbf{R}$.*

PROOF: By applying the chain rule, it is apparent that

$$F_p(\omega) = \left.\frac{dp(s)}{ds}\right|_{s=j\omega} \left.\frac{ds}{d\omega}\right|_{s=j\omega} = jp'(j\omega).$$

Since $p'(s)$ is stable (see Lemma 18.4.5), its angle is monotonically increasing (Lemma 5.7.6). Hence, angle monotonicity for $F_p(\omega)$ follows immediately from the equality above. ∎

NOTATION 18.5.3 (Some Halfplanes):   We now introduce some simple geometric notation to aid in the convexity analysis of frequency response arcs. Indeed, given any polynomial $p(s)$ and a frequency $\omega^* \in \mathbf{R}$ such that $p(j\omega^*) \neq 0$, consider the straight line segment from the origin to $p(j\omega^*)$. The unbounded line containing this segment (its affine hull) defines two halfplanes $\mathcal{R}_p^-(\omega^*)$ and $\mathcal{R}_p^+(\omega^*)$ as follows: Take $\mathcal{R}_p^-(\omega^*)$ to be the closed half plane containing the origin and points $p(j\omega)$ for ω lying in some sufficiently small interval $[\omega^* - \epsilon, \omega^*]$ defined by $\epsilon > 0$, and take $\mathcal{R}_p^+(\omega^*)$ to be the open halfplane which is the complement of $\mathcal{R}_p^-(\omega^*)$. These two halfplanes are depicted below in Figure 18.5.1.

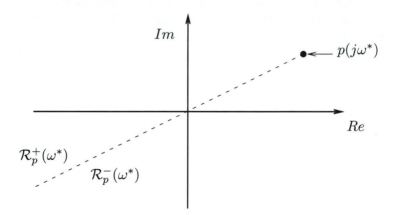

FIGURE 18.5.1   The Halfplanes $\mathcal{R}_p^+$ and $\mathcal{R}_p^-$

Associated with a polynomial $p(s)$ is a second set of halfplanes obtained from chords along an arc. Indeed, if $p(s)$ is a polynomial with values $p(j\omega)$ generating a strictly proper arc for frequencies $\omega \in \Omega = [\omega_1, \omega_2]$, we take $\mathcal{H}_p^-(\omega_1, \omega_2)$ to be the closed halfplane containing the associated chord and the origin of the complex plane, and we take $\mathcal{H}_p^+(\omega_1, \omega_2)$ to be the open halfplane which is the complement of $\mathcal{H}_p^-(\omega_1, \omega_2)$. In addition, we consider the two auxiliary halfplanes $\mathcal{H}_{p,0}^-(\omega_1, \omega_2)$ and $\mathcal{H}_{p,0}^+(\omega_1, \omega_2)$ which are simply parallel translates of $\mathcal{H}_p^-(\omega_1, \omega_2)$ and $\mathcal{H}_p^+(\omega_1, \omega_2)$, with their separating plane passing through the origin; e.g., see Figure 18.5.2.

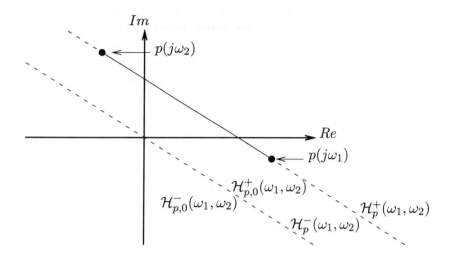

FIGURE 18.5.2   The Halfplanes $\mathcal{H}_p^+, \mathcal{H}_p^-$ and Their Translates

18.6   Machinery 3: The Forbidden Cone

In this section, we establish the existence of a special cone in the complex plane which cannot be penetrated by the flow associated with a strictly proper arc of a stable polynomial.

LEMMA 18.6.1 (Stable Flow Restriction):   *Given a stable polynomial $p(s)$ and a frequency $\omega^* \in \mathbf{R}$, its flow satisfies the condition $F_p(\omega^*) \in \mathcal{R}_p^+(\omega^*)$. Furthermore, there exists an $\epsilon > 0$ such that $p(j\omega) \in \mathcal{R}_p^+(\omega^*)$ for all frequencies $\omega \in (\omega^*, \omega^* + \epsilon]$.*

PROOF: Viewing the flow $F_p(\omega^*)$ as a derivative vector for $p(j\omega)$ at $\omega = \omega^*$ as in Lemma 18.4.5, the monotonically increasing angle of $p(j\omega)$ dictates that $F_p(\omega^*) \in \mathcal{R}_p^+(\omega^*)$. Now, this same flow condition forces $p(j\omega)$ in $\mathcal{R}_p^+(\omega^*)$ for all ω in some sufficiently small right neighborhood of ω^*. ∎

LEMMA 18.6.2 (Stable Flow Further Restricted):   *Suppose that $p(s)$ is a stable polynomial whose values $p(j\omega)$ generate a strictly proper arc for $\omega \in \Omega = [\omega_1, \omega_2]$. Then the associated flow satisfies*

$$F_p(\omega) \in \mathcal{R}_p^+(\omega_1) \cup \mathcal{R}_p^+(\omega_2)$$

for all frequencies $\omega \in \Omega$.

PROOF: By Lemma 18.6.1, the flow must obey the beginning constraint $F_p(\omega_1) \in \mathcal{R}_p^+(\omega_1)$ and endpoint constraint $F_p(\omega_2) \in \mathcal{R}_p^+(\omega_2)$. Furthermore, using Lemma 18.4.5, the angle of the flow is continuous and nondecreasing in ω. Therefore, as ω increases from ω_1 to ω_2, the flow (viewed as a vector) can only rotate continuously counterclockwise. Notice that $F_p(\omega)$ begins in $\mathcal{R}_p^+(\omega_1)$, and by strict properness of the arc, it follows that $F_p(\omega)$ reaches $\mathcal{R}_p^+(\omega_2)$ before exiting $\mathcal{R}_p^+(\omega_1)$. Hence, $F_p(\omega)$ remains within $\mathcal{R}_p^+(\omega_1) \cup \mathcal{R}_p^+(\omega_2)$ for all $\omega \in [\omega_1, \omega_2]$. ∎

REMARKS 18.6.3 (Describing the Forbidden Core):   As a consequence of Lemma 18.6.2, we have established the existence of a *forbidden cone* in the complex plane; i.e., a nonempty cone within which the flow for a strictly proper arc may not reside; see Figure 18.6.1. This concept is summarized in the following lemma. No proof is given because the result follows immediately from Lemma 18.6.2.

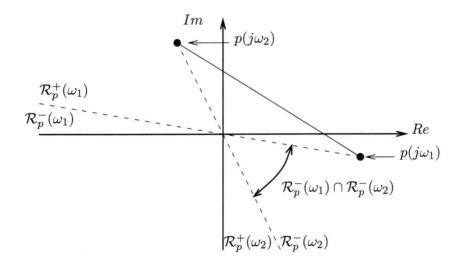

FIGURE 18.6.1   The Forbidden Cone $\mathcal{R}_p^-(\omega_1) \cap \mathcal{R}_p^-(\omega_2)$

LEMMA 18.6.4 (Forbidden Cone):   *Suppose that $p(s)$ is a stable polynomial whose values $p(j\omega)$ generate a strictly proper arc for frequencies $\omega \in \Omega = [\omega_1, \omega_2]$. Then $\mathcal{R}_p^-(\omega_1) \cap \mathcal{R}_p^-(\omega_2) \neq \emptyset$. Moreover, $F_p(\omega) \notin \mathcal{R}_p^-(\omega_1) \cap \mathcal{R}_p^-(\omega_2)$ for all frequencies $\omega \in \Omega$.*

18.7   Proof of the Arc Convexity Theorem

For a polynomial of order $n = 0$, the result is trivial since the frequency response arc is a point. The analysis to follow therefore assumes that $n \geq 1$. Indeed, given a stable polynomial $p(s)$ and a frequency interval $\Omega = [\omega_1, \omega_2] \subseteq \mathbf{R}$ such that the arc described by $p(j\omega)$ for $\omega \in \Omega$ is proper, to establish convexity it suffices to show that the arc lies in $\operatorname{cl} \mathcal{H}_p^+(\omega_1, \omega_2)$. To this end we consider two cases.

Case 1: The arc is proper but not strictly proper. In this case, the net angle increment is exactly 180 degrees, and the chord joining the endpoints of the arc passes through the origin. Therefore, the arc cannot intersect the chord for any $\omega^* \in (\omega_1, \omega_2)$; i.e., if $p(j\omega^*)$ intersects the chord for some $\omega^* \in (\omega_1, \omega_2)$, then either $\angle p(j\omega_1) = \angle p(j\omega^*)$ or $\angle p(j\omega_2) = \angle p(j\omega^*)$, which contradicts the increasing phase of $p(j\omega)$.

Case 2: The arc is strictly proper. We now claim that

$$F_p(\omega_1) \in \operatorname{cl} \mathcal{H}_{p,0}^+(\omega_1, \omega_2),$$

where $\operatorname{cl} \mathcal{H}_{p,0}^+(\omega_1, \omega_2)$ denotes the closure of $\mathcal{H}_{p,0}^+(\omega_1, \omega_2)$. Indeed, from Lemma 18.6.1 we know that $F_p(\omega_1) \in \mathcal{R}_p^+(\omega_1)$. Letting $\operatorname{int} \mathcal{Z}$ denote the interior of a set Z, to prove the claim, we must then show that a flow in $\mathcal{R}_p^+(\omega_1) \cap \operatorname{int} \mathcal{H}_{p,0}^-(\omega_1, \omega_2)$ is not possible. Proceeding by contradiction, assume $F_p(\omega_1) \in \mathcal{R}_p^+(\omega_1) \cap \operatorname{int} \mathcal{H}_{p,0}^-(\omega_1, \omega_2)$. Now, by Lemma 18.5.2, the phase of the flow is nondecreasing and hence is capable of only counterclockwise rotation. Notice that the condition $F_p(\omega_1) \in \mathcal{R}_p^+(\omega_1) \cap \operatorname{int} \mathcal{H}_{p,0}^-(\omega_1, \omega_2)$ requires that there exists an $\epsilon > 0$ such that $p(j\omega) \in \mathcal{R}_p^+(\omega_1) \cap \operatorname{int} \mathcal{H}_p^-(\omega_1, \omega_2)$ for $\omega \in (\omega_1, \omega_1 + \epsilon]$. Hence, in order for the arc to return to the chord at $p(j\omega_2)$, the flow must enter $\mathcal{H}_{p,0}^+(\omega_1, \omega_2)$. However, this would require that en route, the flow must enter the forbidden cone $\mathcal{R}_p^-(\omega_1) \cap \mathcal{R}_p^-(\omega_2)$, contradicting Lemma 18.6.4. Thus the claim is established.

We are now ready to complete the proof. We need to prove that $p(j\omega) \in \operatorname{cl} \mathcal{H}_p^+(\omega_1, \omega_2)$ for all $\omega \in [\omega_1, \omega_2]$. Since we already know that $p(j\omega_1) \in \operatorname{cl} \mathcal{H}_p^+(\omega_1, \omega_2)$ and $p(j\omega_2) \in \operatorname{cl} \mathcal{H}_p^+(\omega_1, \omega_2)$, we proceed by contradiction. Suppose that for $\omega^* \in (\omega_1, \omega_2)$, $p(j\omega) \in \operatorname{cl} \mathcal{H}_p^+(\omega_1, \omega_2)$ for all $\omega \in [\omega_1, \omega^*]$ but $p(j(\omega^* + \epsilon)) \in \operatorname{int} \mathcal{H}_p^-(\omega_1, \omega_2)$ for all $\epsilon > 0$ sufficiently small. This implies that there exists some $\epsilon^* > 0$ such that $F_p(\omega) \in \operatorname{int} \mathcal{H}_{p,0}^-(\omega_1, \omega_2)$ for all $\omega \in (\omega^*, \omega^* + \epsilon^*]$. Now, similar to the argument in the claim above, in order for the arc to return to the chord at $p(j\omega_2)$, the flow must enter $\mathcal{H}_{p,0}^+(\omega_1, \omega_2)$. However, this would require that en route, the flow must enter the forbidden cone $\mathcal{R}_p^-(\omega_1) \cap \mathcal{R}_p^-(\omega_2)$, contradicting Lemma 18.6.4. The proof of Theorem 18.3.1 is now complete. ∎

18.8   Robustness Connections

To stimulate further work on the connections between arc convexity and robustness theory, we begin with a definition.

DEFINITION 18.8.1 (Inner Frequency Response Set):   Given a stable polynomial $p(s)$, the *inner frequency response set*, denoted IFRS[p], is defined to be the open connected subset of the complex plane which contains the origin and is bounded by the curve obtained by plotting $p(j\omega)$ with ω varying from $-\infty$ to $+\infty$.

EXAMPLE 18.8.2 (Generation of Inner Frequency Response):   For the high order polynomial

$$p(s) = s^{12} + 2.53s^{11} + 25.404s^{10} + 47.924s^9 + 224.75s^8 + 311.96s^7$$
$$+ 856.72s^6 + 846.82s^5 + 1388.22s^4 + 920.54s^3$$
$$+ 770.10s^2 + 303.08s + 36.713,$$

the plot of $p(j\omega)$ is generated and the inner frequency response set is indicated in Figure 18.8.1. We also use this example to motivate the analysis to follow. The following observation is critical: Imagine a square of radius $r \geq 0$ (centered at the orgin) inscribed inside the inner frequency response set. For small values of r, this square is wholly contained within the set. As r increases, the square eventually contacts the boundary of the inner frequency response set at an extreme point.

REMARKS 18.8.3 (Convexity of the Inner Frequency Response Set): In the example above, it is no coincidence that the inscribed square contacts the inner frequency response set at an extreme point rather than along an edge. This type of contact is a consequence of the Arc Convexity Theorem; i.e., since $p(s)$ is stable, it can be shown that its inner frequency response set is convex; see Hamann and Barmish (1992) for a formal justification of this statement. The next exercise elaborates on these comments.

EXERCISE 18.8.4 (Inscribed Polygon):   Let $p(s)$ be a stable polynomial with a convex inner frequency response set and let $\Lambda \subseteq \mathbf{C}$ be

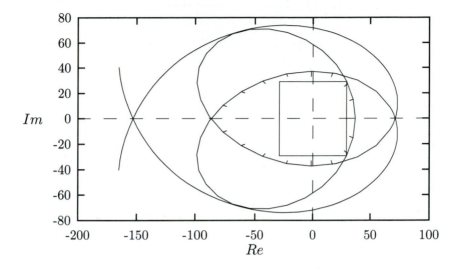

FIGURE 18.8.1   The Inner Frequency Response Set for Example 18.8.2

any (convex) polygon which contains $z = 0$ in its interior. Let

$$\gamma_{max} = \sup\{\gamma : p(j\omega) \notin \gamma\Lambda \text{ for all } \omega \in \mathbf{R}\}.$$

Beginning at $\gamma = 0$, argue that as γ increases, first contact between $p(j\omega)$ and $\gamma\Lambda$ occurs at an extreme point. That is, when $\gamma = \gamma_{max}$, one of the extreme points of the polygon $\gamma_{max}\Lambda$ contacts the arc associated with the polynomial $p(s)$.

EXERCISE 18.8.5 (Stability-Preserving Complex Gains):   For a given rational function $R(s)$, define the notions of arc, proper arc, convex arc and inner frequency response set IFRS[R] by mimicking the definitions given in the polynomial case. Note that $R(s)$ is not necessarily assumed to be proper.
(a) Argue that if $R(s)$ is strictly proper, its inner frequency response set is either empty or the singleton $\{0\}$.
(b) If $R(s)$ is either proper or improper without zeros along the imaginary axis, argue that its inner frequency response set is an open set containing $z = 0$ as an interior point.
(c) Suppose that $P(s)$ is proper and stable and consider the classical interconnection shown in Figure 18.8.2. Allow K to be a complex gain and let $\mathcal{K}_{max} \subseteq \mathbf{C}$ denote the largest pathwise connected set of gains which contains $K = 0$ as an interior point and preserves closed

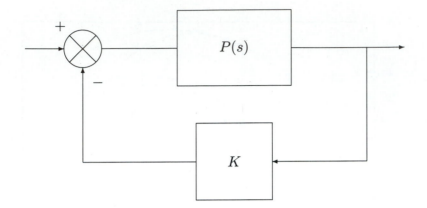

FIGURE 18.8.2   Feedback Interconnection for Exercise 18.8.5

loop stability. Prove that

$$\mathcal{K}_{max} = \text{IFRS}\left[\frac{1}{P}\right].$$

Hint: Use the Zero Exclusion Condition (see Theorem 7.4.2).

REMARKS 18.8.6 (Extreme Point Destabilization):   Let the plant $P(s) = N_P(s)/D_P(s)$ be proper and stable and consider the feedback system in Figure 18.8.2. Taking the set of admissible gains to be the polygon

$$\mathcal{K} = \text{conv}\{K_1, K_2, \ldots, K_\ell\},$$

we define the robustness margin in the usual way; i.e.,

$$r_{max} = \sup\{r : \text{closed loop stability is assumed for all } K \in r\mathcal{K}\}.$$

Now, observe that if the inner frequency response set for $1/P(s)$ is convex, by invoking the result in the exercise above, we arrive at the following conclusion: As r increases from 0 to r_{max}, the first loss of stability occurs at one of the extreme points $r_{max}K_i$ of $r_{max}\mathcal{K}$. In other words, to study closed loop stability, we need only consider the *extreme polynomials* of the form

$$\Psi_{i,r}(s) = rK_iN_P(s) + D_P(s)$$

with $i \in \{1, 2, \ldots, \ell\}$. Letting

$$r_{max,i} = \inf\{r : \Psi_{i,r}(s) \text{ is unstable}\},$$

it follows that

$$r_{max} = \min_{i \le \ell} r_{max,i}.$$

REMARKS 18.8.7 (Unmodeled Dynamics):   The ideas above can be readily extended to deal with classes of unmodeled dynamics. To elaborate, in Figure 18.8.2, suppose that we replace the pure gain K in the feedback path by a proper stable rational function $\Delta(s)$ of the form

$$\Delta(s) = \delta W(s),$$

where $W(s) = N_W(s)/D_W(s)$ is a weighting function describing the shape of the frequency response and δ is a complex gain. Taking $\Delta_{max} \subseteq \mathbf{C}$ to be the largest pathwise connected set of gains which contains $\delta = 0$ as an interior point and preserves closed loop stability, the arguments given above are then easily modified to arrive at the conclusion that

$$\Delta_{max} = \mathrm{IFRS}\left[\frac{1}{WP}\right].$$

Hence, if $1/[W(s)P(s)]$ has a convex inner frequency response set and if the admissible set of gains is a polygon

$$\Delta = \mathrm{conv}\{\delta_1, \delta_2, \ldots, \delta_\ell\},$$

then robust stability can be studied using the extreme polynomials

$$\Psi_{i,r}(s) = r\delta_i N_P(s)N_W(s) + D_P(s)D_W(s)$$

with $i = 1, 2, \ldots, \ell$. For example, if Δ is the unit square, we need only consider the four extreme polynomials

$$\Psi_{1,r}(s) = r(1+j)N_P(s)N_W(s) + D_P(s)D_W(s);$$
$$\Psi_{2,r}(s) = r(1-j)N_P(s)N_W(s) + D_P(s)D_W(s);$$
$$\Psi_{3,r}(s) = r(-1+j)N_P(s)N_W(s) + D_P(s)D_W(s);$$
$$\Psi_{4,r}(s) = r(-1-j)N_P(s)N_W(s) + D_P(s)D_W(s).$$

In view of the discussion above, if $R(s)$ is a proper rational function, it is of interest to give conditions under which its inverse $1/R(s)$ has a convex inner frequency response set. One class of rational functions having this property is described below. This is the class considered by Tesi, Vicino and Zappa (1992); see the notes at the end of the chapter for further discussion.

EXERCISE 18.8.8 (Stable All-Pole Transfer Functions):   For the stable all-pole transfer function described by

$$P(s) = \frac{K}{D_P(s)}$$

with $K \in \mathbf{R}$, prove that IFRS $\left[\frac{1}{P}\right]$ is convex.

18.9   Conclusion

The Arc Convexity Theorem describes a new fundamental property of stable polynomials. However, the ramifications of arc convexity in a robustness context have only been minimally explored. Further exploration of these ramifications appears to be a good area for future research. Our exploitation of arc convexity in a robustness context involved unstructured complex uncertainty; i.e., unmodeled dynamics is implicit in Exercise 18.8.4 and explicit in Example 18.8.8. Thus far, it is unclear if arc convexity is useful in the context of real parametric uncertainty; further research is required.

One possible direction for further work involves making connections between arc convexity and the Tsypkin–Polyak function $G_{TP}(\omega)$; see Section 6.7. If one can identify classes for which the arcs of $G_{TP}(\omega)$ are convex, the attainment of new extreme point results for robust stability becomes possible. In this regard, the issue of frequency scaling is of fundamental importance. More specifically suppose that

$$Z(\omega) = X(\omega) + jY(\omega)$$

is a complex frequency function with convex arcs and $\psi(\omega)$ is a positive function of frequency. Then it is of interest to develop conditions on $\psi(\omega)$ under which the *scaled function*

$$Z_\psi(\omega) = \frac{Z(\omega)}{\psi(\omega)}$$

has convex arcs. For example, if $\psi(\omega)$ is nondecreasing and $X(\omega)$ and $Y(\omega)$ correspond to the real and imaginary parts of a stable polynomial evaluated at $s = j\omega$, under what conditions does $Z_\psi(\omega)$ have convex arcs? Another possibility would be to assume different scale factors $\psi_1(\omega)$ and $\psi_2(\omega)$ for $X(\omega)$ and $Y(\omega)$, respectively.

Notes and Related Literature

NRL 18.1 A more general form of the Arc Convexity Theorem is given in Hamann and Barmish (1992). By sweeping the boundary of a desired root location region $\mathcal{D}$, we obtain *generalized frequency response arcs*. For a $\mathcal{D}$-stable polynomial, these arcs are again convex. A primary source of motivation for this more general framework is the Schur stability problem; i.e., since the unit disc is convex, all proper arcs associated with the frequency response of a Schur polynomial must be convex.

NRL 18.2 In Exercise 18.4.8, the relationship between arc convexity and curvature was explored. In this regard, it is important to mention the body of literature dealing with the *clockwise property* of the Nyquist locus; see Horowitz and Ben-Adam (1989), Bartlett (1990b) and Tesi, Vicino and Zappa (1992). The departure point for comparison with the Arc Convexity Theorem is the interesting result in Tesi, Vicino and Zappa (1992); i.e., if $P(s) = K/D_p(s)$ is a stable all-pole transfer function, the clockwise property for the Nyquist plot of $P(j\omega)$ is guaranteed if the roots of $p(s)$ all lie within a damping cone $\mathcal{C} = \{z \in \mathbf{C} : \pi - \theta < \angle z < \pi + \theta\}$ with $|\theta| \leq \frac{\pi}{4}$. For the same transfer function, the Arc Convexity Theorem leads to a rather different result. Namely, the Nyquist plot of the inverse plant $P^{-1}(s)$ always has the clockwise property—even if the damping cone requirement above is violated. Another important point to note is that arc convexity is a more stringent requirement than the clockwise property; i.e., arc convexity implies that the clockwise property is satisfied but the converse does not hold. One critical difference is the "centering" about $z = 0$ in the arc convexity framework. For this reason, the positive curvature result in Exercise 18.4.8 is not easily modified to yield a proof of Theorem 18.3.1.

Chapter 19

Five Easy Problems

Synopsis

This final chapter provides an overview of five research directions involving robustness of systems with real parametric uncertainty. We immediately make the disclaimer that the five areas being described represent the author's bias. If a later edition of this text emerges, it is quite possible that this chapter will look dramatically different.

19.1 Introduction

The title of this chapter is a deliberate misnomer. A more appropriate chapter title would be "Five Problem Areas of Potential Interest." The choice of chapter title was dictated by two considerations: First, the use of the rather flippant title "Five Easy Problems" is a way of signaling that the style of presentation in this chapter is intended to be less formal than its predecessors. Our goal is to stimulate new research directions rather than providing "mainstream" results which are intended to be cast in stone. In a sense, the reader should view this chapter as a biased survey of a subset of recent developments. The second reason for selection of the chapter title was for nostalgia purposes. The chapter title is intended to conjure up images of a well-known movie of the sixties with which some readers may be familiar. Further details about the movie are provided in the index under the letter N.

19.2   Problem Area 1: Generating Mechanisms

Let $p(s, q)$ be an uncertain polynomial with some complicated uncertainty structure. For example, suppose that $p(s, q)$ has a multilinear or polynomic uncertainty structure as in Chapter 14. Recognizing the fundamental difficulty associated with robustness analysis in this context, the following question arises: Is there a meaningful subclass of uncertainty structures for which "strong" robustness results can be given? In the current literature, there is still not a clear understanding what constitutes a "strong" robustness result versus a "weak" one. For example, is a robustness result considered "strong" if its implementation requires solution of a nonlinear program without a guarantee that a global optimum is attainable? If the answer is no to this question, then we can consider the question again with the word "convex" replacing "nonlinear."

With regard to the questions above, it is felt that an important line of research involves studying how the structure of feedback control systems gives rise to special mathematical properties which facilitate solution of the robustness problem at hand. For example, can we describe rich classes of feedback systems which give rise to an uncertain closed loop polynomial $p(s, q)$ having special properties which can be exploited to facilitate robust stability analysis?

We use the words "generating mechanism" in connection with the discussion above. The topology or physics of the system leads to a closed loop polynomial having special mathematical properties associated with its uncertainty structure. To provide an example, if we restrict our attention to uncertain polynomials and transfer functions obtained via application of Mason's rule for signal flow graphs, we can identify special properties of the resulting characteristic polynomial or closed loop transfer function. In this regard, it is important to note that the uncertain quantities which arise are not arbitrary. To illustrate, if two uncertain branch gains q_1 and q_2 appear in touching loops, then the product term $q_1 q_2$ cannot appear when Mason's rule is applied. We now describe one line of research motivated by these ideas.

The ideas below come from Barmish, Ackermann and Hu (1992). Note that the uncertain parameters q_i can be taken as either real or complex. We now concentrate on a special class of uncertain polynomials which has a special property—polynomials in this class admit a decomposition which can be described via a tree diagram. This idea is illustrated via examples.

19.2.1   The Tree-Structured Decomposition (TSD)

The objective of this subsection is to describe some special properties associated with classes of uncertain polynomials obtained via application of Mason's rule. In the exposition below, if $q \in \mathbf{R}^\ell$ and $I \subseteq \{1, 2, \ldots, \ell\}$, then q^I denotes the subvector obtained from q by deleting components q_i for $i \notin I$.

DEFINITION 19.2.2 (Partitions):   Suppose $I_i \subseteq I = \{1, 2, \ldots, \ell\}$ for $i = 1, 2$. We call $\{I_1, I_2\}$ a *nontrivial partition of the index set* I if I_1 and I_2 are both nonempty, $I_1 \cap I_2 = \emptyset$ and $I_1 \cup I_2 = I$.

DEFINITION 19.2.3 (Decomposability):   The uncertain polynomial $p(s, q)$ is said to be *sum decomposable* if there exists a nontrivial partition $\{I_1, I_2\}$ of $I = \{1, 2, \ldots, \ell\}$ and uncertain polynomials $p(s, q^{I_1})$ and $p(s, q^{I_2})$ such that

$$p(s, q) \equiv p(s, q^{I_1}) + p(s, q^{I_2}).$$

Similarly, $p(s, q)$ is said to be *product decomposable* if there exists a nontrivial partition $\{I_1, I_2\}$ of $I = \{1, 2, \ldots, \ell\}$, uncertain polynomials $p(s, q^{I_1})$ and $p(s, q^{I_2})$ and a fixed polynomial $p_0(s)$ such that

$$p(s, q) \equiv p(s, q^{I_1}) p(s, q^{I_2}) + p_0(s).$$

Finally, $p(s, q)$ is said to be *decomposable* if it is either sum decomposable or product decomposable. In this case, the uncertain polynomials $p(s, q^{I_1})$, $p(s, q^{I_2})$ and $p_0(s)$ involved in the decomposition are called *children* of $p(s, q)$. If $p(s, q)$ is not decomposable, then we say that it is *indecomposable*.

DEFINITION 19.2.4 (Descendents and k-Decomposability):   Suppose that $p(s, q)$ is decomposable and has children $p(s, q^{I_1})$, $p(s, q^{I_2})$ and $p_{i,0}(s)$ with $p(s, q^{I_i})$ being further decomposable for either $i = 1$ or $i = 2$. Let $p(s, q^{I_{i,1}})$, $p(s, q^{I_{i,2}})$ and $p_{i,0}(s)$ denote the children of such $p(s, q^{I_i})$. Then $p(s, q^{I_{i,1}})$, $p(s, q^{I_{i,2}})$ and $p_0(s)$ are called *grandchildren* of $p(s, q)$. By continuing the decomposition process, it may turn out that some of the grandchildren are still further decomposable. If $p(s, q^{I_{i,m}})$ is decomposable, then its children $p(s, q^{I_{i,m,1}})$, $p(s, q^{I_{i,m,2}})$ and possibly $p_{i,m,0}(s)$ are called *great grandchildren* of $p(s, q)$. Continuing this decomposition process, we may generate additional children which we call *descendents* of $p(s, q)$. Finally, we say that $p(s, q)$ is k-*decomposable* if a sequence of decompositions can be applied so that all indecomposable descendents, of $p(s, q)$ depend on at most k components q_i of q. For $k = 1$, we say that $p(s, q)$ is *totally decomposable*.

REMARKS 19.2.5 (Nonuniqueness and Tree Structure):   Notice that the process of generating descendents of $p(s, q)$ is highly nonunique. Although there may be many ways to decompose $p(s, q)$, the definition of k-decomposability requires one sequence of decompositions leading to indecomposable descendents, each depending on at most k components q_i of q. Also note that $p(s, q)$ can have no more than $2\ell - 1$ indecomposable descendents.

There is an obvious tree structure which can be identified with a k-decomposable uncertain polynomial $p(s, q)$. We now illustrate the method of tree construction via an example.

EXAMPLE 19.2.6 (Illustration of TSD): Suppose that $p(s, q)$ is sum decomposable as

$$p(s, q) \equiv p(s, q^{I_1}) + p(s, q^{I_2}),$$

where $p(s, q^{I_1})$ admits a further product decomposition

$$p(s, q^{I_1}) \equiv p(s, q^{I_{1,1}}) p(s, q^{I_{1,2}})$$

and $p(s, q^{I_2})$ admits a further sum decomposition

$$p(s, q^{I_2}) \equiv p(s, q^{I_{2,1}}) + p(s, q^{I_{2,2}}).$$

At the next step, suppose that $p(s, q^{I_{1,1}})$ and $p(s, q^{I_{2,1}})$ are indecomposable but $p(s, q^{I_{1,2}})$ admits a sum decomposition

$$p(s, q^{I_{1,2}}) \equiv p(s, q^{I_{1,2,1}}) + p(s, q^{I_{1,2,2}})$$

and $p(s, q^{I_{2,2}})$ admits a product decomposition

$$p(s, q^{I_{2,2}}) \equiv p(s, q^{I_{2,2,1}}) p(s, q^{I_{2,2,2}}).$$

Then, the associated TSD is seen in Figure 19.2.1.

REMARKS 19.2.7 (Algebra of Sets in the Complex Plane):   The TSD motivates some interesting research problems involving algebraic operations on sets in the complex plane. For example, in the work of Polyak, Scherbakov and Shmulyian (1993), interesting geometrical objects (such as the product of two discs in the complex

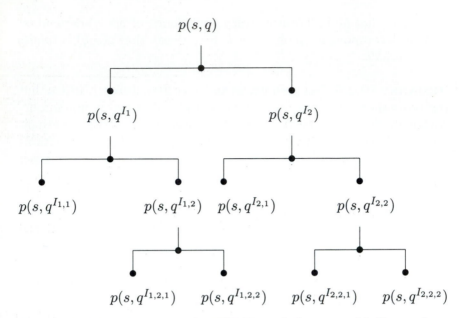

FIGURE 19.2.1   Generation of a TSD Using Indecomposable Descendents

plane) are represented analytically. One appealing aspect of such work is the two-dimensional nature of value sets. This enables us to display results graphically. We now elaborate on this point.

It is felt that an interesting line of research involves the algebra of sets in the complex plane. For example, suppose that $\mathcal{D}_1$, $\mathcal{D}_2$ and $\mathcal{D}_3$ are three discs in the complex plane with $0 \notin \mathcal{D}_3$. Then it is of interest to find ways to characterize or display combinations of $\mathcal{D}_1$, $\mathcal{D}_2$ and $\mathcal{D}_3$; e.g., consider the problem of constructing sets such as

$$\mathcal{D} = \frac{\mathcal{D}_1 \mathcal{D}_2}{\mathcal{D}_3}$$

or

$$\mathcal{D} = \mathcal{D}_1 \mathcal{D}_2 \mathcal{D}_3.$$

Returning to $p(s,q)$ in Figure 19.2.1, notice that the final value set $p(j\omega, Q)$ is described via the elementary operations $\{+, -, \cdot\}$ on pairwise combinations of two-dimensional sets in the complex plane. To further illustrate, suppose that the indecomposable descendents of $p(s,q)$ have independent or affine linear uncertainty structures. Then, associated with the leaves of the TSD in Figure 19.2.1 are the polygonal or rectangular value sets $p(j\omega, Q^{I_{1,2,1}})$, $p(j\omega, Q^{I_{1,2,2}})$, $p(j\omega, Q^{I_{2,2,1}})$, $p(j\omega, Q^{I_{2,2,2}})$, $p(j\omega, Q^{I_{1,1}})$ and $p(j\omega, Q^{I_{2,1}})$, which can easily be computed and stored. Subsequently, the TSD indicates that one can begin at the bottom of the tree and perform pairwise combinations of sets to build up the value set $p(j\omega, Q)$.

EXAMPLE 19.2.8 (Affine Linear Uncertainty Structures):   To see that the TSD concept handles affine linear uncertainty structures as a special case, say $p(s,q) = p_0(s) + \sum_{i=1}^{4} q_i p_i(s)$, where $p_i(s)$ are fixed polynomials. Noting that we can express $p(s,q)$ in the nested form

$$p(s,q) = ((((p_0(s) + q_1 p_1(s)) + q_2 p_2(s)) + q_3 p_3(s)) + q_4 p_4(s)),$$

it follows that $p(s,q)$ is totally decomposable. The associated TSD is illustrated in Figure 19.2.2.

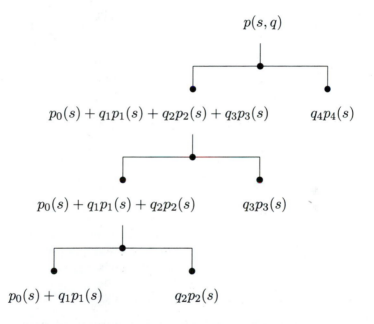

FIGURE 19.2.2   A TSD for Polytopes of Polynomials

EXERCISE 19.2.9 (Uncertainty in Poles and Zeros):   Suppose that the uncertain transfer function

$$P(s,q,r) = \frac{K(s - q_1)(s - q_2) \cdots (s - q_m)}{(s - r_1)(s - r_2) \cdots (s - r_n)}$$

is connected in a unity feedback configuration. Show that the resulting closed loop polynomial is totally decomposable.

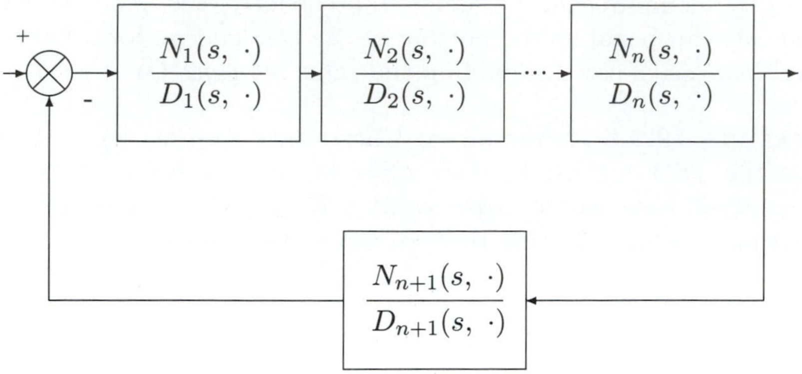

FIGURE 19.2.3   Interconnection for Exercise 19.2.10

EXERCISE 19.2.10 (Cascade Combination with Feedback):   For the system in Figure 19.2.3, assume that each uncertain parameter enters into either one numerator $N_i(s, \cdot)$ or one denominator $D_i(s, \cdot)$. Furthermore, for $i = 1, 2, \ldots, n, n + 1$, assume that $N_i(s, \cdot)$ and $D_i(s, \cdot)$ are k_i-decomposable. Letting

$$k = \max_{i \leq n+1} k_i,$$

prove that the closed loop polynomial is k-decomposable.

19.2.11   The TSD for Rational Functions

The decomposability concept, introduced in the context of polynomials, generalizes to rational functions. In view of the fact that the definitions are nearly identical to those used in the polynomial case, we only sketch the key ideas. Indeed, if $z = (z_1, z_2, \ldots, z_\ell) \in \mathbf{C}^\ell$ and

$$R(z) = R(z_1, z_2, \ldots, z_\ell)$$

is a multivariable polynomial, the various notions of decomposability of $R(z)$ are defined in the natural way; e.g., $R(z) = z_1 + z_2$ is trivially sum decomposable and $R(z) = z_1 z_2$ is trivially product decomposable. This more general TSD framework is useful for rational fractions. For example, suppose that the i-th block in a feedback loop is a transfer function $P_i(s)$ which includes some unmodeled dynamics $\Delta_i(s)$ entering additively. In this case, at fixed frequency $\omega \in \mathbf{R}$ we identify z_i with $P_i(j\omega) + \Delta_i(j\omega)$. Now, if we have apriori bounds for $\Delta_i(j\omega)$, an appropriate choice of $R(z)$ enables us to study value sets for various loop functions of interest.

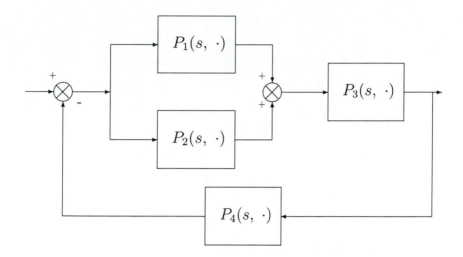

FIGURE 19.2.4   Feedback Interconnection for Example 19.2.12

To illustrate the idea above, we represent a parallel connection of ℓ blocks via

$$R(z) = z_1 + z_2 + z_3 + \cdots + z_\ell.$$

Since $R(z)$ is totally decomposable, we see that a value set description for the overall transfer function is obtained by performing $\ell - 1$ set additions.

EXAMPLE 19.2.12 (Robust Closed Loop Stability):   For the uncertain feedback system in Figure 19.2.4, robust stability is governed by the zeros of the uncertain rational function

$$P(s, \cdot) = 1 + P_1(s, \cdot)P_3(s, \cdot)P_4(s, \cdot) + P_2(s, \cdot)P_3(s, \cdot)P_4(s, \cdot).$$

To make a connection with TSD theory, we take

$$R(z) = 1 + z_1 z_3 z_4 + z_2 z_3 z_4$$

and notice that this function is totally decomposable; i.e.,

$$R(z) = 1 + z_3 z_4 (z_1 + z_2).$$

The resulting TSD is shown in Figure 19.2.5.

REMARKS 19.2.13 (New Directions):   For the example above, the existence of a TSD was demonstrated by performing a factorization.

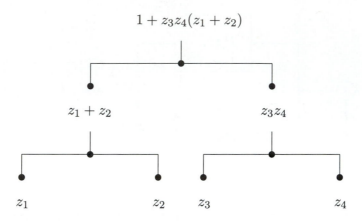

$$1 + z_3 z_4 (z_1 + z_2)$$

$z_1 + z_2$ $z_3 z_4$

z_1 z_2 z_3 z_4

FIGURE 19.2.5   TSD for Example 19.2.12

In some cases, however, such a factorization may not be transparent. For the case when a simple factorization is not available, the theory of unique factorization domains may prove to be useful; e.g., see Lang (1965). Another set of research problems arises by considering the mathematical equations describing the dynamics of the physical system under consideration. In some cases, the interconnection of physical components suggests a natural TSD. A nice illustration of this concept is given by the mass–spring–damper system considered by Ackermann and Sienel (1990).

19.3   Problem Area 2: Conditioning of Margins

When carrying out robustness computations on a digital computer, the following fundamental question arises: Will small changes in the input data lead to small changes in calculated robustness margins? Our focal point in this section is the issue of continuity of computed quantities as a function of the input data. In this regard, we note that there is a fundamental difference between real and complex uncertainties. When the uncertain parameters q_i are complex, discontinuity can be ruled out under rather mild regularity conditions. However, if the q_i are real, then the possibility exists for discontinuous dependence of the robust stability margin on the input data. Matters are further complicated by the fact that at the point of discontinuity in the space of problem data, the robustness margin r_{max} may be much smaller than at neighboring points. This may lead to potentially deceptive conclusions.

Although discontinuity of the robustness margin may be non-generic, motivation for further research is provided by the following fact: In regions of data space close to the discontinuity set, r_{max} can be highly ill-conditioned. It is felt that analysis of conditioning properties of the robust stability problem is an important topic area.

To substantiate many of the remarks above, the remainder of this section proceeds along the lines in Barmish, Khargonekar, Shi and Tempo (1990). The example which we provide illustrating the discontinuity of the robustness margin is based on a unity feedback system—the plant has uncertain parameters entering affine linearly into numerator and denominator coefficients. Using d to represent the data describing the system, the robustness margin is written explicitly as $r_{max}(d)$. Subsequently, we see that there exists a sequence of data $\langle d(k) \rangle_{k=1}^{\infty}$ converging to some d^* such that

$$\lim_{k \to \infty} r_{max}(d(k)) > r_{max}(d^*).$$

That is, if one solves the sequence of robustness margin problems corresponding to $d(k)$, the margins $r_{max}(d(k))$ may differ considerably from $r_{max}(d^*)$. This happens even as $d(k)$ gets arbitrarily close to the limit point d^*.

NOTATION 19.3.1 (Uncertain Polynomials and Data):   We work with the uncertain polynomial

$$p(s, q) = s^n + \sum_{i=0}^{n-1} a_i(q) s^i$$

with coefficient functions $a_i(q)$ which are affine linear in q. Since we are working with a variable uncertainty bound $r > 0$, the dependence on r is emphasized by writing Q_r for the uncertainty bounding set. That is, we consider the box

$$Q_r = \{q \in \mathbf{R}^\ell : |q_i| \leq r \text{ for } i = 1, 2, \ldots, \ell\}$$

and, as usual, the robustness margin is given by

$$r_{max} = \sup\{r : p(s, q) \text{ is stable for all } q \in Q_r\}.$$

Within this framework, the *problem data* consists of integers $\ell = \dim q$ and $n = \deg p(s, q)$ and the set of coefficient functions $a_0(\cdot), a_1(\cdot), \ldots, a_{n-1}(\cdot)$. To illustrate the discontinuity phenomenon, we use a finite-dimensional space for this problem data. That is, each

$a_i(\cdot)$ is viewed as a mapping on *data vectors* $d \in \mathbf{R}^p$ to continuous functions of q. For example, *a family of problems* might be described by $p = 6$, $\ell = 2$, $n = 2$ and

$$p(s, q) = s^2 + (d_1 + d_2 q_1 + d_3 q_2)s + (d_4 + d_5 q_1 + d_6 q_2).$$

A *specific* robustness margin problem is obtained with $d_1 = 2$, $d_2 = 1$, $d_3 = 4$, $d_4 = 3$, $d_5 = 6$ and $d_6 = 12$. This leads to

$$p(s, q) = s^2 + (2 + q_1 + 4q_2)s + (3 + 6q_1 + 12q_2).$$

Within this data space context, two robustness margin problems are deemed to be "close together" if their associated data vectors (call them d^1 and d^2) are close together in some arbitrary but fixed norm on $\mathbf{R}^p$; i.e., $\|d^1 - d^2\|$ is small. To denote dependence on d, we henceforth write $p_d(s, q)$ and $r_{max}(d)$ in lieu of $p(s, q)$ and r_{max}, respectively.

19.3.2   Example Illustrating Discontinuity

Before proceeding, it is important to note that it is easy to construct relatively trivial examples for which discontinuity of r_{max} can easily be demonstrated. Such examples involve cases when there is only one uncertain parameter, cases when the uncertainty structure is highly nonlinear, cases when the limiting polynomial $p_{d^*}(s, q)$ is only marginally stable and cases when $p_{d^*}(s, q)$ is structurally different from $p_{d(k)}(s, q)$; e.g., $p_{d^*}(s, q)$ has lower degree or a smaller number of uncertainties than $p_{d^*}(s, q)$. In contrast, the example below is simple yet nontrivial.

   We consider a unity feedback system with open loop transfer function denoted by

$$P_d(s, q) = K_d \frac{N_d(s, q)}{D_d(s, q)},$$

where $N_d(s, q)$ and $D_d(s, q)$ are uncertain polynomials, K_d is a fixed gain and Q_r is the box given above. In this example, the limiting system is described by $K_{d^*} = a$, $N_{d^*}(s, q) = 4a + 10aq_1$ and

$$D_{d^*}(s, q) = s^4 + (20 - 20q_2)s^3 + (44 + 2a + 10q_1 - 40q_2)s^2$$
$$+ (20 + 8a + 20aq_1 - 20q_2)s + a^2,$$

where $a = 3 + 2\sqrt{2}$. Using our data notation, we write $K_d = d_0$, $N_d(s, q) = d_1 + d_2 q_1$ and

$$D_d(s, q) = s^4 + (d_3 + d_4 q_2)s^3 + (d_5 + d_6 q_1 + d_7 q_2)s^2$$
$$+ (d_8 + d_9 q_1 + d_{10} q_2)s + d_{11}.$$

By comparing the expressions for K_{d^*}, $N_{d^*}(s,q)$ and $D_{d^*}(s,q)$ with K_d, $N_d(s,q)$ and $D_d(s,q)$, respectively, it is clear that the d_i^* are readily available; e.g., $d_0^* = a$, $d_1^* = 4a$, $d_2^* = 10a$, $d_3^* = 20$, etc.

Now, we consider the data sequence $\langle d(k) \rangle_{k=1}^{\infty}$ described by

$$d_i(k) = \begin{cases} d_i^* & \text{for } i \neq 0; \\ a_k & \text{for } i = 0, \end{cases}$$

where $a_k = a - 1/k$. This sequence corresponds to the case where the plant data is fixed and the gain a_k is converging to a.

In order to obtain the robustness margin along the $d(k)$ sequence, we study the closed loop polynomial

$$\begin{aligned} p_{d(k)}(s,q) &= K_{d(k)}N_{d(k)}(s,q) + D_{d(k)}(s,q) \\ &= s^4 + (20 - 20q_2)s^3 + (44 + 2a + 10q_1 - 40q_2)s^2 \\ &\quad + (20 + 8a + 20aq_1 - 20q_2)s + a(5a - \frac{4}{k} + 10(a - \frac{1}{k})q_1) \end{aligned}$$

and for the limiting case, we study the closed loop polynomial

$$\begin{aligned} p_{d^*}(s,q) &= aN_{d^*}(s,q) + D_{d^*}(s,q) \\ &= s^4 + (20 - 20q_2)s^3 + (44 + 2a + 10q_1 - 40q_2)s^2 \\ &\quad + (20 + 8a + 20aq_1 - 20q_2)s + (5a^2 + 10a^2q_1). \end{aligned}$$

By a lengthy computation whose details are described in Barmish, Khargonekar, Shi and Tempo (1990), it can be verified that

$$0.417 \approx 1 - \frac{a}{10} = \lim_{k \to \infty} r_{max}(d(k)) > r_{max}(d^*) = \frac{7-a}{5} \approx 0.234.$$

In other words, $r_{max}(d)$ depends discontinuously on the data d.

REMARKS 19.3.3 (Interpretation and Further Research): The example above illustrates the "false sense of security" associated with the robustness margin. To further elaborate, if $q_1^* = q_2^* \approx 0.234$, two of the roots of the closed loop polynomial $p_{d(k)}(s, q^*)$ approach the imaginary axis as $k \to \infty$. That is, $p_{d(k)}(s, q^*)$ is "nearly" destabilized by an uncertainty vector q^* whose norm is 0.234 despite the fact that the predicted margin is approximately 0.417.

For the case of affine linear uncertainty structures, the papers by Tesi and Vicino (1991) and Rantzer (1992c) provide conditions under which the robustness margin depends continuously on the data; the paper by Packard and Pandey (1991) aims to regularize the computation of $r_{max}(d)$ by adding "small" fictitious complex perturbations.

At the heart of the discontinuity problem is a certain rank-dropping phenomenon. That is, with $q \in \mathbf{R}^\ell$, we can always write

$$\begin{bmatrix} Re\ p(j\omega, q) \\ Im\ p(j\omega, q) \end{bmatrix} = A(\omega)q + b(\omega)$$

with $A(\omega) \in \mathbf{R}^{2 \times \ell}$ and $b(\omega) \in \mathbf{R}^{2 \times 1}$. It turns out that discontinuity in the margin is accompanied by $A(\omega_0)$ losing rank at some $\omega_0 \geq 0$.

The example by Ackermann, Hu and Kaesbauer (1990), however, indicates that the problem of regularity of robust stability computations is a lot more subtle than simply detecting whether $r_{max}(d)$ is continuous; i.e., for the uncertain polynomial

$$p(s, q) = s^3 + (q_1 + q_2 + 1)s^2 + (q_1 + q_2 + 3)s + (1 + \epsilon^2 + 6q_1 + 6q_2 + 2q_1 q_2),$$

it is straightforward to verify that for small $\epsilon > 0$, there is an "island" of instability described by

$$(q_1 - 1)^2 + (q_2 - 1)^2 \leq \epsilon^2.$$

However, for any fixed $\epsilon > 0$, the robust stability computations are continuous with respect to the data. Many methods of computation will "miss" the instability when ϵ is suitably small.

19.4   Problem Area 3: Parametric Lyapunov Theory

Throughout this text, we have worked almost exclusively with uncertain polynomials rather than uncertain matrices. For example, in a robust stability analysis involving an uncertain state space system

$$\dot{x}(t) = A(q)x(t),$$

we indicated that it can be addressed using the uncertain characteristic polynomial

$$p(s, q) = \det(sI - A(q)).$$

In this section, we raise the possibility that in many cases there may often be an advantage to working directly with matrices. To motivate further research at the matrix level, we survey a number of existing results. Unless otherwise stated, throughout this section, the uncertainty bounding set Q is taken to be a box.

To begin, we first mention some matrix results which can be trivially obtained by "lifting" the polynomial theory to the matrix level. For example, suppose that $A(q)$ is the companion canonical form

$$A(q) = \begin{bmatrix} 0 & 1 & 0 & \cdots & \cdots \\ 0 & 0 & 1 & 0 & \cdots \\ \vdots & \vdots & \vdots & \vdots & \vdots \\ 0 & 0 & 0 & \cdots & 1 \\ q_0 & q_1 & q_2 & \cdots & q_{n-1} \end{bmatrix}$$

Then we obtain a simple matrix analogue of Kharitonov's Theorem: $\mathcal{A}$ *is robustly stable if and only if four distinguished extreme matrices are stable.* Of course, the four distinguished matrices to which we refer are obtained from the Kharitonov polynomials associated with the interval polynomial family with the characteristic given by polynomial $p(s, q) = \det(sI - A(q))$.

There are also some trivial matrix analogues of robust stability results for polytopes of polynomials. For example, if q enters affine linearly into only a single row or column of $A(q)$, then the resulting characteristic polynomial $p(s, q)$ turns out to have an affine linear uncertainty structure and the many results in this text on polytopes of polynomials are applicable. More generally, if $A(q)$ has rank one dependence on q, the same result holds.

In fact, in the dissertation of El Ghaoui (1990), there is discussion of the class of matrix uncertainty structures which permit linkage with the theory for polytopes of polynomials: Indeed, if

$$A(q) = A_0 + \sum_{i=0}^{\ell} A_i q_i$$

with $A_i \in \mathbf{R}^{n \times n}$ fixed for $i = 0, 1, 2, \ldots, \ell$, it is of interest to provide conditions on the matrices A_0, A_1, A_2, ..., A_ℓ for which the characteristic polynomial $p(s, q) = \det(sI - A(q))$ has affine linear uncertainty structure.

19.4.1 Polytopes of Matrices and Lyapunov Functions

If the entries of $A(q)$ depend affine linearly on q, then the matrix family $\mathcal{A} = \{A(q) : q \in Q\}$ is called a *polytope of matrices* or a *polytopic matrix family*. The fact that $\mathcal{A}$ is polytopic is explained by noting that if q^i is the i-th extreme point of Q and we take $A_i = A(q^i)$, then

$$\mathcal{A} = \text{conv}\{A_i\}.$$

In Chapter 14, we already discussed a special subclass of matrix polytopes. Namely, we considered the robust stability problem for an interval matrix family. In view of the limited results available for this special case, we expect the more general matrix polytope problem to be difficult at the level of 4×4 and above. We now mention some approaches to this problem and special cases which have been solved.

There are a number of papers in the literature which aim to establish robust stability using a so-called *common Lyapunov function*. For example, in the paper by Horisberger and Belanger (1976), the following idea is central: Suppose that $P = P^T > 0$ is such that

$$A_i^T P + P A_i < 0$$

for all generators A_i of $\mathcal{A}$. Then it follows that

$$A^T P + P A < 0$$

for all $A \in \mathcal{A}$. To see that this conclusion is correct, notice that any $A \in \mathcal{A}$ can be expressed as a convex combination $A = \sum_i \lambda_i A_i$ with all $\lambda_i \geq 0$ and $\sum_i \lambda_i = 1$. Now, using the fact that positively weighted sums of negative-definite matrices are still negative-definite, we obtain the desired result by noting that

$$A^T P + P A = \sum_i \lambda_i (A_i^T P + P A_i).$$

An interesting generalization of the ideas above is given by Garofalo, Celentano and Glielmo (1992): *If each entry of $A(q)$ is a quotient of multilinear functions of q, then once again, a common Lyapunov matrix for the extremes of Q serves as a common Lyapunov matrix for all of $\mathcal{A}$.*

The work of Shi and Gao (1986) provides a concrete example for which a common Lyapunov function is readily available: *If $\mathcal{A}$ is a polytope of symmetric matrices, then stability of the set of generators $\{A_i\}$ is equivalent to robust stability of $\mathcal{A}$.* This is easily explained by noting that $A_i + A_i^T < 0$ implies that $P = I$ serves as a common Lyapunov function.

More generally, the problem of finding a common Lyapunov function does not appear to admit a simple analytical solution. From a practical point of view, however, this presents no major obstacle because this problem can be cast in a convex programming framework. To see this, we provide a convexity argument in the style of Boyd and Barratt (1990). Indeed, if we let $\mathcal{P}$ denote the set of $n \times n$ positive-definite symmetric matrices and define

$$J_i(P) = \lambda_{max}[A_i^T P + PA_i]$$

for each generator A_i and

$$J(P) = \max_i \ J_i(P),$$

it is easy to see that the existence of a common Lyapunov function is equivalent to

$$\inf_{P \in \mathcal{P}} \ J(P) < 0.$$

We now claim that this infimum problem is a convex program. Indeed, convexity of the set $\mathcal{P}$ is immediate; in fact, $\mathcal{P}$ is a convex cone. To see that $J(P)$ is a convex function, we use the well-known fact (for example, see Rockafellar (1970)) that the pointwise supremum of an indexed collection of convex functions $\{J_x : x \in X\}$ is still convex. Subsequently, to establish convexity of $J_i(P)$, we take

$$J_x(P) = x^T[A_i^T P + PA]x,$$

$$X = \{x : \|x\| \leq 1\}$$

and observe that $J_x(P)$ is linear (hence convex) in the entries of the matrix P. Hence, by viewing $J(P)$ as the pointwise supremum of the finite collection of the $J_i(P)$, we conclude that $J(P)$ is convex.

REMARKS 19.4.2 (Parameterized Lyapunov Function):   Further research is motivated by the following simple fact: It is easy to construct polytopes of matrices which are robustly stable but do not admit a common Lyapunov function. In other words, solution of robust stability problems via the common Lyapunov function approach is inherently conservative. Roughly speaking, if the "spread" of the uncertainty is large, it is unreasonable to expect the same Lyapunov function to work for all $A \in \mathcal{A}$. These comments motivate the search for *parameterized Lyapunov functions*. The takeoff point for work along these lines is the following obvious fact: *The polytopic family of $n \times n$ matrices $\mathcal{A} = \{A(q) : q \in Q\}$ is robustly stable if* *and only if there exists a positive-definite symmetric matrix function* $P : Q \to \mathbf{R}^{n \times n}$ *such that*

$$A^T(q)P(q) + P(q)A(q) < 0$$

for all $q \in Q$. A fundamental research problem involves identification of classes of polytopes $\mathcal{A}$ for which the existence of an appropriate $P(q)$ can be ascertained. Only a few papers have been written along those lines. We mention a sampling of the rather specialized results obtained to date.

In the paper by Barmish and DeMarco (1986), the uncertain parameter vector q is expressed as the convex combination

$$q = \sum_i \lambda_i q^i,$$

where q^i denotes the i-th extreme point of q. Subsequently, if P_i is a Lyapunov matrix for the i-th generator A_i, a parameterization

$$P(\lambda) = \sum_i \lambda_i P_i$$

is proposed and conditions are given under which $P(\lambda)$ proves the stability of $\mathcal{A}$. We refer to $P(\lambda)$ above as *affinely parameterized* and note that a different class of affine parameterizations is pursued in Leal and Gibson (1990). In both papers, stringent side conditions must be satisfied in order to prove robust stability of $\mathcal{A}$.

For the important special case when $A(q)$ arises by embedding a stable interval polynomial family into a matrix companion form, the paper by Mansour and Anderson (1992) establishes that there is a bilinearly parameterized Lyapunov matrix $P(q)$ which can be used to prove robust stability. In conclusion, the theory of parametric Lyapunov functions is only in its infancy—some of the most fundamental questions are yet to be answered. For example, with the same setup as in Mansour and Anderson (1992), can one prove the robust stability using a parametric Lyapunov function with $P(q)$ having affine linear dependence on q? Another interesting area of research involves the relationship between the classical Popov criterion and the parametric Lyapunov function; for example, see the paper by Haddad and Bernstein (1992).

19.4.3   A Conjecture

To further emphasize the fact that there is a wealth of open research problems involving parametric Lyapunov functions, we conclude this section with a conjecture involving one of the most basic problems which one might address.

We concentrate on the special case when $\mathcal{A}$ is simply the convex hull of two real $n \times n$ matrices A_0 and A_1. Assuming that $\mathcal{A}$ is robustly stable, we conjecture that $\mathcal{A}$ admits an affine linearly parameterized quadratic Lyapunov function. That is, if

$$A(\lambda) = (1 - \lambda)A_0 + \lambda A_1$$

is stable for all $\lambda \in [0, 1]$, there exist $n \times n$ symmetric matrices P_0 and P_1 such that

$$P(\lambda) = P_0 + \lambda P_1$$

is positive-definite and

$$A^T(\lambda)P(\lambda) + P(\lambda)A(\lambda) < 0$$

for all $\lambda \in [0, 1]$.

19.5   Problem Area 4: Polytopes of Matrices

In this section, we continue to focus on the robust stability problem for a polytope of matrices. In contrast to the preceding section, we no longer concentrate on Lyapunov theory. Instead, we consider problems whose solutions shed light on the computational complexity of the robust stability problem.

Motivated by the many robust stability results available for polytopes of polynomials, it is natural to ask whether results along the same lines are possible for a polytope of matrices. Recalling the example of Barmish, Fu and Saleh (1988) in Exercise 14.5.1, we already know that even for the special case of interval matrices, neither extreme point results nor edge results are possible. The interesting result of Cobb and DeMarco (1989) tells us even more: *If $\mathcal{A}$ is a polytope of $n \times n$ matrices with $n \geq 3$, then stability of all faces of dimension $2n - 4$ is sufficient to guarantee robust stability of $\mathcal{A}$. Moreover, if the dimension of $\mathcal{A}$ (viewed as a subset of $\mathcal{R}^{n^2}$) is $2n - 4$ or greater, there are examples of matrix polytopes which are unstable but have the property that all faces of dimension $2n - 5$ are stable.*

One pathway to the study of computational complexity is motivated by a certain relationship between robust stability and *robust nonsingularity*. To clearly explain this linkage, we consider the family $\mathcal{A} = \{A(q) : q \in Q\}$ of $n \times n$ matrices with $A(q)$ depending continuously on q. Under rather mild conditions, there is a family of linear transformations (for example, see Bialas (1985) and Fu and Barmish (1988)) mapping $\mathcal{A}$ into a new family $\bar{\mathcal{A}}$ having the following property: *The family $\mathcal{A}$ is robustly stable if and only if the family $\bar{\mathcal{A}}$ is robustly nonsingular;* see Section 4.12 and also note that this idea is used many times in Chapter 17. As an example, using the linear map

$$TA = A \oplus A,$$

the robust stability problem is readily transformed into a robust nonsingularity problem.

An important point to note is that any linear transformation on $\mathcal{A}$ preserves the affine linear dependence of matrix entries on the uncertain parameters. Hence, the resulting nonsingularity problem also involves a polytope of matrices. However, it is also important to mention that an independent uncertainty structure in $A(q)$ gets transformed into an affine linear uncertainty for $A(q) \oplus A(q)$. Said another way, a linear transformation taking the robust stability problem to a robust nonsingularity problem does not preserve the interval matrix structure; i.e., we begin with an interval matrix family $\mathcal{A}$ and end up with a polytope of matrices $T\mathcal{A}$. To illustrate this, observe that the transformation $TA(q) = A(q) \oplus A(q)$ maps

$$A(q) = \begin{bmatrix} q_{11} & 1 \\ 3 & q_{22} \end{bmatrix}$$

into

$$TA(q) = \begin{bmatrix} 2q_{11} & 1 & 1 & 0 \\ 3 & q_{11} + q_{22} & 0 & 1 \\ 3 & 0 & q_{11} + q_{22} & 1 \\ 0 & 3 & 3 & 2q_{22} \end{bmatrix}$$

which no longer has an independent uncertainty structure. The independent uncertainty structure has been transformed into an affine linear uncertainty structure. Hence, a solution to the robust stability problem via this approach involves solution of the robust nonsingularity problem for a class of matrix polytopes.

19.5.1   Robust Nonsingularity

Motivated by the discussion above, we now concentrate on the robust nonsingularity problem for a polytope of matrices. For the special case of an interval matrix family $\mathcal{A} = \{A(q) : q \in Q\}$, we recall Exercise 14.5.6: $\mathcal{A}$ *is robustly nonsingular if and only if for each extreme point q_i of Q, det $A(q^i)$ has the same sign.* From an application point of view, a weakness of this result is that as the dimension of matrix $A(q)$ in $\mathcal{A}$ increases, there is a combinatoric explosion in the number of extreme points. Notice that if $\mathcal{A}$ is an $n \times n$ interval matrix family, there can be as many as 2^{n^2} extreme points. This immediately suggests a number of basic questions about the computational complexity of the robust stability and robust nonsingularity problems.

Under strengthened hypotheses, however, special classes of robust stability problems can be solved. For example, suppose that A is an $n \times n$ matrix with nonnegative off-diagonal entries. Then, letting A_k denote the upper $k \times k$ block of A for $k = 1, 2, \ldots, n$, in accordance with classical results from matrix algebra (for example, see Gantmacher (1959)), it follows that A is stable if and only if

$$(-1)^k \det(I_k - A_k) > 0$$

for $k = 1, 2, \ldots, n$, where I_k denotes the $k \times k$ identity matrix. A similar result holds for Schur stability with the added restriction that the diagonal entries of A are nonnegative and the determinant condition above is replaced by $\det(I_k - A_k) > 0$. If we consider an $n \times n$ interval matrix family $\mathcal{A} = \{A(q) : q \in Q\}$ with nonnegative off-diagonal entries, we can exploit the result above in combination with the fact that a multilinear function on a box achieves both its minimum and maximum at an extreme point; e.g., see Lemma 14.5.5. Now, we arrive at the following extreme point result: $\mathcal{A}$ *is robustly stable if and only if*

$$(-1)^k \det(I_k - A_k(q^i)) > 0$$

for all extreme points q^i of Q and all $k \in \{1, 2, \ldots, n\}$. For the case of robust Schur stability, minor modifications of the arguments above lead to a similar result given by Shafai, Perev, Cowley and Chehab (1991). For further extensions involving irreducible interval matrices, see Mayer (1984).

To reduce the number of extreme points to be tested, the important paper by Rohn (1989) begins with a nonsingular $n \times n$ matrix A_0 and given bounds $r_{ij} \geq 0$ for the entries q_{ij} of an interval matrix $\Delta A(q)$. Defining the family of matrices

$$\mathcal{A}_r = \{A_0 + \epsilon \Delta A(q) : 0 \leq \epsilon \leq r \text{ and } q \in Q\}$$

with variable *magnification factor* $r \geq 0$, the objective is to obtain the robustness margin

$$r_{max} = \sup\{r : \mathcal{A}_r \text{ is robustly nonsingular}\}.$$

Taking R to be the $n \times n$ matrix having (i,j)-th entry r_{ij}, we provide some standard terminology which is needed in order to describe Rohn's result; see also Demmel (1988). Indeed, if M is a square matrix, let

$$\rho_0(M) = \max\{|\lambda| : \lambda \text{ is a real eigenvalue of } M\}$$

with $\rho_0(M) = 0$ if no eigenvalues of M are real. A square matrix S is said to be a *signature matrix* if it is diagonal with all diagonal entries equal to either $+1$ or -1. Now, we let $\mathcal{S}$ be the set of $n \times n$ signature matrices and observe that $\mathcal{S}$ has 2^n members. We are now prepared to present Rohn's Theorem.

THEOREM 19.5.2 (Rohn (1989)):   *Given $r \geq 0$, the interval matrix family $\mathcal{A}_r$ is robustly nonsingular if and only if $\det A_0$ and the 4^n determinants $\det(A_0 + rS_1RS_2)$, obtained with $S_1, S_2 \in \mathcal{S}$, have the same nonzero sign. Moreover,*

$$r_{max} = \frac{1}{\max_{S_i \in \mathcal{S}} \rho_0(S_1 A_0^{-1} S_2 R)}.$$

REMARKS 19.5.3 (Connections with μ Theory and Complexity): It is interesting to note that the theorem above has a bearing on the computation of the real structured singular value (real μ) for the case when the ℓ^∞ norm is used for the uncertain parameter vector. To this end, we begin with a real square rank one matrix M and the goal is to compute

$$\frac{1}{\mu_\infty(M)} = \inf\{\|q\|_\infty : \det(I + M\Delta(q)) = 0\},$$

where $\|q\|_\infty$ denotes the ℓ^∞ norm of q and

$$\Delta(q) = \text{diag}\{q_1, q_2, \ldots, q_\ell\}.$$

In the numerical computation of $\mu_\infty(M)$, the following lower bound has traditionally been used:

$$\mu_\infty(M) \geq \max_{S \in \mathcal{S}} \rho_0(SM).$$

In view of the extreme point theory of Rohn (1989), it can be shown that this lower bound is sharp. For further discussion of extreme point results in this framework, see also Holohan and Safonov (1989), El Ghaoui (1990) and Chen, Fan and Nett (1992).

To conclude this section, we note that the line of research above has resulted in control system researchers devoting attention to issues of computational complexity as defined in the field of computer science; e.g., see Garey and Johnson (1979). Some initial results in this direction are given in the paper by Rohn and Poljak (1992) where a class of robust nonsingularity problems are shown to be NP–hard. Coxson and DeMarco (1992) establish that the problem of deciding if the real structured singular value of a matrix is bounded above by a given constant is NP–hard, and Nemirovskii (1992) addresses the computational complexity of a class of robust stability problems.

19.6   Problem Area 5: Robust Performance

Given the degree to which this text has emphasized extreme point results for robust stability, it is natural ask: What type of robust performance criteria can be addressed in an extreme point context? Over the last few years, we see the beginning of a new line of research in this direction. In the subsections to follow, we overview some of the recent developments and briefly mention some of the interesting open problems.

19.6.1   Parametric H^∞ Norm

The first result which we mention involves H^∞ analysis with structured real uncertainty. For completeness, recall that if $P(s)$ is proper, stable and rational, then the H^∞ norm is given by

$$\|P\|_\infty = \sup_\omega |P(j\omega)|.$$

In the theorem below, we see that for a stable interval plant family, the worst-case H^∞ norm is attained by one of the sixteen Kharitonov plants; see Chapter 11 for further details.

THEOREM 19.6.2 (See Mori and Barnett (1988) and Chapellat, Dahleh and Bhattacharyya (1990)):   *Consider a robustly stable proper interval plant family* $\mathcal{P} = \{P(s,q,r) : q \in Q, r \in R\}$ *with monic denominator and sixteen associated Kharitonov plants* $P_{i_1,i_2}(s)$

for $i_1, i_2 = 1, 2, 3, 4$. Then it follows that

$$\max_{q \in Q, r \in R} \|P(\cdot, q, r)\|_\infty = \max_{i_1, i_2} \|P_{i_1, i_2}\|_\infty.$$

REMARKS 19.6.3 (Extensions):   The ideas central to the proof of Theorem 19.6.2 enter into the proof of a number of closely related results. For example, in Chapellat, Dahleh and Bhattacharyya (1990), a real parameter version of the Small Gain Theorem is established and in Mori and Barnett (1988), a robust Popov-like criterion is provided in the interval plant context. These results are extended in papers by Dahleh, Tesi and Vicino (1991), Vicino and Tesi (1991) and Rantzer (1992a). It is also interesting to point out that for the frequency-weighted version of the problem above, severe restrictions on the weighting function are required in order to obtain an extreme point result; see Hollot, Tempo and Blondel (1992).

19.6.4   Positive-Realness

The issue of positive-realness (addressed in the classical Popov theory) is studied in an extreme point context in the paper by Dasgupta and Bhagwat (1987). If

$$P(s, q) = \frac{N(s, q)}{D(s, q)},$$

is an uncertain plant, then for fixed q, we recall that $P(s, q)$ is strictly positive-real (SPR) if both $N(s, q)$ and $D(s, q)$ are stable and $Re\, P(j\omega, q) > 0$ for all $\omega \in \mathbf{R}$. We say that a family of plants $\mathcal{P} = \{P(\cdot, q) : q \in Q\}$ is *robustly SPR* if $P(s, q)$ is SPR for all $q \in Q$. In the theorem below, the problem of guaranteeing that an interval plant family is robustly SPR is addressed; the exercise following the theorem allows for a more general class of plants.

THEOREM 19.6.5 (Dasgupta and Bhagwat (1987)):   *Consider the interval plant family $\mathcal{P}$ with fixed numerator $N(s, q) \equiv N(s)$. Letting $D_1(s), D_2(s), D_3(s)$ and $D_4(s)$ denote the four Kharitonov denominator polynomials, it follows that $\mathcal{P}$ is robustly SPR if and only if $N(s)/D_i(s)$ is SPR for $i = 1, 2, 3, 4$.*

EXERCISE 19.6.6 (More General Interval Plant):   Consider an interval plant family $\mathcal{P}$ with its four Kharitonov numerators $N_1(s)$, $N_2(s)$, $N_3(s)$ and $N_4(s)$ and its four denominators $D_1(s)$, $D_2(s)$,

$D_3(s)$ and $D_4(s)$. Prove that $\mathcal{P}$ is robustly SPR if and only if each of the extreme plants

$$P_{i_1, i_2}(s) = \frac{N_{i_1}(s)}{D_{i_2}(s)}$$

is SPR. *Hint*: With $P(s, q, r) = N(s, q)/D(s, r)$ and fixed $\omega \in \mathbf{R}$, instead of studying the condition

$$\min_{q \in Q, r \in R} Re\ P(j\omega, q, r) > 0,$$

work with

$$\min_{q \in Q, r \in R} Re\ N(j\omega, q)D^*(j\omega, r) > 0,$$

where $D^*(j\omega, r)$ denotes the complex conjugate of $D(j\omega, r)$. Now, this problem can be analyzed by noting that it involves minimization of a multilinear function on a box.

REMARKS 19.6.7 (Extensions):   In their paper, Dasgupta and Bhagwat (1987) also point out the applicability of robust SPR results in an adaptive output error identification context. There are also a number of papers dealing with extensions and variations on the SPR theme. In papers by Dasgupta (1987), Bose and Delansky (1989), Chapellat, Dahleh and Bhattacharyya (1991) and Shi (1991), the SPR problem is studied under the weaker hypothesis that the numerator $N(s, q)$ can be uncertain as in the exercise above; in some cases, plants with complex coefficients are considered. It is also worth noting that the line of proof used in Exercise 19.6.6 above also works for SPR problems with multilinear uncertainty structures. However, instead of a four-plant result as in Theorem 19.6.5, the plants associated with all of the extreme q^i come into play; see the recent papers by Dasgupta, Parker, Anderson, Kraus and Mansour (1991), Bose and Delansky (1989) and Shi (1991). Finally, we mention extreme point results for the case when the plant is not necessarily SPR but can be rendered SPR by suitable addition of a positive constant; see Chapellat, Dahleh and Bhattacharyya (1991).

19.6.8   Steady State, Overshoot and Nyquist

In this final subsection, we review a number of results which seem to raise at least as many questions as they answer. In the work of Bartlett (1990c), an uncertain plant $P(s, q)$ with multilinear uncertainty structure is considered and steady-state error for a unit step input is the prime consideration. A robust stability assumption is imposed and it is shown that *both the maximum and the minimum of the steady-state error occur on one of the extreme plants* $P(s, q^i)$. Bartlett then goes on to provide a counterexample to the tempting conjecture that the maximal peak overshoot is also attained at an extreme. To this end, he considers the family of plants with affine linear uncertainty structure described by

$$P(s, q) = \frac{1}{(3.4q + 0.1)s^2 + (1.7q + 0.8)s + 1}$$

and $q \in [0, 1]$. It turns out that the peak overshoot corresponding to a step input is not maximized at the extremes $q = 0$ or $q = 1$; e.g., $q = 0.5$ leads to a higher overshoot value. A similar example is also given for discrete-time systems.

In the work of Hollot and Tempo (1991), the frequency response of an interval plant is considered. To briefly overview some of their results, suppose that $\mathcal{P} = \{P(s, q, r) : q \in Q, r \in R\}$ is a strictly proper interval plant family with monic denominator and, to keep the exposition simple, we also assume that for all $r \in R$ the plant denominator $D(s, r)$ has no roots on the imaginary axis. Then the *Nyquist set* associated with this family of plants is defined by

$$\mathcal{N} = \{P(j\omega, q, r) : \omega \in \mathbf{R}; q \in Q; r \in R\}.$$

The boundary $\partial \mathcal{N}$ of the Nyquist set, called the *Nyquist envelope*, is the focal point and the following question is addressed: What points $z \in \partial \mathcal{N}$ on the Nyquist envelope are Kharitonov points? By this, we mean points $z \in \partial \mathcal{N}$ such that $z = P_{i_1, i_2}(j\omega)$ for one of the sixteen Kharitonov plants $P_{i_1, i_2}(s)$ and some $\omega \in \mathbf{R}$.

In this regard, it is already known from the work of Fu (1991) that all points on the Nyquist envelope come from the edges of the uncertainty bounding set $Q \times R$. In fact, one can restrict attention to the thirty-two edges identified by Chapellat and Bhattacharyya (1989). If $\mathcal{P}$ is robustly stabilized by unity feedback, then the Kharitonov points on the envelope include the minimal gain margin points and minimal phase margin points. If $P(s, q)$ is (open loop) stable for all $q \in Q$, the points associated with the maximum H^∞ norm also lie on the Nyquist envelope. Such a point is identified with a $q^* \in Q$ and an $\omega^* \in \mathbf{R}$ such that

$$|P(j\omega^*, q^*)| = \max_{q, \omega} |P(j\omega, q)|.$$

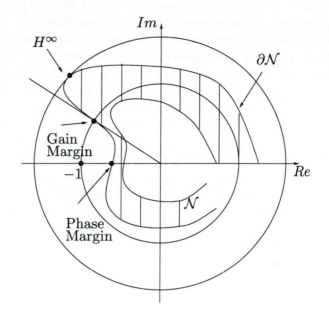

FIGURE 19.6.1   Kharitonov Points on the Nyquist Envelope

Hollot and Tempo (1991) also establish that there are many other interesting Kharitonov points. For example, they prove that the critical points associated with minimum sensitivity and complementary sensitivity are Kharitonov points. Along these same lines, we also mention the paper by Kimura and Hara (1991) where a more restrictive class of interval plants with fixed numerator is considered. To further clarify the geometry associated with the discussion above, we refer to Figure 19.6.1 where three distinguished points on the Nyquist envelope are depicted.

19.7   Conclusion

One of the objectives of this chapter was to bombard the reader with a large number of results in the hope that a subset of them will suggest topics for further research. At a number of selected points within the chapter, rather specific research problems were suggested.

From a control theoretic point of view, it can be argued that the "most important problem" was omitted from the list of five given. That is, this chapter made no mention of the robust synthesis problem with structured real parametric uncertainty. Although the recent literature already contains some interesting new synthesis results (for example, see the paper by Rantzer and Magretsky (1992)), a number of basic questions remain unanswered. From the viewpoint of this author, one of the most important of these questions involves the hope for synthesis results in the spirit of LQR theory. That is, for plants with real parametric uncertainty, can we reduce classes of robust synthesis problems to solution of standard Riccati equations? For classes of complex uncertainty problems, we already know of many cases for which a Riccati equation solution is possible; e.g., see Doyle, Glover, Khargonekar and Francis (1989). Of course, in asking questions of this nature, we are ruling out conservative solution techniques which involve overbounding real uncertainty intervals by discs in the complex plane; for such cases only sufficient conditions are obtained. The objective is to obtain "tight nonconservative" conditions which are both necessary and sufficient.

Appendix and Bibliography

Appendix A

Symbolic Computation for Fiat Dedra

The closed loop polynomial for the Fiat Dedra case study (see Chapter 3) was obtained using a symbolic manipulator. Note that the formulas given below have been simplified; before combining like terms, the formulas for the $a_i(q)$ are much more complicated.

$$a_0(q) = 6.82079 \times 10^{-5} q_1 q_3 q_4^2 + 6.82079 \times 10^{-5} q_1 q_2 q_4 q_5;$$

$$\begin{aligned}
a_1(q) = \ & 7.61760 \times 10^{-4} q_2^2 q_5^2 + 7.61760 \times 10^{-4} q_3^2 q_4^2 \\
& + 4.02141 \times 10^{-4} q_1 q_2 q_5^2 + 0.00336706 q_1 q_3 q_4^2 \\
& + 6.82079 \times 10^{-5} q_1 q_4 q_5 + 5.16120 \times 10^{-4} q_2^2 q_5 q_6 \\
& + 0.00336706 q_1 q_2 q_4 q_5 + 6.82079 \times 10^{-5} q_1 q_2 q_4 q_7 \\
& + 6.28987 \times 10^{-5} q_1 q_2 q_5 q_6 + 4.02141 \times 10^{-4} q_1 q_3 q_4 q_5 \\
& + 6.28987 \times 10^{-5} q_1 q_3 q_4 q_6 + 0.00152352 q_2 q_3 q_4 q_5 \\
& + 5.16120 \times 10^{-4} q_2 q_3 q_4 q_6;
\end{aligned}$$

$$\begin{aligned}
a_2(q) = \ & 4.02141 \times 10^{-4} q_1 q_5^2 + 0.00152352 q_2 q_5^2 \\
& + 0.0552 q_2^2 q_5^2 + 0.0552 q_3^2 q_4^2 \\
& + 0.0189477 q_1 q_2 q_5^2 + 0.034862 q_1 q_3 q_4^2 \\
& + 0.00336706 q_1 q_4 q_5 + 6.82079 \times 10^{-5} q_1 q_4 q_7 \\
& + 6.28987 \times 10^{-5} q_1 q_5 q_6 + 0.00152352 q_3 q_4 q_5 \\
& + 5.16120 \times 10^{-4} q_3 q_4 q_6 - 0.00234048 q_3^2 q_4 q_6
\end{aligned}$$

$$+\ 0.034862q_1q_2q_4q_5 + 0.0237398q_2^2q_5q_6$$
$$+\ 0.00152352q_2^2q_5q_7 + 5.16120 \times 10^{-4}q_2^2q_6q_7$$
$$+\ 0.00336706q_1q_2q_4q_7 + 0.00287416q_1q_2q_5q_6$$
$$+\ 8.04282 \times 10^{-4}q_1q_2q_5q_7 + 6.28987 \times 10^{-5}q_1q_2q_6q_7$$
$$+\ 0.0189477q_1q_3q_4q_5 + 0.00287416q_1q_3q_4q_6$$
$$+\ 4.02141 \times 10^{-4}q_1q_3q_4q_7 + 0.1104q_2q_3q_4q_5$$
$$+\ 0.0237398q_2q_3q_4q_6 + 0.00152352q_2q_3q_4q_7$$
$$-\ 0.00234048q_2q_3q_5q_6 + 0.00103224q_2q_5q_6;$$

$$a_3(q) = 0.0189477q_1q_5^2 + 0.1104q_2q_5^2$$
$$+\ 5.16120 \times 10^{-4}q_5q_6 + q_2^2q_5^2 + 7.61760 \times 10^{-4}q_2^2q_7^2$$
$$+\ q_3^2q_4^2 + 0.1586q_1q_2q_5^2 + 4.02141 \times 10^{-4}q_1q_2q_7^2$$
$$+\ 0.0872q_1q_3q_4^2 + 0.034862q_1q_4q_5$$
$$+\ 0.00336706q_1q_4q_7 + 0.00287416q_1q_5q_6$$
$$+\ 6.28987 \times 10^{-5}q_1q_6q_7 + 0.00103224q_2q_6q_7$$
$$+\ 0.1104q_3q_4q_5 + 0.0237398q_3q_4q_6$$
$$+\ 0.00152352q_3q_4q_7 - 0.00234048q_3q_5q_6$$
$$+\ 0.1826q_2^2q_5q_6 + 0.1104q_2^2q_5q_7$$
$$+\ 0.0237398q_2^2q_6q_7 - 0.0848q_3^2q_4q_6$$
$$+\ 0.0872q_1q_2q_4q_5 + 0.034862q_1q_2q_4q_7$$
$$+\ 0.0215658q_1q_2q_5q_6 + 0.0378954q_1q_2q_5q_7$$
$$+\ 0.00287416q_1q_2q_6q_7 + 0.1586q_1q_3q_4q_5$$
$$+\ 0.0215658q_1q_3q_4q_6 + 0.0189477q_1q_3q_4q_7$$
$$+\ 2q_2q_3q_4q_5 + 0.1826q_2q_3q_4q_6$$
$$+\ 0.1104q_2q_3q_4q_7 - 0.0848q_2q_3q_5q_6$$
$$-\ 0.00234048q_2q_3q_6q_7 + 7.61760 \times 10^{-4}q_5^2$$
$$+\ 0.0474795q_2q_5q_6 + 8.04282 \times 10^{-4}q_1q_5q_7$$
$$+\ 0.00304704q_2q_5q_7;$$

$$a_4(q) = 0.1586q_1q_5^2 + 4.02141 \times 10^{-4}q_1q_7^2$$
$$+\ 2q_2q_5^2 + 0.00152352q_2q_7^2 + 0.0237398q_5q_6$$
$$+\ 0.00152352q_5q_7 + 5.16120 \times 10^{-4}q_6q_7$$
$$+\ 0.0552q_2^2q_7^2 + 0.0189477q_1q_2q_7^2$$
$$+\ 0.0872q_1q_4q_5 + 0.034862q_1q_4q_7$$

$$+ 0.0215658q_1q_5q_6 + 0.00287416q_1q_6q_7$$
$$+ 0.0474795q_2q_6q_7 + 2q_3q_4q_5 + 0.1826q_3q_4q_6$$
$$+ 0.1104q_3q_4q_7 - 0.0848q_3q_5q_6 - 0.00234048q_3q_6q_7$$
$$+ 2q_2^2q_5q_7 + 0.1826q_2^2q_6q_7 + 0.0872q_1q_2q_4q_7$$
$$+ 0.3172q_1q_2q_5q_7 + 0.0215658q_1q_2q_6q_7$$
$$+ 0.1586q_1q_3q_4q_7 + 2q_2q_3q_4q_7 - 0.0848q_2q_3q_6q_7$$
$$+ 0.0552q_5^2 + 0.3652q_2q_5q_6 + 0.0378954q_1q_5q_7$$
$$+ 0.2208q_2q_5q_7;$$

$$a_5(q) = 0.0189477q_1q_7^2 + 0.1104q_2q_7^2 + 0.1826q_5q_6$$
$$+ 0.1104q_5q_7 + 0.0237398q_6q_7$$
$$+ q_2^2q_7^2 + 0.1586q_1q_2q_7^2 + 0.0872q_1q_4q_7$$
$$+ 0.0215658q_1q_6q_7 + 0.3652q_2q_6q_7$$
$$+ 2q_3q_4q_7 - 0.0848q_3q_6q_7 + q_5^2 + 7.61760 \times 10^{-4}q_7^2$$
$$+ 0.3172q_1q_5q_7 + 4q_2q_5q_7;$$

$$a_6(q) = 0.1586q_1q_7^2 + 2q_2q_7^2 + 2q_5q_7 + 0.1826q_6q_7$$
$$+ 0.0552q_7^2;$$

$$a_7(q) = q_7^2.$$

Bibliography

Abate, M. and V. Di Nunzio, "Idle speed control using optimal regulation," *XXIII FISITA Congress*, Associazione Tecnica dell'Automobile, Technical paper, vol. 1, no. 905008, Torino, Italy, 1990.

Abate, M., B. R. Barmish, C. E. Murillo–Sanchez and R. Tempo, "Robust performance analysis of the idle speed control of a spark ignition engine," *Proceedings of the American Control Conference*, Chicago, Ill., 1992.

Ackermann, J. E., "Parameter space design of robust control systems," *IEEE Transactions on Automatic Control*, vol. AC-25, pp. 1058–1072, 1980.

Ackermann, J. E. and B. R. Barmish, "Robust Schur stability of a polytope of polynomials," *IEEE Transactions on Automatic Control*, vol. AC-33, pp. 984–986, 1988.

Ackermann, J. E., H. Z. Hu and D. Kaesbauer, "Robustness analysis: A case study, " *IEEE Transactions on Automatic Control*, vol. AC-35, pp. 352–356, 1990.

Ackermann, J. E. and W. Sienel, "What is a "large" number of parameters in robust systems," *Proceedings of the IEEE Conference of Decision and Control*, Honolulu, Hawaii, 1990.

Anderson, B. D. O., N. K. Bose and E. I. Jury, "Output feedback stabilization and related problems–Solution via decision methods," *IEEE Transactions on Automatic Control,* vol. AC-20, pp. 53–66, 1975.

Anderson, B. D. O., E. I. Jury and M. Mansour, "On robust Hurwitz polynomials," *IEEE Transactions on Automatic Control,* vol. AC-32, pp. 909–913, 1987.

Ando, H. and M. Motomochi, "Contribution of fuel transport lag and statistical perturbation in combustion to oscillation of SI engine speed at idle," *Society of Automotive Engineers Technical Paper Series,* no. 870545, 1987.

Argoun, M. B., "Allowable coefficient perturbations with preserved stability of a Hurwitz polynomial," *International Journal of Control,* vol. 44, pp. 927–934, 1986.

Argoun, M. B., "Stability of a Hurwitz polynomial under coefficient perturbations: Necessary and sufficient conditions, *International Journal of Control,* vol. 45, pp. 739–744, 1987.

Aström, K. J. and B. Wittenmark, *Adaptive Control,* Addison–Wesley, Reading, Mass., 1989.

Aubin, J. P. and R. B. Vinter (editors), *Convex Analysis and Optimization,* (Research notes in mathematics; 57), Pitman Advanced Publishing Program, Boston, Mass., 1980.

Bailey, F. N., D. Panzer and G. Gu, "Two algorithms for frequency domain design of robust control systems," *International Journal of Control,* vol. 48, pp. 1787–1806, 1988.

Barmish, B. R., "Invariance of the strict Hurwitz property for polynomials with perturbed coefficients," *Proceedings of the IEEE Conference on Decision and Control,* San Antonio, Tex., 1983; see also *IEEE Transactions on Automatic Control,* vol. AC-29, pp. 935–936, 1984.

Barmish, B. R., "Necessary and sufficient conditions for quadratic stabilizability of an uncertain system," *Journal of Optimization Theory and Applications,* vol. 46, pp. 399–408, 1985.

Barmish, B. R., "New tools for robustness analysis," *Proceedings of the IEEE Conference on Decision and Control,* Austin, Tex., 1988.

Barmish, B. R., "A generalization of Kharitonov's four polynomial concept for robust stability problems with linearly dependent coefficient perturbations," *IEEE Transactions on Automatic Control*, vol. AC-34, pp. 157–165, 1989.

Barmish, B. R., J. E. Ackermann and H. Z. Hu, "The tree structured decomposition: A new approach to robust stability analysis," *Proceedings of the Conference on Information Sciences and Systems*, Johns Hopkins University, Baltimore, Md., 1989.

Barmish, B. R. and C. L. DeMarco, "A new method for improvement of robustness bounds for linear state equations," *Proceedings of the Conference on Information Sciences and Systems*, Johns Hopkins University, Baltimore, Md., 1986.

Barmish, B. R., M. Fu and S. Saleh, "Stability of a polytope of matrices; Counterexamples," *IEEE Transactions on Automatic Control*, vol. AC-33, pp. 569–572, 1988.

Barmish, B. R. and C. V. Hollot, "Counter-example to a recent result on the stability of interval matrices by S. Bialas," *International Journal of Control*, vol. 39, pp. 1103–1104, 1984.

Barmish, B. R., C. V. Hollot, F. J. Kraus and R. Tempo, "Extreme point results for robust stabilization of interval plants with first order compensators," *IEEE Transactions on Automatic Control*, vol. AC-37, pp. 707–714, 1992.

Barmish, B. R. and P. P. Khargonekar, "Robust stability of feedback control systems with uncertain parameters and unmodelled dynamics," *Mathematics of Control, Signals and Systems*, vol. 3, pp. 197–210, 1990.

Barmish, B. R., P. P. Khargonekar, Z. Shi and R. Tempo, "Robustness margin need not be a continuous function of the problem data," *Systems and Control Letters*, vol. 15, pp. 91–98, 1990.

Barmish, B. R. and Z. Shi, "Robust stability of perturbed systems with time delays," *Automatica*, vol. 25, pp. 371–381, 1989.

Barmish, B. R. and Z. Shi, "Robust stability of a class of polynomials with coefficients depending multilinearly on perturbations," *IEEE Transactions on Automatic Control*, vol. AC-35, pp. 1040–1043, 1990.

Barmish, B. R. and R. Tempo, "The robust root locus," *Automatica*, vol. 26, pp. 283–292, 1990.

Barmish, B. R. and R. Tempo, "On the spectral set for a family of polynomials," *IEEE Transactions on Automatic Control*, vol. AC-36, pp. 111–115, 1991.

Barmish, B. R., R. Tempo, C. V. Hollot and H. I. Kang, "An extreme point result for robust stability of a diamond of polynomials," *IEEE Transactions on Automatic Control*, vol. AC-37, pp. 1460–1462, 1992.

Barmish, B. R. and K. H. Wei, "Simultaneous stabilizability of single input–single output systems," *Proceedings of the International Symposium on Mathematical Theory of Networks and Systems*, Stockholm, 1985.

Bartlett, A. C., "Vertex and Edge Theorems Which Simplify Classical Analyses of Linear Systems with Uncertain Parameters," Ph.D. Dissertation, Department of Electrical and Computer Engineering, University of Massachusetts, Amherst, Mass., 1990a.

Bartlett, A. C., "Counter–example to 'Clockwise nature of Nyquist locus of stable transfer functions,'" *International Journal of Control*, vol. 51, pp. 1479–1483, 1990b.

Bartlett, A. C., "Vertex results for the steady state analysis of uncertain systems," *Proceedings of the IEEE Conference of Decision and Control*, Honolulu, Hawaii, 1990c.

Bartlett, A. C. and C. V. Hollot, "A necessary and sufficient condition for Schur invariance and generalized stability of polytopes of polynomials," *IEEE Transactions on Automatic Control*, vol. AC-33, pp. 575–578, 1988.

Bartlett, A. C., C. V. Hollot and L. Huang, "Root locations of an entire polytope of polynomials: It suffices to check the edges," *Mathematics of Control, Signals and Systems*, vol. 1, pp. 61–71, 1988.

Berge, C., *Topological Spaces*, Oliver and Boyd, London, England, 1963.

Bialas, S., "A necessary and sufficient condition for stability of interval matrices," *International Journal of Control*, vol. 37, pp. 717–722, 1983.

Bialas, S., "A necessary and sufficient condition for the stability of convex combinations of stable polynomials or matrices," *Bulletin of the Polish Academy of Sciences, Technical Sciences,* vol. 33, pp. 473–480, 1985.

Bialas, S. and J. Garloff, "Convex combinations of stable polynomials," *Journal of the Franklin Institute,* vol. 319, pp. 373–377, 1985.

Biernacki, R. M., H. Hwang and S. P. Bhattacharyya, "Robust stability with structured real parameter perturbations," *IEEE Transactions on Automatic Control,* vol. AC-32, pp. 495–506, 1987.

Bose, N. K., *Applied Multidimensional Systems Theory,* Van Nostrand Reinhold, New York, 1982.

Bose, N. K. and J. F. Delansky, "Boundary implications for interval positive rational functions," *IEEE Transactions on Circuits and Systems,* vol. CAS-36, pp. 454–458, 1989.

Bose, N. K., E. I. Jury and E. Zeheb, "On robust Hurwitz and Schur polynomials," *Proceedings of the IEEE Conference on Decision and Control,* Athens, Greece, 1986.

Bose, N. K. and Y. Q. Shi, "A simple general proof of Kharitonov's generalized stability criterion," *IEEE Transactions on Circuits and Systems,* vol. CAS-34, pp. 1233–1237, 1987.

Bose, N. K. and E. Zeheb, "Kharitonov's theorem and stability test of multidimensional digital filters," *IEE Proceedings,* Part G, vol. 133, pp. 187–190, 1986.

Boyd, S. P. and C. H. Barratt, *Linear Controller Design: Limits of Performance,* Prentice–Hall, Englewood Cliffs, N.J., 1990.

Brewer, J. W., "Kronecker products and matrix calculus in system theory," *IEEE Transactions on Circuits and Systems,* vol. CAS-25, pp. 772–781, 1978.

Brockett, R. W., "Lie algebras and Lie groups in control theory," in *Geometric Methods in System Theory,* D. Q. Mayne and R. W. Brockett, eds., Reidel, Dordrecht, Netherlands, pp. 32–82, 1973.

Chang, B. C. and O. Ekdal, "Calculation of the real structured singular value via polytopic polynomials," *Proceedings of the American Control Conference,* Pittsburgh, Pa., 1989.

Chang, B. C. and J. B. Pearson, Jr., "Optimal disturbance reduction in linear multivariable systems," *IEEE Transactions on Automatic Control*, vol. AC-29, pp. 880–887, 1984.

Chapellat, H. and S. P. Bhattacharyya, "A generalization of Kharitonov's theorem: Robust stability of interval plants," *IEEE Transactions on Automatic Control*, vol. AC-34, pp. 306–311, 1989.

Chapellat, H., M. Dahleh and S. P. Bhattacharyya, "Robust stability under structured and unstructured perturbations," *IEEE Transactions on Automatic Control*, vol. AC-35, pp. 1100–1108, 1990.

Chapellat, H., M. Dahleh and S. P. Bhattacharyya, "On robust non-linear stability of interval control systems," *IEEE Transactions on Automatic Control*, vol. AC-36, pp. 59–67, 1991.

Chen, C. T., *Linear System Theory and Design*, Holt, Rinehart and Winston, New York, 1984.

Chen, J., M. K. H. Fan and C. N. Nett, "The structured singular value and stability of uncertain polynomials: A missing link," *Proccedings of the American Control Conference*, Chicago, Ill., 1992.

Cieslik, J., "On possibilities of the extension of Kharitonov's stability test for interval polynomials to the discrete–time case," *IEEE Transactions on Automatic Control*, vol. AC-32, pp. 237–238, 1987.

Cobb, J. D. and C. L. DeMarco, "The minimal dimension of stable faces to guarantee stability of a matrix polytope," *IEEE Transactions of Automatic Control*, vol. AC-34, pp. 990–992, 1989.

Coxson, G. E. and C. L. DeMarco, "Computing the structured real singular value is NP–hard," Technical Report ECE-92-4, Department of Electrical and Computer Engineering, University of Wisconsin, Madison, Wis., 1992.

Dahleh, M., A. Tesi and A. Vicino, "On the robust Popov criterion for interval Lur'e systems," Technical Report RT 24/91, Dipartimento di Sistemi e Informatica, Universita di Firenze, Italy, 1991.

Dasgupta, S., "A Kharitonov-like theorem for systems under non-linear parameter feedback," *Proceedings of the IEEE Conference on Decision and Control*, Los Angeles, Calif., 1987.

Dasgupta, S., "Kharitonov's theorem revisited," *Systems and Control Letters*, vol. 11, pp. 381–384, 1988.

Dasgupta, S. and A. S. Bhagwat, "Conditions for designing strictly positive real transfer functions for adaptive output error identification," *IEEE Transactions on Circuits and Systems*, vol. CAS-34, pp. 731–736, 1987.

Dasgupta, S., P. J. Parker, B. D. O. Anderson, F. J. Kraus and M. Mansour, "Frequency domain conditions for robust stability verification of linear and nonlinear dynamical systems," *IEEE Transactions on Circuits and Systems*, vol. CAS-38, pp. 389–397, 1991.

Davison, E. J., "The robust control of a servomechanism problem for linear time-invariant multivariable systems," *Proceedings of the Allerton Conference*, 1973.

de Gaston, R. R. E. and M. G. Safonov, "Exact calculation of the multiloop stability margin," *IEEE Transactions on Automatic Control*, vol. AC-33, pp. 156–171, 1988.

Demmel, J., "On structured singular values," *Proceedings of the IEEE Conference on Decision and Control*, Austin, Tex., 1988.

Desoer, C. A. and M. Vidyasagar, *Feedback Systems: Input–Output Properties*, Academic Press, New York, 1975.

Djaferis, T. E., "Shaping conditions for the robust stability of polynomials with multilinear parameter uncertainty," *Proceedings of the Conference on Decision and Control*, Los Angeles, Calif., 1988.

Djaferis, T. E., "To stabilize an interval plant family, it suffices to stabilize 64 polynomials," Technical Report ECE-DEC-90-1, Department of Electrical and Computer Engineering, University of Massachusetts, Mass., 1991.

Djaferis, T. E. and C. V. Hollot, "The stability of a family of polynomials can be deduced from a finite number of $O(k^3)$ of frequency checks," *IEEE Transactions on Automatic Control*, vol. AC-34, pp. 982–986, 1989a.

Djaferis, T. E. and C. V. Hollot, "Parameter partitioning via shaping conditions for the stability of families of polynomials," *IEEE Transactions of Automatic Control*, vol. AC-34, pp. 1205–1209, 1989b.

Dobner, D. J., "A mathematical engine model for development of dynamic engine control," *Transactions of the Society of Automotive Engineers*, vol. 89, no. 800054, 1980.

Dobner, D. J. and R. D. Fruechte, "An engine model for dynamic engine control development," *Proceedings of the American Control Conference*, San Francisco, Calif., 1983.

Dorf, R. C., *Modern Control Systems*, Addison–Wesley, Reading, Mass., 1974.

Doyle, J. C., "Analysis of feedback system with structured uncertainty," *IEE Proceedings*, vol. 129, Part D, pp. 242–250, 1982.

Doyle, J. C., K. Glover, P. P. Khargonekar and B. A. Francis, "State space solutions to standard H_2 and H_∞ control problems," *IEEE Transactions on Automatic Control*, vol. AC-34, pp. 831–847, 1989.

Eggleston, H. G., *Convexity*, (Cambridge Tracts in Mathematics and Mathematical Physics; No. 47), Cambridge University Press, Cambridge, England, 1958.

El Ghaoui, L., "Robustness of Linear Systems," Ph.D. dissertation, Department of Aeronautics and Astronautics, Stanford University, Stanford, Calif., 1990.

Faedo, S., "A new stability problem for polynomials with real coefficients," *Ann. Scuola Norm. Sup. Pisa Sci. Fis. Mat. Ser. 3-7*, pp. 53–63, 1953.

Fam, A. T. and J. S. Meditch, "A canonical parameter space for linear systems," *IEEE Transactions on Automatic Control*, vol. AC-23, pp. 454–458, 1978.

Fogel, E. and Y. F. Huang, "On the value of information in system identification," *Automatica*, vol. 18, pp. 229–238, 1982.

Francis, B. A., *A Course in H_∞ Control Theory* (Lecture notes in control and information sciences, no. 88), Springer-Verlag, New York, 1987.

Francis, B. A., J. W. Helton and G. Zames, "H^∞ optimal feedback controllers for linear multivariable systems," *IEEE Transactions on Automatic Control*, vol. AC-29, pp. 888–900, 1984.

Franklin, G. F., J. D. Powell and A. Emani-Naeini, *Feedback Control of Dynamic Systems*, Addison-Wesley, New York, 1986.

Frazer, R. A. and W. J. Duncan, "On the criteria for the stability of small motion," *Proceedings of the Royal Society*, A, vol. 124, pp. 642–654, London, England, 1929.

Fruchter, G., U. Srebro and E. Zeheb, "On several variable zero sets and application to MIMO robust feedback stabilization," *IEEE Transactions on Circuits and Systems*, vol. CAS-34, pp. 1210–1220, 1987.

Fu, M., "Comments on 'A necessary and sufficient condition for the positive-definiteness of interval symmetric matrices,' " *International Journal of Control*, vol. 46, p. 1485, 1987.

Fu, M., "A Class of weak Kharitonov regions for robust stability of linear uncertain systems," *IEEE Transactions on Automatic Control*, vol. AC-36, pp. 975–978, 1991.

Fu, M. and B. R. Barmish, "Maximal unidirectional perturbation bounds for stability of polynomials and matrices," *Systems and Control Letters*, vol. 11, pp. 173–179, 1988.

Fu, M. and B. R. Barmish, "Polytopes of polynomials with zeros in a prescribed set," *IEEE Transactions on Automatic Control*, vol. AC-34, pp. 544–546, 1989.

Fu, M., A. W. Olbrot and M. P. Polis, " Robust stability for time delay systems: The edge theorem and graphical tests," *IEEE Transactions on Automatic Control*, vol. AC-34, pp. 813–820, 1989.

Gantmacher, F. R., *The Theory of Matrices*, vols. I and II, Chelsea, New York, 1959.

Garey, M. R. and D. S. Johnson, *Computers and Intractability: A Guide to the Theory of NP–Completeness*, W. H. Freeman, San Francisco, Calif., 1979.

Garofalo, F., G. Celentano and L. Glielmo, "Stability robustness of interval matrices via Lyapunov quadratic forms," to appear in *IEEE Transactions on Automatic Control*, 1992.

Ghosh, B. K., "Some new results on the simultaneous stabilizability of a family of single input, single output systems," *Systems and Control Letters*, vol. 6, pp. 39–45, 1985.

Guiver, J. P. and N. K. Bose, "Strictly Hurwitz property invariance of quartics under coefficient perturbation," *IEEE Transactions on Automatic Control*, vol. AC-28, pp. 106–107, 1983.

Gutman, S., *Root Clustering in Parameter Space*, Springer-Verlag, New York, 1990.

Gutman, S. and E. I. Jury, "A general theory for matrix root clustering in subregions of the complex plane," *IEEE Transactions on Automatic Control*, vol. AC-26, pp. 853–863, 1981.

Haddad, W. M. and D. S. Bernstein, "Parameter-dependent Lyapunov functions and the discrete-time Popov criterion for robust analysis and synthesis," *Proceedings of the American Control Conference*, Chicago, Ill., 1992.

Hamann, J. C. and B. R. Barmish, "Convexity of frequency response arcs associated with a stable polynomial," *Proceedings of the American Control Conference*, Chicago, Ill., 1992.

Hara, S., T. Kimura and R. Kondo, "H_∞ control system design by a parameter space approach," *Proceedings of the Symposium on the Mathematical Theory of Networks and Systems*, Kobe, Japan, 1991.

Hazell, P. A. and J. O. Flower, "Sampled-data theory applied to the modelling and control analysis of compression ignition engines–Part I," *International Journal of Control*, vol. 13, pp. 549–562, 1971.

Hertz, D., E. I. Jury and E. Zeheb, "Root exclusion from complex polydomains and some of its applications," *Automatica*, vol. 23, pp. 399–404, 1987.

Hinrichsen, D. and A. J. Pritchard, "A robustness measure for linear systems under structured real parameter perturbations," Report No. 184, Institut fur Dynamische Systeme, Bremen, Germany, 1988.

Hinrichsen, D. and A. J. Pritchard, "An application of state space methods to obtain explicit formulae for robustness measures of polynomials," in *Robustness in Identification and Control*, M. Milanese, R. Tempo and A. Vicino, eds., Plenum, New York, pp. 183–206, 1989.

Hollot, C. V., "Kharitonov-like results in the space of Markov parameters," *IEEE Transactions on Automatic Control*, vol. AC-34, pp. 536–538, 1989.

Hollot, C. V. and A. C. Bartlett, "Some discrete-time counterparts to Kharitonov's stability criterion for uncertain systems," *IEEE Transactions on Automatic Control*, vol. AC-31, pp. 355–356, 1986.

Hollot, C. V. and R. Tempo, "On the Nyquist envelope of an interval plant family," *Proceedings of the American Control Conference*, Boston, Mass., 1991.

Hollot, C. V., R. Tempo and V. Blondel, "H_∞ performance of interval plants and interval feedback systems," in *Robustness of Dynamic Systems with Parameter Uncertainty*, M. Mansour, S. Balemi and W. Truöl, eds., Birkhauser, Basel, Switzerland, 1992.

Hollot, C. V. and F. Yang, "Robust stabilization of interval plants using lead or lag compensators," *Systems and Control Letters*, vol. 14, pp. 9–12, 1990.

Hollot, C. V. and Z. L. Xu, "When is the image of a multilinear function a polytope? A conjecture," *Proceedings of the IEEE Conference on Decision and Control*, Tampa, Fla., 1989.

Holohan, A. M. and M. G. Safonov, "On computing the MIMO real structured stability margin," *Proceedings of the IEEE Conference on Decision and Control*, Tampa, Fla., 1989.

Horisberger, H. P. and P. R. Belanger, "Regulators for linear time invariant plants with uncertain parameters," *IEEE Transactions on Automatic Control*, vol. AC-21, pp. 705–708, 1976.

Horn, R. and C. Johnson, *Matrix Analysis*, Cambridge University Press, Cambridge, England, 1988.

Horowitz, I. M., *Synthesis of Feedback Systems*, Academic Press, New York, 1963.

Horowitz, I. M., "Feedback systems with nonlinear uncertain plants," *International Journal of Control*, vol. 36, pp. 155–171, 1982.

Horowitz, I. M. and S. Ben-Adam, "Clockwise nature of Nyquist locus of stable transfer functions," *International Journal of Control*, vol. 49, pp. 1433–1436, 1989.

Horowitz, I. M. and M. Sidi, "Synthesis of feedback systems with large plant ignorance for prescribed time-domain tolerances," *International Journal of Control*, vol. 16, pp. 287–309, 1972.

Isidori, A., *Nonlinear Control Systems,* Springer-Verlag, New York, 1985.

Jury, E. I., *Inners and Stability of Dynamic Systems*, Wiley, New York, 1974.

Jury, E. I. and T. Pavlidis, "Stability and aperiodicity constraints for systems design," *IEEE Transactions on Circuit Theory*, vol. 10, pp. 137–141, 1963.

Kang, H. I., "Extreme Point Results for Robustness of Control Systems," Ph.D. Dissertation, Department of Electrical and Computer Engineering, University of Wisconsin, Madison, Wis., 1992.

Karl, W. C., J. P. Greschak and G. C. Verghese, "Comments on 'A necessary and sufficient condition for the stability of interval matrices,' " *International Journal of Control*, vol. 39, pp. 849–851, 1984.

Keel, L. H., S. P. Bhattacharyya and J. W. Howze, "Robust control with structured perturbations," *IEEE Transactions on Automatic Control*, vol. AC-33, pp. 68–78, 1988.

Khalil, H. K., *Nonlinear Systems*, Macmillan, New York, 1992.

Kharitonov, V. L., "Asymptotic stability of an equilibrium position of a family of systems of linear differential equations," *Differentsial'nye Uravneniya*, vol. 14, pp. 2086–2088, 1978a.

Kharitonov, V. L., "On a generalization of a stability criterion," *Izvestiya Akademii Nauk Kazakhskoi SSR Seriya Fizika Matematika*, no. 1, pp. 33–57, 1978b.

Kharitonov, V. L., "The Routh–Hurwitz problem for families of polynomials and quasipolynomials," *Izvetiy Akademii Nauk Kazakhskoi SSR, Seria fizikomatematicheskaia*, no 26, pp. 69–79, 1979.

Kharitonov, V. L. and A. P. Zhabko, "Stability of convex hull of quasi polynomials," in *Robustness of Dynamic Systems with Parameter Uncertainty*, M. Mansour, S. Balemi and W. Truöl, eds., Birkhauser, Basel, Switzerland, 1992.

Kiendl, H., "Totale stabilitat von linearen regelungssystemen beiungenau bekannten parametern der regelstrecke," *Automatisierungstechnik*, vol. 33, pp. 379–386, 1985.

Kiendl, H., "Robustheitanalyse von regelungssystemen mit der methode der konvexen zerlegung," *Automatisierungstechnik*, vol. 35, pp. 192–202, 1987.

Kim, K. D. and N. K. Bose, "Invariance of the strict Hurwitz property for bivariate polynomials under coefficient perturbations," *IEEE Transactions on Automatic Control*, vol. AC-33, pp. 1172–1174, 1988.

Kimura, T. and S. Hara, "A robust control system design by a parameter space approach based on sign definite condition," *Proceedings of the Korean Automatic Control Conference*, Seoul, Korea, 1991.

Kraus, F. J., B. D. O. Anderson, E. I. Jury and M. Mansour, "On the robustness of low order Schur polynomials," *IEEE Transactions on Circuits and Systems*, vol. CAS-35, pp. 570–577, 1988.

Kraus, F. J., B. D. O. Anderson and M. Mansour, "Robust Schur polynomial stability and Kharitonov's theorem," *International Journal of Control*, vol. 47, pp. 1213–1225, 1988.

Kraus, F. J. and W. Truöl, "Robust stability of control systems with polytopical uncertainty: A Nyquist approach," *International Journal of Control*, vol. 53, pp. 967–983, 1991.

Kumar, P. R. and P. Varaiya, *Stochastic Systems: Estimation, Identification and Adaptive Control*, Prentice–Hall, Englewood Cliffs, N.J., 1986.

Kurzhanskii, A. B., "Dynamic control system identification under uncertainty conditions," *Problems of Control and Information Theory*, vol. 9, no. 6, pp. 395–406, 1980.

Kwakernaak, H., "A condition for robust stabilizability," *Systems and Control Letters*, vol. 2, pp. 1–5, 1982.

Lang, S., *Algebra*, Addison-Wesley, New York, 1965.

Leal, M. A. and J. S. Gibson, "A first-order Lyapunov robustness method for linear systems with uncertain parameters," *IEEE Transactions on Automatic Control*, vol. AC-35, pp. 1068–1070, 1990.

Leitmann, G., "Guaranteed asymptotic stability for some linear systems with bounded uncertainties," *Journal of Dynamic Systems, Measurement and Control*, vol. 101, pp. 212–216, 1979.

Luenberger, D. G., *Introduction to Linear and Nonlinear Programming*, Addison–Wesley, Reading, Mass, 1973.

Mansour, M., "Robust stability of interval matrices," *Proceedings of the IEEE Conference on Decision and Control*, Tampa, Fla., 1989.

Mansour, M. and B. D. O. Anderson, "Kharitonov's Theorem and the second method of Lyapunov," in *Robustness of Dynamic Systems with Parameter Uncertainty* M. Mansour, S. Balemi and W. Truöl, eds., Birkhauser, Basel, Switzerland, 1992.

Mansour, M., F. J. Kraus and B. D. O. Anderson, "Strong Kharitonov theorem for discrete systems," *Proceedings of the IEEE Conference on Decision and Control*, Austin, Tex., 1988.

Marden, M., *Geometry of Polynomials*, American Mathematical Society, Providence, R.I., 1966.

Mayer, G., "On the convergence of powers of interval matrices," *Linear Algebra and Its Applications*, vol. 58, pp. 201–216, 1984.

Meerov, M. V., "Automatic control systems with indefinitely large gains," *Avtomatika i Telemekhanika,* vol. 8, pp. 152–167, 1947.

Middleton, R. H. and G. C. Goodwin, "Improved finite word length characteristics in digital control using delta operators," *IEEE Transactions on Automatic Control*, vol. AC-31, pp. 1015–1021, 1986.

Mikhailov, A. W., "Method of harmonic analysis in control theory," *Avtomatika i Telemekhanika*, vol. 3, pp. 27–81, 1938.

Minnichelli, R. J., J. J. Anagnost and C. A. Desoer, "An elementary proof of Kharitonov's theorem with extensions," *IEEE Transactions on Automatic Control*, vol. AC-34, pp. 995–998, 1989.

Monov, V. V., "On the spectral set of a family of matrices," to appear, 1992.

Mori, T. and S. Barnett, "On stability tests for some classes of dynamical systems with perturbed coefficients," *IMA Journal of Mathematical Control and Information*, vol. 5, pp. 117–123, 1988.

Mori, T. and H. Kokame, "Convergence property of interval matrices and interval polynomials," *International Journal of Control*, vol. 45, pp. 481-484, 1987.

Mori, T. and H. Kokame, "Stability of interval polynomials with vanishing extreme coefficients," *Recent Advances in Mathematical Theory of Systems, Control, Networks and Signal Processing I*, pp. 409–414, Mita Press, Tokyo, Japan, 1992.

Neimark, Y. I., "On the problem of the distribution of the roots of polynomials," *Dokl. Akad. Nauk*, vol. 58, 1947.

Neimark, Y. I., *Stability of Linearized Systems*, Leningrad Aeronautical Engineering Academy, Leningrad, USSR, 1949.

Nemirokvskii, A., "Several NP–hard problem arise in robust stability analysis," to appear, 1992.

Nise, N. S., *Control Systems Engineering*, Benjamin Cummings, New York, 1992.

Nishimura, Y. and K. Ishii, "Engine idle stability analysis and control," *Society of Automotive Engineers Technical Paper Series*, no. 860415, 1986.

Ossadnik, H. and H. Kiendl, "Robuste quadratische Ljapunovfunktionen," *Automatisierungstechnik*, vol. 38, pp. 174–182, 1990.

Olbrot, A. W. and B. K. Powell, " Robust design and analysis of third and fourth order time delay systems with applications to automotive idle speed control," *Proceedings of the American Control Conference*, Pittsburgh, Pa., 1989.

Packard, A. A., *What's New with μ?*, Department of Mechanical Engineering, University of California, Berkeley, Calif., 1987.

Packard, A. A. and A. P. Pandey, "Continuity properties of the real/complex structured singular value," Technical report, Department of Mechanical Engineering, University of California, Berkeley, Calif., 1991.

Panier, E. R., M. K. H. Fan and A. L. Tits, "On the robust stability of polynomials with no cross-coupling between the perturbations in the coefficients of even and odd powers," *Systems and Control Letters*, vol. 12, pp. 291–299, 1989.

Perez, F., D. Decampo and C. Abdallah, "New extreme-point robust stability results for discrete-time polynomials," *Proceedings of the American Control Conference*, Chicago, Ill., 1992.

Petersen, I. R., "A collection of results on the stability of families of polynomials with multilinear parameter dependence," Technical Report EE8801, Department of Electrical Engineering, University College, University of New South Wales, Australian Defence Force Academy, Canberra, Australia, 1988.

Petersen, I. R., "A class of stability regions for which a Kharitonov-like theorem holds," *IEEE Transactions on Automatic Control*, vol. AC-34, pp. 1111–1115, 1989.

Petersen, I. R., "A new extension to Kharitonov's theorem," *IEEE Transactions on Automatic Control*, vol. AC-35, pp. 825–828, 1990.

Polyak, B. T., "Robustness analysis for multilinear perturbations," in *Robustness of Dynamic Systems with Parameter Uncertainty*, M. Mansour, S. Balemi and W. Truöl, eds., Birkhauser, Basel, Switzerland, 1992.

Polyak, B. T., P. S. Scherbakov and S. B. Shmulyian, "Construction of value set for robustness analysis via circular arithmetic," to appear in *International Journal of Robust and Nonlinear Control*, 1993.

Polyak, B. T. and Y. Z. Tsypkin, "Frequency criteria of robust stability and aperiodicity of linear systems," *Avtomatika i Telemekhanika*, pp. 45–54, 1990.

Powell, B. K., "A dynamic model for automotive engine control analysis," *Proceedings of the IEEE Conference on Decision and Control*, Fort Lauderdale, Fla., 1979.

Powell, B. K., J. A. Cook and J. W. Grizzle, "Modeling and analysis of an inherently multi–rate sampling fuel injected engine idle speed control loop," *Proceedings of the American Control Conference*, Minneapolis, Minn., 1987.

Pujara, L. R., "On the stability of uncertain polynomials with dependent coefficients," *IEEE Transactions on Automatic Control*, vol. AC-35, pp. 756–759, 1990.

Qiu, L. and E. J. Davison, "Bounds on the real stability radius," in *Robustness of Dynamic Systems with Parameter Uncertainties*, M. Mansour, S. Balemi and W. Truöl, eds., Birkhauser, Basel, Switzerland, 1992.

Rantzer, A., "Hurwitz testing sets for parallel polytopes of polynomials," *Systems and Control Letters*, vol. 15, pp. 99–104, 1990.

Rantzer, A., "Stability conditions for polytopes of polynomials," *IEEE Transactions on Automatic Control*, vol. AC-37, pp. 79–89, 1992a.

Rantzer, A., "Kharitonov's weak theorem holds if and only if the stability region and its reciprocal are convex," to appear in *International Journal of Nonlinear and Robust Control*, 1992b.

Rantzer, A., "Continuity properties of the parameter stability margin," *Proceedings of the American Control Conference*, Chicago, Ill., 1992c.

Rantzer, A. and A. Magretsky, "A convex parameterization of robustly stabilizing controllers," in *Robustness of Dynamic Systems with Parameter Uncertainty*, M. Mansour, S. Balemi and W. Truöl, eds., Birkhauser, Basel, Switzerland, 1992.

Rockafellar, R. T., *Convex Analysis*, Princeton University Press, Princeton, N.J., 1970.

Rohn, J., "Systems of linear interval equations," *Linear Algebra and Applications*, vol. 126, pp. 39–78, 1989.

Rohn, J. and S. Poljak, "Checking robust nonsingularity is NP–hard," to appear in *Systems and Control Letters*, 1992.

Rudin, W., *Real and Complex Analysis*, McGraw-Hill, New York, 1968.

Rugh, W. J., *Nonlinear System Theory*, Johns Hopkins University Press, Baltimore, Md., 1981.

Saeki, M., "Method of robust stability analysis with highly structured uncertainties," *IEEE Transactions on Automatic Control*, vol. AC-31, pp. 935–940, 1986.

Saeks, R. and J. Murray, "Fractional representation, algebraic geometry and the simultaneous stabilization problem," *IEEE Transactions on Automatic Control*, vol. AC-27, pp. 859–903, 1982.

Safonov, M. G., *Stability and Robustness of Multivariable Feedback Systems*, MIT Press, Cambridge, Mass., 1980.

Safonov, M. G., "Stability margins of diagonally perturbed multivariable feedback systems," *IEE Proceedings*, vol. 129, Part D, pp. 251–256, 1982.

Saridereli, M. K. and F. J. Kern, "The stability of polynomials under correlated coefficient perturbations," *Proceedings of the IEEE Conference on Decision and Control*, Los Angeles, Calif., 1987.

Saydy, L., A. L. Tits and E. H. Abed, "Guardian maps and the generalized stability of parametrized families of matrices and polynomials," *Mathematics of Control, Signals and Systems*, vol. 3, pp. 345–371, 1990.

Schweppe, F. C. *Uncertainty Dynamical Systems*, Prentice–Hall, Englewood Cliffs, N.J., 1973.

Shafai, B., K. Perev, D. Cowley and Y. Chehab, "A necessary and sufficient condition for the stability of nonnegative interval discrete systems," *IEEE Transactions on Automatic Control*, vol. AC-36, pp. 742–746, 1991.

Shi, Y. Q., "Robust (strictly) positive interval rational functions," *IEEE Transactions on Circuits and Systems*, vol. CAS-38, pp. 552–554, 1991.

Shi, Z. and W. B. Gao, "A necessary and sufficient condition for the positive–definiteness of interval symmetric matrices," *International Journal of Control*, vol. 43, pp. 325–328, 1986.

Sideris, A., "An efficient algorithm for checking the robust stability of a polytope of polynomials," *Mathematics of Control, Signals, and Systems*, vol. 4, pp. 315–337, 1991.

Sideris, A. and B. R. Barmish, "An edge theorem for polytopes of polynomials which can drop in degree," *Systems and Control Letters*, vol. 13, pp. 233–238, 1989.

Sideris, A. and R. S. Sanchez Pena, "Fast computation of the multivariable stability margin for real interrelated uncertain parameters," *IEEE Transactions on Automatic Control*, vol. AC-34, pp. 1272–1276, 1989.

Siljak, D. D., *Nonlinear Systems*, Wiley, New York, 1969.

Soh, C. B., "Robust stability of discrete-time systems using delta operators," *IEEE Transactions on Automatic Control*, vol. AC-36, pp. 377–380, 1991.

Soh, C. B. and C. S. Berger, "Damping margins of polynomials with perturbed coefficients," *IEEE Transactions on Automatic Control*, vol. AC-33, pp. 509–511, 1988.

Soh, C. B., C. S. Berger and K. P. Dabke, "On the stability properties of polynomials with perturbed coefficients," *IEEE Transactions on Automatic Control*, vol. AC-30, pp. 1033–1036, 1985.

Soh, Y. C. and Y. K. Foo, "Generalized edge theorem," *Systems and Control Letters*, vol. 12, pp. 219–224, 1989.

Soh, Y. C. and Y. K. Foo, "A note on the edge theorem," *Systems and Control Letters*, vol. 15, pp. 41–43, 1990.

Sondergeld, K. P., "A generalization of the Routh–Hurwitz stability criteria and an application to a problem in robust controller design," *IEEE Transactions on Automatic Control*, vol. AC-28, pp. 965–970, 1983.

Stöer, J. and C. Witzgall, *Convexity and Optimization in Finite Dimensions*, Springer-Verlag, Berlin, Germany, 1970.

Tarski, A., *A Decision Method for Elementary Algebra and Geometry*, University of California Press, Berkeley, Calif., 1951.

Tempo, R., "A dual result to Kharitonov's theorem," *IEEE Transactions on Automatic Control*, vol. AC-35, pp. 195–198, 1990.

Tesi A. and A. Vicino, "Design of optimally robust controllers with few degrees of freedom," Technical Report RT 5/91, Dipartimento di Sistemi e Informatica, Universita di Firenze, Firenze, Italy, 1991.

Tesi, A., A. Vicino and G. Zappa, "Clockwise property of the Nyquist plot with implications for absolute stability," *Automatica*, vol. 28, pp. 71–80, 1992.

Tits, A., "Comments on 'Polytopes of polynomials with zeros in a prescribed set' by M. Fu and B. R. Barmish," *IEEE Transactions on Automatic Control*, vol. AC-35, pp. 1276–1277, 1990.

Tsing, N. K. and A. L. Tits, "On the multilinear image of a cube," in *Robustness of Dynamic Systems with Parameter Uncertainty,* M. Mansour, S. Balemi and W. Truöl, eds., Birkhauser, Basel, Switzerland, 1992.

Truxal, J. G., *Automatic Feedback Control System Synthesis,* McGraw-Hill, New York, 1955.

Tsypkin, Y. Z. and B. T. Polyak, "Frequency domain criterion for the ℓ^p–robust stability of continuous linear systems," *IEEE Transactions on Automatic Control,* vol. AC-36, pp. 1464–1469, 1991.

Utkin, V. I., "Variable structure systems with sliding modes," *IEEE Transactions on Automatic Control,* vol. AC-22, pp. 212–222, 1977.

Vaidyanathan, P. P., "Derivation of new and existing discrete-time Kharitonov theorems based on discrete-time reactances," *IEEE Transactions on Acoustics, Speech and Signal Processing,* vol. ASSP-38, pp. 277–285, 1990.

Vicino, A., "Robustness of pole locations in perturbed systems," *Automatica,* vol. 25, pp. 109–114, 1989.

Vicino, A., A. Tesi and M. Milanese, "Computation of nonconservative stability perturbation bounds for systems with nonlinearly correlated uncertainties," *IEEE Transactions on Automatic Control,* vol. AC-35, pp. 835–841, 1990.

Vicino, A. and A. Tesi, "Strict positive realness: New results for interval plants plus controller families," *Proceedings of the IEEE Conference on Decision and Control,* Brighton, England, 1991.

Vidyasagar, M., *Control System Synthesis: A Factorization Approach,* MIT Press, Boston, Mass., 1985.

Vidyasagar, M. and N. Viswanadham, "Algebraic design techniques for reliable stabilization," *IEEE Transactions on Automatic Control,* vol. AC-27, pp. 1085–1095, 1982.

Washino, S., R. Nishiyama and S. Ohkubo, "A fundamental study for the control of periodic oscillations of SI engine revolutions," *Society of Automotive Engineers Technical Paper Series,* no. 860411, 1986.

Wei, K. and R. K. Yedavalli, "Robust stabilizability for linear systems with both parameter variation and unstructured uncertainty,"

IEEE Transactions on Automatic Control, vol. AC-34, pp. 149–156, 1989.

Weinmann, A., *Uncertain Models and Robust Control*, Springer-Verlag, New York, 1991.

Yamagushi, H., S. Takizawa, H. Sanbuichi and K. Ikeura, "Analysis on idle speed stability in port fuel injection engines," *Society of Automotive Engineers Technical Paper Series*, no. 861389, 1986.

Youla, D. C., J. J. Bongiorno, Jr. and C. N. Lu, "Single-loop feedback-stabilization of linear multivariable dynamical plants," *Automatica*, vol. 10, pp. 159–173, 1974.

Youla, D. C., H. A. Jabr and J. J. Bongiorno, Jr., "Modern Wiener–Hopf design of optimal controllers, Part II: The multivariable case," *IEEE Transactions on Automatic Control*, vol. AC-21, pp. 319–338, 1976.

Zadeh, L. A. and C. A. Desoer, *Linear System Theory—A State-Space Approach*, McGraw-Hill, New York, 1963.

Zames, G., "Feedback and optimal sensitivity: Model reference transformations, multiplicative seminorms and approximate inverses," *IEEE Transactions on Automatic Control*, vol. AC-26, pp. 301–320, 1981.

Zames, G. and B. A. Francis, "A new approach to classical frequency methods: Feedback and minimax sensitivity," *Proceedings of the IEEE Conference on Decision and Control*, San Diego, Calif., 1981.

Zeheb, E. and E. Walach, "2-Parameter Root-Loci Concept and Some Applications," *IEEE International Journal of Circuit Theory and Applications*, vol. 5, pp. 305–315, 1977.

Zeheb, E. and E. Walach, "Zero sets of multidimensional functions and stability of multidimensional systems," *IEEE Transactions on Acoustics, Speech and Signal Processing*, vol. ASSP-29, pp. 197–206, 1981.

Zong, S. Q., "Robust Stability Analysis of Control Systems with Structured Parametric Uncertainties and an Application to a Robot Control System," Ph.D. dissertation, Department of Physics and Electrical Engineering, University of Bremen, Germany, 1990.

Index

Index